Undergraduate Topics in Computer Science

Undergraduate Topics in Computer Science (UTiCS) delivers high-quality instructional content for undergraduates studying in all areas of computing and information science. From core foundational and theoretical material to final-year topics and applications, UTiCS books take a fresh, concise, and modern approach and are ideal for self-study or for a one- or two-semester course. The texts are all authored by established experts in their fields, reviewed by an international advisory board, and contain numerous examples and problems. Many include fully worked solutions.

For further volumes:
http://www.springer.com/series/7592

Boris Mirkin

Core Concepts in Data Analysis: Summarization, Correlation and Visualization

 Springer

Boris Mirkin
Research University – Higher School
 of Economics
School of Applied Mathematics
 and Informatics
11 Pokrovsky Boulevard
Moscow RF
Russia
and
Department of Computer Science
Birkbeck University of London
Malet Street, London
UK

QA
76.9
.D35
M57
2011

Series editor
Ian Mackie

Advisory board
Samson Abramsky, University of Oxford, Oxford, UK
Karin Breitman, Pontifical Catholic University of Rio de Janeiro, Rio de Janeiro, Brazil
Chris Hankin, Imperial College London, London, UK
Dexter Kozen, Cornell University, Ithaca, USA
Andrew Pitts, University of Cambridge, Cambridge, UK
Hanne Riis Nielson, Technical University of Denmark, Kongens Lyngby, Denmark
Steven Skiena, Stony Brook University, Stony Brook, USA
Iain Stewart, University of Durham, Durham, UK

ISSN 1863-7310
ISBN 978-0-85729-286-5   e-ISBN 978-0-85729-287-2
DOI 10.1007/978-0-85729-287-2
Springer London Dordrecht Heidelberg New York

British Library Cataloguing in Publication Data
A catalogue record for this book is available from the British Library

Library of Congress Control Number: 2011922052

© Springer-Verlag London Limited 2011
Apart from any fair dealing for the purposes of research or private study, or criticism or review, as permitted under the Copyright, Designs and Patents Act 1988, this publication may only be reproduced, stored or transmitted, in any form or by any means, with the prior permission in writing of the publishers, or in the case of reprographic reproduction in accordance with the terms of licenses issued by the Copyright Licensing Agency. Enquiries concerning reproduction outside those terms should be sent to the publishers.
The use of registered names, trademarks, etc., in this publication does not imply, even in the absence of a specific statement, that such names are exempt from the relevant laws and regulations and therefore free for general use.
The publisher makes no representation, express or implied, with regard to the accuracy of the information contained in this book and cannot accept any legal responsibility or liability for any errors or omissions that may be made.

Printed on acid-free paper

Springer is part of Springer Science+Business Media (www.springer.com)

Preface

In this textbook, I take an unconventional approach to data analysis. Its contents are heavily influenced by the idea that data analysis should help in enhancing and augmenting knowledge of the domain as represented by the concepts and statements of relation between them. According to this view, two main pathways for data analysis are summarization, for developing and augmenting concepts, and correlation, for enhancing and establishing relations. Visualization, in this context, is a way of presenting results in a cognitively comfortable way. The term *summarization* is understood quite broadly here to embrace not only simple summaries like totals and means, but also more complex summaries such as the principal components of a set of features or cluster structures in a set of entities.

The material presented in this perspective makes a unique mix of subjects from the fields of statistical data analysis, data mining, and computational intelligence, which follow different systems of presentation.

Another feature of the text is that its main thrust is to give an in-depth understanding of a few basic techniques rather than to cover a broad spectrum of approaches developed so far. Most of the described methods fall under the same least-squares paradigm for mapping an "idealized" structure to the data. This allows me to bring forward a number of relations between methods that are usually overlooked. Just one example: a relation between the choice of a scoring function for classification trees and normalization options for dummies representing the target categories.

Although the in-depth study approach involves a great deal of technical details, these are encapsulated in specific fragments of the text termed "formulation" parts. The main, "presentation", part is written in a very different style. The presentation involves no mathematical formulas and explains a method by actually applying it to a small real-world dataset – this part can be read and studied with no concern for the formulation at all. There is one more part, "computation", targeted at a computer-oriented reader. This part describes the computational implementation of the methods, illustrated using the MatLab computing environment. I have arrived at this three-way narrative style as a result of my experiences in teaching data analysis and computational intelligence to students in Computer Science. Some students might be mainly interested in just one of the parts, whereas others might try to get to grips with two or even all three of them.

One more device to stimulate the reader's interest is a multi-layer system of proactive learning materials for class- and self-study:

- *Worked examples* provided to show how specific methods apply to particular datasets;
- More complex problems solved, *case studies*, possibly involving a rule for data generation, rather than a pre-specified dataset, or an informal way of analyzing results;
- Even more complex problems, *projects*, possibly involving uncharted terrain and a small-scale investigation; these should be used as models for similar self-study projects on other data;
- A number of computational or theoretical problems, *questions*, formulated as self-study exercises; answers are provided for most of them.

The text is based on my courses for full-time and part-time students in the MS program in Computer Science at Birkbeck, University of London (2003–2010), in the BS and MS programs in Applied Mathematics and Informatics at Higher School of Economics, Moscow (2008–2010), and post-graduate School of Data Analysis at Yandex, a popular Russian search engine, Moscow (2009–2010). The material covers lectures and labs for about 35–40 lecture hours in advanced BS programs or MS programs in Computer Science or Engineering. It can also be used in application-oriented courses such as Bioinformatics or Methods in Marketing Research.

No prerequisite beyond a conventional school background for reading through the presentation part is required, yet some training in reading academic material is expected. The reader interested in studying the formulation part should have some background in: (a) basic calculus including the concepts of function, derivative and the first-order optimality conditions, (b) basic linear algebra including vectors, inner products, Euclidean distances and matrices (these are reviewed in the Appendix), and (c) basic set theory notation such as the symbols for inclusion and membership. The computation part is oriented towards those interested in coding for computer implementation, specifically focusing on working with MatLab as a user-friendly environment.

Acknowledgments

Too many people contributed to the material of the book to list all their names. First of all, my gratitude goes to Springer's editors who were instrumental in bringing forth the very idea of writing such a book and in channeling my efforts by providing good critical reviews. Then, of course, I thank the students at my classes in MS programs in Computer Science at Birkbeck and, more recently, in BS and MS programs in Applied Mathematics and Informatics at HSE. Here is a list of people who directly contributed to this book with advice, and sometimes with computation: I. Muchnik (Rutgers University), M. Levin (Higher School of Economics Moscow), T. Fenner (Birkbeck University of London), S. Nascimento (New University of Lisbon), T. Krauze (Hofstra University), I. Mandel (Telmar Inc), I. Mirkin (Yext), V. Sulimova (Tula Technical University), and V. Topinsky (Higher School of Economics Moscow). The HSE students J. Askarova, K. Chernyak, O. Chugunova, K. Kovaleva, and A Kramarenko helped in debugging the final version.

Contents

1 **Introduction: What Is Core** 1
   1.1 Summarization and Correlation: Two Main Goals
        of Data Analysis 1
   1.2 Case Study Problems 9
   1.3 An Account of Data Visualization 21
        1.3.1 General 21
        1.3.2 Highlighting 22
        1.3.3 Integrating Different Aspects 25
        1.3.4 Narrating a Story 28
   1.4 Summary 28
   References 29

2 **1D Analysis: Summarization and Visualization
of a Single Feature** 31
   2.1 Quantitative Feature: Distribution and Histogram 31
        P2.1.1 Presentation 31
        F2.1.2 Formulation 33
        C2.1.3 Computation 35
   2.2 Further Summarization: Centers and Spreads 36
        P2.2.1 Centers and Spreads: Presentation 36
        F2.2.2 Centers and Spreads: Formulation 39
        C2.2.3 Centers and Spreads: Computation 43
   2.3 Binary and Categorical Features 43
        P2.3.1 Presentation 43
        F2.3.2 Formulation 46
        C2.3.3 Computation 49
   2.4 Modeling Uncertainty: Intervals and Fuzzy Sets 49
        2.4.1 Individual Membership Functions 49
        2.4.2 Central Fuzzy Set 52
   2.5 Summary 64
   References 65

3 2D Analysis: Correlation and Visualization of Two Features 67
- 3.1 General 67
- 3.2 Two Quantitative Features Case 68
  - P3.2.1 Scatter-Plot, Linear Regression and Correlation Coefficients 68
  - P3.2.2 Validity of the Regression 70
  - F3.2.3 Linear Regression: Formulation 74
  - C3.2.4 Linear Regression: Computation 78
- 3.3 Mixed Scale Case: Nominal Feature Versus a Quantitative One . 89
  - P3.3.1 Box-Plot, Tabular Regression and Correlation Ratio ... 89
  - F3.3.2 Tabular Regression: Formulation 93
  - 3.3.3 Nominal Target 95
- 3.4 Two Nominal Features Case 100
  - P3.4.1 Analysis of Contingency Tables: Presentation 100
  - F3.4.2 Analysis of Contingency Tables: Formulation 107
- 3.5 Summary 111
- References 112

4 Learning Multivariate Correlations in Data 113
- 4.1 General: Decision Rules, Fitting Criteria, and Learning Protocols 113
- 4.2 Naïve Bayes Approach 118
  - 4.2.1 Bayes Decision Rule 118
  - 4.2.2 Naïve Bayes Classifier 120
  - 4.2.3 Metrics of Accuracy 123
- 4.3 Linear Regression 128
  - P4.3.1 Linear Regression: Presentation 128
  - F4.3.2 Linear Regression: Formulation 131
- 4.4 Linear Discrimination and SVM 133
  - P4.4.1 Linear Discrimination and SVM: Presentation 133
  - F4.4.2 Linear Discrimination and SVM: Formulation 137
- 4.5 Decision Trees 141
  - P4.5.1 General: Presentation 141
  - F4.5.2 General: Formulation 142
  - 4.5.3 Measuring Correlation for Classification Trees 145
  - 4.5.4 Building Classification Trees 152
  - C4.5.5 Building Classification Trees: Computation 157
- 4.6 Learning Correlation with Neural Networks 159
  - 4.6.1 General 159
  - 4.6.2 Learning a Multi-layer Network 163
- 4.7 Summary 171
- References 171

5 Principal Component Analysis and SVD 173
- 5.1 Decoder Based Data Summarization 173
  - 5.1.1 Structure of a Summarization Problem with Decoder ... 173

		P5.1.2 Data Recovery Criterion: Presentation	174
		F5.1.3 Data Recovery Criterion: Formulation	176
		5.1.4 Data Standardization	177
		C5.1.5 Data Standardization: Computation	182
	5.2	Principal Component Analysis: Model, Method, Usage	188
		P5.2.1 SVD Based PCA and Its Usage: Presentation	188
		F5.2.2 Mathematical Model of PCA-SVD and Its Properties: Formulation	198
		C5.2.3 Computing Principal Components	205
	5.3	Application: Latent Semantic Analysis	207
		P5.3.1 Latent Semantic Analysis: Presentation	207
		F5.3.2 Latent Semantic Analysis: Formulation	210
		C5.3.3 Latent Semantic Analysis: Computation	211
	5.4	Application: Correspondence Analysis	212
		P5.4.1 Correspondence Analysis: Presentation	212
		F5.4.2 Correspondence Analysis: Formulation	213
		C5.4.3 Correspondence Analysis: Computation	216
	5.5	Summary	218
	References		219
6	**K-Means and Related Clustering Methods**		221
	6.1	General	221
	6.2	K-Means Clustering	222
		P6.2.1 Batch K-Means Partitioning	222
		F6.2.2 Batch K-Means and Its Criterion: Formulation	229
		C6.2.3 A Pseudo-Code for Batch K-Means: Computation	235
		6.2.4 Incremental K-Means	237
		6.2.5 Nature Inspired Algorithms for K-Means	238
		6.2.6 Partition Around Medoids PAM	245
		6.2.7 Initialization of K-Means	246
		6.2.8 Anomalous Pattern and Intelligent K-Means	253
	6.3	Cluster Interpretation Aids	260
		P6.3.1 Cluster Interpretation Aids: Presentation	260
		F6.3.2 Cluster Interpretation Aids: Formulation	268
	6.4	Extension of K-Means to Different Cluster Structures	271
		6.4.1 Fuzzy K-Means Clustering	271
		6.4.2 Mixture of Distributions and EM Algorithm	275
		6.4.3 Kohonen's Self-Organizing Maps SOM	278
	6.5	Summary	280
	References		280
7	**Hierarchical Clustering**		283
	7.1	General	283
	7.2	Agglomerative Clustering and Ward's Criterion	285
		P7.2.1 Agglomerative Clustering: Presentation	285
		F7.2.2 Square-Error Criterion and Ward Distance: Formulation .	289

		C7.2.3 Agglomerative Clustering: Computation	291
	7.3	Divisive and Conceptual Clustering	292
		P7.3.1 Divisive Clustering: Presentation	292
		F7.3.2 Divisive and Conceptual Clustering: Formulation	299
		C7.3.3 Divisive and Conceptual Clustering: Computation	300
	7.4	Single Linkage Clustering, Connected Components and Maximum Spanning Tree	302
		P7.4.1 Maximum Spanning Tree and Clusters: Presentation	302
		F7.4.2 MST, Connected Components and Single Link Clustering: Formulation	309
		C7.4.3 Building a Maximum Spanning Tree: Computation	311
	7.5	Summary	312
	References		312
8	**Approximate and Spectral Clustering for Network and Affinity Data**		315
	8.1	One Cluster Summary Similarity with Background Subtracted	316
		P8.1.1 Summary Similarity and Two Types of Background: Presentation	316
		F8.1.2 One Cluster Summary Criterion and Its Properties: Formulation	324
		C8.1.3 Local Algorithms for One Cluster Summary Criterion: Computation	328
	8.2	Two Cluster Case: Cut, Normalized Cut and Spectral Clustering	329
		8.2.1 Minimum Cut and Spectral Clustering	329
		8.2.2 Normalized Cut and Laplace Transformation	334
	8.3	Additive Clusters	341
		P8.3.1 Decomposing a Similarity Matrix over Clusters: Presentation	341
		F8.3.2 Additive Clusters One-by-One: Formulation	348
		C8.3.3 Finding (Sub)Optimal Additive Clusters: Computation	353
	8.4	Summary	356
	References		356
Appendix			357
	A1	Basic Linear Algebra	357
		A1.1 Inner Product and Distance	357
		A1.2 Matrix Algebra	360
	A2	Basic Optimization	362
	A3	Basic MatLab	364
		A3.1 Introduction	364
		A3.2 Loading and Storing Files	365
		A3.3 Using Subsets of Entities and Features	368
	A4	MatLab Program Codes	370
		A4.1 Minkowski's Center: Evolutionary Algorithm	370

	A4.2	Fitting Power Law: Non-linear Evolutionary and	
		Linearization	372
	A4.3	Training Neuron Network with One Hidden Layer	378
	A4.4	Building Classification Trees	380
A5	Two Random Samples		383
	A5.1	Short.dat	383
	A5.2	A Sample of 280 N(0,10) Values	384
Index			387

List of Projects

2.1	Computing Minkowski metric's center	52
2.2	Analysis of a multimodal distribution	55
2.3	Computational validation of the mean by bootstrapping	57
2.4	K-fold cross validation	60
3.1	2D analysis, linear regression and bootstrapping	78
3.2	Non-linear and linearized regression: a nature-inspired algorithm	84
4.1	Prediction of learning outcome at Student data	153
5.1	Standardization of mixed scale data and its effect	182
6.1	Using contributions to determine the number of clusters	257
6.2	Does PCA clean the data structure indeed: K-Means after PCA	258
8.1	Analysis of structure of amino acid substitution rates	345

List of Case Study

3.1	Growth of Investment	86
3.2	Correlation Between Iris Sepal Length and Width	88
3.3	Trimming Contingency Data: A Bad Option	102
3.4	Has There Been a Bias in S'nS' Policy?	104
4.1	Prevalence and Quetelet coefficients	125
4.2	Linear regression for Market town data	128
4.3	Using feature weights standardized	130
6.1	Dependence of K-Means on Initialization: A Drawback and Advantage	226
6.2	Uniform Clusters Can Be Too Costly	228
6.3	Robustness of K-Means Criterion with Data Normalization	228
6.4	Hartigan's Index for Choosing the Number of Clusters	247
6.5	iK-Means Clustering of a Normally Distributed 1D Dataset	256
6.6	2D Analysis of Most Contributing Features	265
7.1	Divisive Clustering of Companies with Two-Splitting	294
7.2	Anomalous Cluster Versus Two-Split Cluster	295
7.3	Conceptual Clustering of Digit Data as Related to Ward Clustering	296
7.4	Difference Between K-Means and Single Link Clustering	307
8.1	Repeated One-Cluster Clustering with Repeated Removal of Background	319
8.2	Summary Clusters at Ordinary Network Data	321
8.3	Circular Cluster Exposed by Lapin Transformation	337

List of Worked Example

2.1	Mean	36
2.2	Median	36
2.3	P-quantile (percentile)	38
2.4	Mode	38
2.5	Entropy and Gini index of a distribution	46
3.1	Determination coefficient	70
3.2	Bootstrap validity testing	71
3.3	Prediction error of the regression equation	73
3.4	Tabular regression of Age (quantitative target) over Occupation (categorical predictor) in Students data	91
3.5	Box-plot of Age at Occupation categories at Students data	92
3.6	Correlation ratio	93
3.7	Nearest neighbor classifier	97
3.8	Category contributions for interval predicate productions	99
3.9	Contingency table on Market towns data	100
3.10	Equivalence and implication from a contingency table	101
3.11	Quetelet index in a contingency table	103
3.12	Visualization of contingency table using weighted Quetelet coefficients	105
3.13	A conventional decomposition of chi-square coefficient	106
4.1	A failure of Fisher discrimination criterion	134
4.2	SVM for Iris dataset	136
4.3	Classification tree for Iris dataset	152
4.4	Learning Iris petal sizes	164
4.5	Predicting marks at Student dataset	165
5.1	Standardizing Iris dataset	179
5.2	Explained proportion of data scatter in Equation (5.8)	191
5.3	Principal components after feature centering	193
5.4	Rescaling the talent score from Worked example 5.3	194
5.5	Visualization of a fragment of Students dataset	195
5.6	Interpretation of principal components at the standardized Student data	197
5.7	SVD for Six Students dataset	205

5.8	Standardized Student data visualized	205
5.9	Evaluation of the quality of visualization of the standardized Student data	206
5.10	Latent semantic space for article-to-term data	209
5.11	Drawing Figure 5.11	211
5.12	Correspondence analysis of Protocol/Attack contingency table	213
6.1	K-Means clustering of Company data	224
6.2	PAM applied to Company data	245
6.3	Selection of initial medoids in Company data	249
6.4	Anomalous pattern in Market towns	251
6.5	Iterated Anomalous patterns in Market towns	254
6.6	Centroids of Market town clusters	261
6.7	Representatives of Company clusters	262
6.8	Contributions of features to Market town clusters	263
6.9	Contributions and relative contributions of features at Company clusters	264
6.10	Describing Market town clusters conceptually	267
6.11	Describing Company clusters conceptually	267
7.1	Agglomerative clustering of Company dataset	285
7.2	Ward algorithm with distances only	287
7.3	Concept of MST	304
7.4	Building an MST on Confusion data	304
7.5	MST and connected components	305
7.6	Single link hierarchy corresponding to an MST	306
7.7	MST and single linkage clusters for Company dataset	308
8.1	Summary similarity clusters at a genuine similarity dataset	317
8.2	Similarity clusters at affinity data	323
8.3	Spectral clusters for Confusion dataset	330
8.4	Spectral clusters for Cockroach network	330
8.5	Spectral clustering of affinity data	332
8.6	Normalized cut for Company data: Laplacian and Lapin matrices	335
8.7	Failure of spectral clustering at Cockroach network	336
8.8	Additive clusters at Confusion dataset	344

Chapter 1
Introduction: What Is Core

1.1 Summarization and Correlation: Two Main Goals of Data Analysis

The term Data Analysis has been used for quite a while, even before the advent of computer era, as an extension of mathematical statistics, starting from developments in cluster analysis and other multivariate techniques before WWII and eventually bringing forth the concepts of "exploratory" data analysis and "confirmatory" data analysis in statistics (see, for example, Tukey 1977). The former was supposed to cover a set of techniques for finding patterns in data, and the latter to cover more conventional mathematical statistics approaches for hypothesis testing. "A possible definition of data analysis is the process of computing various summaries and derived values from the given collection of data" and, moreover, the process may become more intelligent if attempts are made to automate some of the reasoning of skilled data analysts and/or to utilize approaches developed in the Artificial Intelligence areas (Berthold and Hand 2003, p. 3). Overall, the term Data Analysis is usually applied as an umbrella to cover all the various activities mentioned above, with an emphasis on mathematical statistics and its extensions (see, for example, Lohninger 1999, Hair et al. 2010, Lebart et al. 1995).

The situation can be looked at as follows. Classical statistics takes the view of data as a vehicle to fit and test mathematical models of the phenomena the data refer to. The data mining and knowledge discovery discipline uses data to add new knowledge in any format. It should be sensible then to look at those methods that relate to an intermediate level and contribute to the theoretical – rather than any – knowledge of the phenomenon. These would focus on ways of augmenting or enhancing theoretical knowledge of the specific domain which the data being analyzed refer to. The term "knowledge" encompasses many a diverse layer or form of information, starting from individual facts to those of literary characters to major scientific laws. But when focusing on a particular domain the dataset in question comes from, its "theoretical" knowledge structure can be considered as comprised of just two types of elements: (i) concepts and (ii) statements relating them. *Concepts* are terms referring to aggregations of similar entities, such as apples or plums, or similar categories such as fruit comprising both apples and plums, among others. When created over data objects or features, these are referred to, in data analysis, as clusters or factors, respectively. *Statements of relation* between concepts express regularities relating different categories. Two features are said to correlate when a co-occurrence of specific patterns in their values is observed as, for instance, when a feature's value tends to be the square of the other feature. The observance of a correlation pattern can lead sometimes to investigation of a broader structure behind the pattern, which may further lead to finding or developing a theoretical framework for the phenomenon in question from which the correlation follows. It is useful to distinguish between quantitative correlations such as functional dependencies between features and categorical ones expressed conceptually, for example, as logical production rules or more complex structures such as decision trees. Correlations may be used for both understanding and prediction. In applications, the latter is by far more important. Moreover, the prediction problem is much easier to make sense of operationally so that the sciences so far have paid much attention to this.

What is said above suggests that there are two main pathways for augmenting knowledge: (i) developing new concepts by "summarizing" data and (ii) deriving new relations between concepts by analyzing "correlation" between various aspects of the data. The quotation marks are used here to point out that each of the terms, summarization and correlation, much extends its conventional meaning. Indeed, while everybody would agree that the average mark does summarize the marking scores on test papers, it would be more daring to see in the same light derivation of students' hidden talent scores by approximating their test marks on various subjects or finding a cluster of similarly performing students. Still, the mathematical structures behind each of these three activities – calculating the average, finding a hidden factor, and designing a cluster structure – are analogous, which suggests that classing them all under the "summarization" umbrella may be reasonable. Similarly, term "correlation" which is conventionally utilized in statistics to only express the extent of linear relationship between two or more variables, is understood here in its generic sense, as a supposed affinity between two or more aspects of the same data that can be variously expressed, not necessarily by a linear equation or by a quantitative expression at all.

It would be useful to spell out that view of the data as a subject of computational data analysis that is adhered to here. Typically, in sciences and in statistics, a problem comes first, and then the investigator turns to data that might be useful in advancing towards a solution. In computational data analysis, it may also be the case sometimes. Yet the situation is reversed frequently. Typical questions then would be: Take a look at this data set – what sense can be made out of it? – Is there any structure in the data set? Can these features help in predicting those? This is more reminiscent to a traveler's view of the world rather than that of a scientist. The scientist sits at his desk, gets reproducible signals from the universe and tries to accommodate them into the great model of the universe that the science has been developing. The traveler deals with what comes on their way. Helping the traveler in making sense of data is the task of data analysis. It should be pointed out that this view much differs from the conventional scientific method in which the main goal is to identify a pre-specified model of the world, and data is but a vehicle in achieving

1.1 Summarization and Correlation: Two Main Goals of Data Analysis

this goal. It is that view that underlies the development of data mining, though the aspect of data being available as a database, quite important in data mining, is rather tangential to data analysis.

Any data set comprises two parts, data and metadata entries. Data entries are the set of measurements taken, whereas metadata is a most straightforward relation between knowledge and measurements. Metadata usually involves names for the entities and features as well as indications of the measurement scales for the latter. Depending on the data domain, entities may be alternatively but synonymously referred to as individuals, objects, cases, instances, patterns, or observations. Data features may be synonymously referred to as variables, attributes, states, or characters. Depending on the way they are assigned to entities, the features can be of elementary structure [e.g., age, sex, or income of individuals] or complex structure [e.g., an image or a statement or a cardiogram]. Metadata may involve relations between entities and other relevant information.

The two fold goal clearly delineates the place of the data analysis core within the set of approaches involving various data analysis tasks. Here is a list of some popular approaches:

- *Classification* – this term applies to denote either a meta-scientific area of organizing the knowledge of a phenomenon into a set of separate classes to structure the phenomenon and relate different aspects of it to each other, or a discipline of supervised classification, that is, developing rules for assigning class labels to a set of entities under consideration. Data analysis can be utilized as a tool for designing the former, whereas the latter can be thought of as a problem in data analysis (see Duda et al. 2001, Vapnik 2006).
- *Cluster analysis* – is a discipline for obtaining (sets of) separate subsets of similar entities or features or both from the data, one of the most generic activities in data analysis (see Hartigan 1975, Murtagh 1985, Mirkin 2005).
- *Computational intelligence* – a discipline utilizing fuzzy sets, nature-inspired algorithms, neural nets and the like to computationally imitate human intelligence, which does overlap other areas of data analysis (see Engelbrecht 2002).
- *Data mining* – a discipline for finding interesting patterns in data stored in databases, which is considered part of the process of knowledge discovery. This has a significant overlap with computational data analysis. Yet data mining is structured somewhat differently by putting more emphasis on fast computations in large databases and finding "interesting" associations and patterns (see Han and Kamber 2006).
- *Document retrieval* – a discipline developing algorithms and criteria for query-based retrieval of as many relevant documents as possible, from a document base, which is similar to establishing a classification rule in data analysis. This area has become most popular with the development of search engines over the internet (see Manning et al. 2008).
- *Factor analysis* – a discipline emerged in psychology for modeling and finding hidden factors in data, which can be considered part of quantitative summarization in data analysis (see Hair et al. 2010).

- *Genetic algorithms* – an approach to globally search through the solution space in complex optimization problems by representing solutions as a population of "chromosomes" that evolves in iterations by mimicking micro-evolutionary events such as "cross-over" and "mutation". This can play a role in solving optimization problems in data analysis.
- *Knowledge discovery* – a set of techniques for deriving quantitative formulas and categorical productions to associate different features and feature sets, which hugely overlaps with the corresponding parts of data analysis (see Berthold and Hand 2003).
- *Mathematical statistics* – a discipline of data analysis based on the assumption of a probabilistic model underlying the data generation and/or decision making so that data or decision results are used for fitting or testing the models. This obviously has a lot to do with data analysis, including the idea that an adequate mathematical model is a finest knowledge format (see Kendall and Stewart 1973).
- *Machine learning* – a discipline in data analysis oriented at producing classification rules for predicting unknown class labels at entities usually arriving one by one in a random sequence (see Vapnik 2006, Schölkopf and Smola 2005, Mitchell 2005).
- *Neural networks* – a technique for modeling relations between (sets of) features utilizing structures of interconnected artificial neurons; the parameters of a neural network are learned from the data (Haykin 1999, Abdi et al. 1999).
- *Nature-inspired algorithms* – a set of contemporary techniques for optimization of complex functions such as the squared error of a data fitting model, using a population of admissible solutions evolving in iterations mimicking a natural process such as genetic recombination or ant colony or particle swarm search for foods (see Engelbrecht 2002).
- *Optimization* – a discipline for analyzing and solving problems in finding optima of a function such as the difference between observed values and those produced by a model whose parameters are being fitted (error) (see Polyak 1987).
- *Pattern recognition* – a discipline for deriving classification rules (supervised learning) and clusters (unsupervised learning) from observed data (see Duda et al. 2001, Webb 2002).
- *Social statistics* – a discipline for measuring social and economic indexes using observation or sampling techniques.
- *Text analysis* – a set of techniques and approaches for the analysis of unstructured text documents such as establishing similarity between texts, text categorization, deriving synopses and abstracts, etc (Weiss et al. 2005).

The text describes methods for enhancing knowledge by finding in data either

(a) Correlation among features (Cor) or
(b) Summarization of entities or features (Sum),

in either of two ways, quantitative (Q) or categorical (C). Combining these two bases makes four major groups of methods: CorQ, CorC, SumQ, and SumC that form the core of data analysis. It should be pointed out that currently different categorizations of tasks related to data analysis prevail: the classical mathematical

1.1 Summarization and Correlation: Two Main Goals of Data Analysis

statistics focuses mostly on mathematically treatable models (see, for example, Hair et al. 2010), whereas the system of machine learning and data mining expressed by the popular account by Duda et al. (2001) concentrates on the problem of learning categories of objects, thus leaving such important problems as quantitative summarization outside.

A correlation or summarization problem typically involves the following five ingredients:

- Stock of mathematical structures sought in data
- Computational model relating the data and the mathematical structure
- Criterion to score the match between the data and structure (fitting criterion)
- Method for optimizing the criterion
- Visualization of the results.

Here is a brief outline of those described in this text:

Mathematical structures:

– *linear* combination of features;
– *neural network* mapping a set of input features into a set of target features;
– *decision* tree built over a set of features;
– *cluster* of entities;
– *partition* of the entity set into a number of non-overlapping clusters.

When the type of mathematical structure to be used has been chosen, its parameters are to be learnt from the data.

A fitting method relies on a computational model involving a function *scoring* the adequacy of the mathematical structure underlying the rule – a criterion, and, usually, visualization aids. The data *visualization* is a way to represent the found structure to human eye. In this capacity, it is an indispensible part of the data analysis, which explains why this term is raised into the title. We briefly outline some aspects of visualization within the data analysis approach in Section 1.3.

The *criterion* measures either the deviation from the target (to be minimized) or goodness of fit to the target (to be maximized).

Currently available *computational methods* to optimize the criterion encompass three major groups:

– *global* optimization, that is, finding the best possible solution, computationally feasible sometimes for linear quantitative and simple discrete structures;
– *local* improvement using such general approaches as:

  - gradient ascent and descent
  - alternating optimization
  - greedy neighborhood search (hill climbing)

– *nature-inspired approaches* involving a population of admissible solutions and its iterative evolution, an approach involving relatively recent advancements in computing capabilities, of which the following will be used in some problems:

- genetic algorithms
- evolutionary algorithms
- particle swarm optimization

It should be pointed out that currently there is no systematic description of all possible combinations of problems, data types, mathematical structures, criteria, and fitting methods available. Here we rather focus on the generic and better explored problems in each of the four data analysis groups that can be safely claimed as being prototypical within the groups:

Sum	Quant	**Principal component analysis**
	Categ	**Cluster analysis**
Cor	Quant	**Regression analysis**
	Categ	**Supervised classification**

The four approaches on the right have emerged in different frameworks and usually are considered as unrelated. However, they are related in the context of data analysis. Moreover, they can be unified by the so-called data-driven modeling together with the least-squares criterion that will be adopted for all main methods described in this text. In fact, the criterion is part of a unifying data-recovery perspective that has been developed in mathematical statistics for fitting probabilistic models and then was extended to data analysis. In data analysis, this perspective is useful not only for supplying a nice fitting criterion but also because it involves the decomposition of the data scatter into "explained" and "unexplained" parts in all four methods. The data recovery approach takes in a type of mathematical structure to model the data and proceeds in three stages:

(1) fitting a model representing the structure to the data (this can be referred to as "coding"),
(2) deriving data from the model in the format of the data used to build the model (this can be referred to as "decoding"), and
(3) looking at the discrepancies between the observed data and those recovered from the model. The smaller are the discrepancies, the better the fit – this is a principle underlying the data-driven modeling approach.

Using the data recovery approach provided me with tools to develop and describe a number of innovative relations bringing together popular concepts conventionally considered as being worlds apart (Mirkin 1996, 2005). Among them:

(a) Reinterpretation and visualization of Pearson chi-square contingency coefficient as a summary association index rather than a statistical independence criterion;
(b) Use of anomalous patterns, an extension of principal component analysis to clustering, for both initializing K-Means and setting the number of clusters;
(c) A multitude of different reformulations of the square-error clustering criterion potentially leading to different clustering strategies;

1.1 Summarization and Correlation: Two Main Goals of Data Analysis

(d) Interrelation between association measures utilized for building decision trees and normalization of dummies representing categorical data, and
(e) A unified framework for network clustering including:
   (i) a number of combinatorial clustering criteria,
   (ii) spectral clustering, a recent very popular approach,
   (iii) additive clustering, a less popular yet powerful paradigm.

There can be distinguished at least three different levels of studying a computational data analysis method. A reader can be interested in learning of the approach on the level of concepts only – what a concept is for, why it should be applied at all, etc. A somewhat more practically oriented tackle would be of an information system/tool that can be utilized without any knowledge beyond the structure of its input and output. A more technically oriented way would be studying the method involved and its properties. Comparable advantages (pro) and disadvantages (contra) of these three levels can be stated as follows.

	Pro	Con
Concepts	Awareness	Superficial
Systems	Usable now	Short-term
	Simple	Stupid
Techniques	Workable	Technical
	Extendable	Boring

Many in Computer Sciences rely on the Systems approach assuming that good methods have been developed and put in there already. Although it is largely true for well defined mathematical problems, the situation is by far different in data analysis because there are no well posed problems here – basic formulations are intuitive and rarely supported by sound theoretical results. This is why, in many aspects, intelligence of currently popular "intelligent methods" may be rather superficial potentially leading to wrong results and decisions.

> Consider, for instance, a very popular concept, the power law – many say that in unconstrained social processes, such as those on the Web networks, this law, expressed with formula $y=ax^{-b}$ where x and y are some features and a and b are constant coefficients, dominates. Here are a few examples: the decay in the numbers of people who read a news story on the web over time; the distribution of page requests on a web-site according to their popularity; the distribution of website connections, etc. According to a very popular recipe, to fit a power law (that is, to estimate a and b from the data), one needs to fit the logarithm of the power-law equation, that is, $\log(y)=c-b*\log(x)$ where $c=\log(a)$, which is much easier to fit because it is linear. Therefore, this recipe advises: take logarithms of the x and y first and then use any popular linear regression program to find the constants. The recipe works well when the regularity is observed with no noise, which cannot be in real world social processes. With the real-world noise, this recipe may lead to big errors. For example, if x is generated between 0–10 and y is related to x by the power law $y=2*x^{1.07}$, which can be interpreted as the growth with the rate of approximately 7% per time unit, with an added Gaussian noise $N(0,2)$ of the zero mean and the standard deviation equal to 2, the recipe can lead to disastrous results. With the parameters above the linear transformation led to estimates of $a=3.08$ and $b=0.8$ to suggest that the process does not grow with x but rather decays. In contrast, when an evolutionary optimization method was applied to the original non-linear problem, the estimates were realistic: $a=2.03$ and $b=1.076$.

This is a relatively simple data analysis example, at which a correct procedure can be used. However, in more complex situations of clustering or categorization, the very idea of a correct method seems rather debatable; at least, methods in the existing systems can be of a rather poor quality.

One may compare the usage of an unsound data analysis method with that of getting services of an untrained medical doctor or car driver – the results can be as devastating. This is why it is important to study not only How's but What's and Why's, which are addressed in this course by focusing on Concepts and Techniques rather than Systems. Another, perhaps even more important, reason for studying concepts and techniques is the constant emergence of new data types (see, for example, recent books by Gama 2010, Mitsa 2010, Zhang and Zhang 2009), such as related to internet networks or medicine, that cannot be tackled by existing systems, yet the concepts and methods are readily extensible to cover them.

This text is oriented towards a student in Computer Sciences or related disciplines and reflects my experiences in teaching students of this type. Most of them prefer a hands-on rather than mathematical style of presentation. This is why almost all of the narrative is divided in three streams: presentation, formulation, and computation. The presentation states the problem and approach taken to tackle it, and it illustrates the solution at some data. The formulation provides a mathematical description of the problem as well as a method or two to solve it. The computation shows how to do that computationally with basic MatLab. Each of the streams can be read independently. In this way, the reader can choose the way of using the book and adjust it to their individual style.

This three-way narrative corresponds to the three typical roles in a successful work team in engineering. One role is of general grasp of things, a visionary. Another role is of a designer who translates the general picture into a technically sound project. Yet one more role is needed to implement the project into a product. The reader can choose either role or combine two or all three of them, even if having preferences for a specific type of narrative.

To help the reader to study the material actively, the text is interlaced with problems along with their solutions. Many of the problems are put as "worked examples" to show how a specific method applies to a specific dataset. More complex problems, "case studies", may involve a rule for data generation rather than a pre-specified data set or an informed way for looking at the results. Yet more complex problems may involve uncharted terrain and an investigation, however small, – these are referred to as "projects".

There is a bias in the volumes of material devoted to correlation and summarization subjects – the latter prevails rather considerably. This can be explained by both personal and objective reasons. The personal reason is that my main research area lies in clustering, that is, summarization. The objective reason is that the correlation problems, and their theoretical underpinnings, have been already subjects of a multitude of monographs and texts in statistics, data analysis, machine learning, data mining, and computational intelligence. In contrast, neither clustering nor principal component analysis – the main constituents of summarization efforts – has received a proper theoretical foundation; in the available books both are treated as heuristics, however useful. This text presents these two as based on a model of data, which raises a number of issues that are addressed here, including that of the theoretical structure of a summarization problem. The concept of coder-decoder is borrowed from the data processing area to draw a theoretical framework in which summarization is considered as a pair of coding/decoding activities so that the quality of the coding part is evaluated by the quality of decoding. Luckily, the theory of singular value decomposition of matrices (SVD) can be safely utilized as a framework for explaining the principal component analysis, and extension of the SVD equations to binary scoring vectors provides a base for K-Means clustering and the like. This raises an important question of mathematical proficiency the reader should have as a prerequisite. There is no prerequisite for reading through the presentation and computation parts. Yet an assumed background of the reader interested in studying formulation parts should include: (a) basics of calculus including the concepts of function, derivative and the first-order optimality condition; (b) basic linear algebra including vectors, inner products, Euclidean distances and matrices (these are reviewed in the Appendix), and (c) basic set theory notation such as symbols for relations of inclusion and membership.

1.2 Case Study Problems

To be more specific, the presentation is illustrated using a number of small datasets – the sizes allow the reader to see the data by a naked eye, which is always a good idea to do before engaging into the analysis. The datasets and problems are selected in such a way that methods further described could be immediately illustrated by using a relevant dataset from the collection.

Case 1.2.1: Company

There are eight companies and five features in Table 1.1.:

(1) Income, $ Mln;
(2) SharP – share price, $;
(3) NSup – the number of principal suppliers;
(4) EC – Yes or No depending on the usage of e-commerce in the company;
(5) Sector – which sector of the economy: (a) Retail, (b) Utility, and (c) Industrial.

These features should not be taken too seriously – they are purely illustrative. For example, term "share price" can be easily changed for "market cap". Examples of computational data analysis problems related to this data set:

– How to map companies to the screen with their similarity reflected in distances on the plane? (Summarization)
– Would clustering of companies reflect the product? What features would be involved then? (Summarization)

Table 1.1 Company: A set of eight companies characterized by mixed scale features. The company names reflect the fact not present in the data – product affinities: first three companies mostly adhere to product group A, the next three to product group B, and the last two to product group C

Company name	Income, $mln	SharP $	NSup	EC	Sector
Aversi	19.0	43.7	2	No	Utility
Antyos	29.4	36.0	3	No	Utility
Astonite	23.9	38.0	3	No	Industrial
Bayermart	18.4	27.9	2	Yes	Utility
Breaktops	25.7	22.3	3	Yes	Industrial
Bumchist	12.1	16.9	2	Yes	Industrial
Civok	23.9	30.2	4	Yes	Retail
Cyberdam	27.2	58.0	5	Yes	Retail

– Can rules be derived to make an attribution of the product for another company, coming outside of the table? (Correlation)
– Is there any relation between the structural features, such as NSup, and market related features, such as Income? (Correlation)

Q.1.1. Is the following statement true? "There is no information on the company products within the table". **A.** Yes: no "Product" feature is present in the table; the separating lines are not part of the data.

An issue related to Table 1.1 is that not all of its entries are quantitative. Specifically, there are three conventional types of features in it:

– Quantitative, that is, such that the averaging of its values is considered meaningful. In the Table 1.1, these are: Income, ShareP and NSup;
– Binary, that is, admitting one of two answers, Yes or No: this is EC;
– Nominal, that is, with a few disjoint not ordered categories, such as Sector in Table 1.1.

Most models and methods presented in this text relate to quantitative data formats only – which does not mean that categorical data are left on their own, just the opposite. The two non-quantitative feature types, binary and nominal, can be pre-processed into a quantitative format too – which is the subject treated at length in Sections 2.3, 4.5 and 7.3, among others.

A binary feature can be recoded into 1/0 format by substituting 1 for "Yes" and 0 for "No". In the author's, rather unconventional, view the recoded feature can be considered quantitative, because its averaging is meaningful: the average value is equal to the proportion of unities, that is, the frequency of "Yes" in the original feature.

A nominal feature is first enveloped into a set of binary "Yes"/"No" features corresponding to individual categories. In Table 1.1, binary features yielded by categories of feature "Sector" are:

1.2 Case Study Problems

Is it Retail?
Is it Utility?
Is it Industrial?

They are put as questions to make "Yes" or "No" as answers to them. These binary features now can be converted to the quantitative format advised above, by recoding 1 for "Yes" and 0 for "No". The 1/0 version is frequently referred to as a dummy.

These are applied to convert the Company data into a quantitative format (see Table 1.2).

Table 1.2 Company data from Table 1.1 converted to the quantitative format

Code	Income	SharP	NSup	EC	Util	Indu	Retail
1	19.0	43.7	2	0	1	0	0
2	29.4	36.0	3	0	1	0	0
3	23.9	38.0	3	0	0	1	0
4	18.4	27.9	2	1	1	0	0
5	25.7	22.3	3	1	0	1	0
6	12.1	16.9	2	1	0	1	0
7	23.9	30.2	4	1	0	0	1
8	27.2	58.0	5	1	0	0	1

Case 1.2.2: Iris

Iris is a popular dataset collected by botanist E. Anderson and presented by R. Fisher in his founding paper on discriminant analysis (1936), see Table 1.3. It presents 150 Iris specimens, representing three taxa of Iris flowers, I *Iris setosa* (diploid), II *Iris versicolor* (tetraploid) and III *Iris virginica* (hexaploid), 50 specimens from each. Each specimen is measured over four morphological variables: sepal length (w1), sepal width (w2), petal length (w3), and petal width (w4) (see Fig. 1.1).

The taxa are defined by the genotype whereas the features are of the appearance (phenotype). The question arises whether the taxa can be described, and indeed predicted, in terms of the features or not. It is well known from previous studies that taxa II and III are not well separated in the variable space. Some non-linear machine learning techniques such as Neural Nets (Haykin 1999 and Section 4.6 further on) can tackle the problem and produce a decent decision rule involving non-linear transformation of the features. Unfortunately, rules derived with Neural Nets are typically not comprehensible to the human. The human mind needs a somewhat less artificial logic that is capable of reproducing and extending botanists' observations such as that the petal area, roughly expressed by the product of w3 and w4, provides for much better resolution than the original linear sizes. Other problems that are of interest: (a) visualize the data; (b) build a predictor of sepal sizes from the petal sizes.

Table 1.3 *Iris* data: 150 Iris specimens measured over four features each

	I Iris setosa				II Iris versicolor				III Iris virginica			
#	w1	w2	w3	w4	w1	w2	w3	w4	w1	w2	w3	w4
1	5.1	3.5	1.4	0.3	6.4	3.2	4.5	1.5	6.3	3.3	6.0	2.5
2	4.4	3.2	1.3	0.2	5.5	2.4	3.8	1.1	6.7	3.3	5.7	2.1
3	4.4	3.0	1.3	0.2	5.7	2.9	4.2	1.3	7.2	3.6	6.1	2.5
4	5.0	3.5	1.6	0.6	5.7	3.0	4.2	1.2	7.7	3.8	6.7	2.2
5	5.1	3.8	1.6	0.2	5.6	2.9	3.6	1.3	7.2	3.0	5.8	1.6
6	4.9	3.1	1.5	0.2	7.0	3.2	4.7	1.4	7.4	2.8	6.1	1.9
7	5.0	3.2	1.2	0.2	6.8	2.8	4.8	1.4	7.6	3.0	6.6	2.1
8	4.6	3.2	1.4	0.2	6.1	2.8	4.7	1.2	7.7	2.8	6.7	2.0
9	5.0	3.3	1.4	0.2	4.9	2.4	3.3	1.0	6.2	3.4	5.4	2.3
10	4.8	3.4	1.9	0.2	5.8	2.7	3.9	1.2	7.7	3.0	6.1	2.3
11	4.8	3.0	1.4	0.1	5.8	2.6	4.0	1.2	6.8	3.0	5.5	2.1
12	5.0	3.5	1.3	0.3	5.5	2.4	3.7	1.0	6.4	2.7	5.3	1.9
13	5.1	3.3	1.7	0.5	6.7	3.0	5.0	1.7	5.7	2.5	5.0	2.0
14	5.0	3.4	1.5	0.2	5.7	2.8	4.1	1.3	6.9	3.1	5.1	2.3
15	5.1	3.8	1.9	0.4	6.7	3.1	4.4	1.4	5.9	3.0	5.1	1.8
16	4.9	3.0	1.4	0.2	5.5	2.3	4.0	1.3	6.3	3.4	5.6	2.4
17	5.3	3.7	1.5	0.2	5.1	2.5	3.0	1.1	5.8	2.7	5.1	1.9
18	4.3	3.0	1.1	0.1	6.6	2.9	4.6	1.3	6.3	2.7	4.9	1.8
19	5.5	3.5	1.3	0.2	5.0	2.3	3.3	1.0	6.0	3.0	4.8	1.8
20	4.8	3.4	1.6	0.2	6.9	3.1	4.9	1.5	7.2	3.2	6.0	1.8
21	5.2	3.4	1.4	0.2	5.0	2.0	3.5	1.0	6.2	2.8	4.8	1.8
22	4.8	3.1	1.6	0.2	5.6	3.0	4.5	1.5	6.9	3.1	5.4	2.1
23	4.9	3.6	1.4	0.1	5.6	3.0	4.1	1.3	6.7	3.1	5.6	2.4
24	4.6	3.1	1.5	0.2	5.8	2.7	4.1	1.0	6.4	3.1	5.5	1.8
25	5.7	4.4	1.5	0.4	6.3	2.3	4.4	1.3	5.8	2.7	5.1	1.9
26	5.7	3.8	1.7	0.3	6.1	3.0	4.6	1.4	6.1	3.0	4.9	1.8
27	4.8	3.0	1.4	0.3	5.9	3.0	4.2	1.5	6.0	2.2	5.0	1.5
28	5.2	4.1	1.5	0.1	6.0	2.7	5.1	1.6	6.4	3.2	5.3	2.3
29	4.7	3.2	1.6	0.2	5.6	2.5	3.9	1.1	5.8	2.8	5.1	2.4
30	4.5	2.3	1.3	0.3	6.7	3.1	4.7	1.5	6.9	3.2	5.7	2.3
31	5.4	3.4	1.7	0.2	6.2	2.2	4.5	1.5	6.7	3.0	5.2	2.3
32	5.0	3.0	1.6	0.2	5.9	3.2	4.8	1.8	7.7	2.6	6.9	2.3
33	4.6	3.4	1.4	0.3	6.3	2.5	4.9	1.5	6.3	2.8	5.1	1.5
34	5.4	3.9	1.3	0.4	6.0	2.9	4.5	1.5	6.5	3.0	5.2	2.0
35	5.0	3.6	1.4	0.2	5.6	2.7	4.2	1.3	7.9	3.8	6.4	2.0
36	5.4	3.9	1.7	0.4	6.2	2.9	4.3	1.3	6.1	2.6	5.6	1.4
37	4.6	3.6	1.0	0.2	6.0	3.4	4.5	1.6	6.4	2.8	5.6	2.1
38	5.1	3.8	1.5	0.3	6.5	2.8	4.6	1.5	6.3	2.5	5.0	1.9
39	5.8	4.0	1.2	0.2	5.7	2.8	4.5	1.3	4.9	2.5	4.5	1.7
40	5.4	3.7	1.5	0.2	6.1	2.9	4.7	1.4	6.8	3.2	5.9	2.3
41	5.0	3.4	1.6	0.4	5.5	2.5	4.0	1.3	7.1	3.0	5.9	2.1
42	5.4	3.4	1.5	0.4	5.5	2.6	4.4	1.2	6.7	3.3	5.7	2.5
43	5.1	3.7	1.5	0.4	5.4	3.0	4.5	1.5	6.3	2.9	5.6	1.8
44	4.4	2.9	1.4	0.2	6.3	3.3	4.7	1.6	6.5	3.0	5.5	1.8
45	5.5	4.2	1.4	0.2	5.2	2.7	3.9	1.4	6.5	3.0	5.8	2.2
46	5.1	3.4	1.5	0.2	6.4	2.9	4.3	1.3	7.3	2.9	6.3	1.8
47	4.7	3.2	1.3	0.2	6.6	3.0	4.4	1.4	6.7	2.5	5.8	1.8
48	4.9	3.1	1.5	0.1	5.7	2.6	3.5	1.0	5.6	2.8	4.9	2.0
49	5.2	3.5	1.5	0.2	6.1	2.8	4.0	1.3	6.4	2.8	5.6	2.2
50	5.1	3.5	1.4	0.2	6.0	2.2	4.0	1.0	6.5	3.2	5.1	2.0

Fig. 1.1 Sepal and petal in an Iris flower

Case 1.2.3: Market Towns

In Table 1.4 a set of Market towns in West Country, England is presented along with features characterizing population and social infrastructure according to census 1991. For the purposes of social planning, it would be good to monitor a smaller number of towns, each representing a cluster of similar towns. In the table, the towns are sorted according to their population size. One can see that 21 towns have less than 4,000 residents. The value 4,000 is taken as a divider since it is round and, more importantly, there is a gap of more than thirteen hundred residents between Kingskerswell (3,672 inhabitants) and next in the list Looe (5,022 inhabitants). Next big gap occurs after Liskeard (7,044 inhabitants) separating the nine middle sized towns from two larger town groups containing six and nine towns respectively. The divider between the latter groups is taken between Tavistock (10,222) and Bodmin (12,553). In this way, we get three or four groups of towns for the purposes of social monitoring. Is this enough, regarding the other features available? Are the groups, defined in terms of population size only, homogeneous enough for the purposes of monitoring?

As further computations will show, the numbers of services on average do follow the town sizes, but this set (as well as the complete set of about thirteen hundred England Market towns) is much better represented with seven somewhat different clusters: large towns of about 17–20,000 inhabitants, two clusters of medium sized towns (8–10,000 inhabitants), three clusters of small towns (about 5,000 inhabitants), and a cluster of very small settlements with about 2,500 inhabitants. Each of the three small town clusters is characterized by the presence of a facility, which is absent in two others: a Farm market or a Hospital or a Swimming pool, respectively.

One may suggest that the only difference between these seven clusters and the grouping over the town resident numbers would be just difference in the dividing points, but both are expressed in terms of the population size only. However, one should not forget that the number of residents for the seven clusters is a posterior selection – because of our knowledge of the clusters not prior to that.

Table 1.4 Data of West Country England Market Towns 1991

Town	Pop	PS	D	Hos	Ba	Sst	Pet	DIY	Swi	Po	CAB	FM
Mullion	2040	1	0	0	2	0	1	0	0	1	0	0
So Brent	2087	1	1	0	1	1	0	0	0	1	0	0
St Just	2092	1	0	0	2	1	1	0	0	1	0	0
St Columb	2119	1	0	0	2	1	1	0	0	1	1	0
Nanpean	2230	2	1	0	0	0	0	0	0	2	0	0
Gunnislake	2236	2	1	0	1	0	1	0	0	3	0	0
Mevagissey	2272	1	1	0	1	0	0	0	0	1	0	0
Ipplepen	2275	1	1	0	0	0	1	0	0	1	0	0
Be Alston	2362	1	0	0	1	1	0	0	0	1	0	0
Lostwithiel	2452	2	1	0	2	0	1	0	0	1	0	1
St Columb	2458	1	0	0	0	1	3	0	0	2	0	0
Padstow	2460	1	0	0	3	0	0	0	0	1	1	0
Perranporth	2611	1	1	0	1	1	2	0	0	2	0	0
Bugle	2695	2	0	0	0	0	1	0	0	2	0	0
Buckfastle	2786	2	1	0	1	2	2	0	1	1	1	1
St Agnes	2899	1	1	0	2	1	1	0	0	2	0	0
Porthleven	3123	1	0	0	1	1	0	0	0	1	0	0
Callington	3511	1	1	0	3	1	1	0	1	1	0	0
Horrabridge	3609	1	1	0	2	1	1	0	0	2	0	0
Ashburton	3660	1	0	1	2	1	2	0	1	1	1	0
Kingskers	3672	1	0	0	0	1	2	0	0	1	0	0
Looe	5022	1	1	0	2	1	1	0	1	3	1	0
Kingsbridge	5258	2	1	1	7	1	2	0	0	1	1	1
Wadebridge	5291	1	1	0	5	3	1	0	1	1	1	0
Dartmouth	5676	2	0	0	4	4	1	0	0	2	1	1
Launceston	6466	4	1	0	8	4	4	0	1	3	1	0
Totnes	6929	2	1	1	7	2	1	0	1	4	0	1
Penryn	7027	3	1	0	2	4	1	0	0	3	1	0
Hayle	7034	4	0	1	2	2	2	0	0	2	1	0
Liskeard	7044	2	2	2	6	2	3	0	1	2	2	0
Torpoint	8238	2	3	0	3	2	1	0	0	2	1	0
Helston	8505	3	1	1	7	2	3	0	1	1	1	1
St Blazey	8837	5	2	0	1	1	4	0	0	4	0	0
Ivybridge	9179	5	1	0	3	1	4	0	0	1	1	0
St Ives	10092	4	3	0	7	2	2	0	0	4	1	0
Tavistock	10222	5	3	1	7	3	3	1	2	3	1	1
Bodmin	12553	5	2	1	6	3	5	1	1	2	1	0
Saltash	14139	4	2	1	4	2	3	1	1	3	1	0
Brixham	15865	7	3	1	5	5	3	0	2	5	1	0
Newquay	17390	4	4	1	12	5	4	0	1	5	1	0
Truro	18966	9	3	1	19	4	5	2	2	7	1	1
Penzance	19709	10	4	1	12	7	5	1	1	7	2	0
Falmouth	20297	6	4	1	11	3	2	0	1	9	1	0
St Austell	21622	7	4	2	14	6	4	3	1	8	1	1
Newton Abb	23801	13	4	1	13	4	7	1	1	7	2	0

1.2   Case Study Problems

The data in Table 1.4 involve the counts of the following 12 features surveyed in the census 1991:

Pop	– Population resident	Pet	– Petrol stations
PS	– Primary schools	DIY	– Do It Yourself shops
D	– General Practitioners	Swi	– Swimming pools
Hos	– Hospitals	Po	– Post offices
Ba	– Banks	CAB	– Citizen Advice Bureaus
Sst	– Superstores	FM	– Farmer markets

Case 1.2.4: Student

In Table 1.5, a fictitious dataset is presented as imitating a typical set up for a group of Birkbeck University of London part-time students pursuing Master's degree in Computer Sciences.

This dataset refers to a hundred students along with six features, three of which are personal characteristics (1. Occupation (Oc): either Information technology (IT) or Business Administration (BA) or anything else (AN); 2. Age, in years; 3. Number of children (Ch)) and three are their marks over courses in 4. Software and Programming (SE), 5. Object-Oriented Programming (OO), and 6. Computational Intelligence (CI).

Related questions are:

– Whether the students' marks are affected by the personal features;
– Are there any patterns in marks, especially in relation to occupation?

Case 1.2.5: Intrusion

With the growing range and scope of computer networks, their security becomes an urgent issue. An attack on a network results in its malfunctioning. The simplest kind of attack is a denial of service (DoS). A DoS is caused by an intruder who makes some resource – in computing, memory, or I/O such as network or disk – too busy or too full to handle legitimate requests. Two of the DoS attacks are denoted as "apache2" and "smurf" in the data. Other types of attack include user-to-root attacks and remote-to-local attacks.

The "apache2" attack targets a popular open source web server, the Apache HTTP Server, and results in denying services to a client by sending a request with a large number of http headers, triggering buffer overflow vulnerabilities. A "smurf" attack works by sending forged ICMP echo messages to a host. An ICMP echo, also known as ping, is a message to a computer attached to an IP network. On receipt of this message, the receiving computer will respond with an ICMP echo reply back to the computer that sent the echo, as determined by the source IP address of the echo request. However, there are techniques that can be used to forge source IP

Table 1.5 *Student* data in two columns

Oc	Age	Ch	SE	OO	CI	Oc	Age	Ch	SE	OO	CI
IT	28	0	41	66	90	BA	51	2	75	73	57
IT	35	0	57	56	60	BA	44	3	53	43	60
IT	25	0	61	72	79	BA	49	3	86	39	62
IT	29	1	69	73	72	BA	27	2	93	58	62
IT	39	0	63	52	88	BA	30	1	75	74	70
IT	34	0	62	83	80	BA	47	0	46	36	36
IT	24	0	53	86	60	BA	38	2	86	70	47
IT	37	1	59	65	69	BA	49	1	76	36	66
IT	33	1	64	64	58	BA	45	0	80	56	47
IT	23	1	43	85	90	BA	44	2	50	43	72
IT	24	1	68	89	65	BA	36	3	66	64	62
IT	32	0	67	98	53	BA	31	2	64	45	38
IT	33	0	58	74	81	BA	31	3	53	72	38
IT	27	1	48	94	87	BA	32	3	87	40	35
IT	32	1	66	73	62	BA	38	0	87	56	44
IT	29	0	55	90	61	BA	48	1	68	71	56
IT	21	0	62	91	88	BA	39	2	93	73	53
IT	21	0	53	59	56	BA	47	1	52	48	63
IT	26	1	69	70	89	BA	39	2	88	52	58
IT	20	1	42	76	79	AN	23	0	54	50	41
IT	28	1	57	85	85	AN	34	0	46	33	25
IT	34	1	49	78	59	AN	33	0	51	38	51
IT	22	0	66	73	69	AN	31	0	59	45	35
IT	21	1	50	72	54	AN	25	0	51	41	53
IT	32	1	60	55	85	AN	40	0	41	61	22
IT	32	0	42	72	73	AN	41	0	44	43	44
IT	20	1	51	69	64	AN	42	0	40	56	58
IT	20	1	55	66	66	AN	34	0	47	69	32
IT	24	1	53	92	86	AN	37	0	45	50	56
IT	32	0	57	87	66	AN	24	0	47	68	24
IT	21	1	58	97	54	AN	34	0	50	63	23
IT	27	1	43	78	59	AN	41	0	37	67	29
IT	33	0	67	52	53	AN	47	1	43	35	57
IT	34	1	63	80	74	AN	28	0	50	62	23
IT	34	0	64	90	56	AN	28	0	39	66	31
BA	36	2	86	54	68	AN	46	0	51	36	60
BA	35	2	79	72	60	AN	27	0	41	35	28
BA	36	1	55	44	57	AN	44	0	50	61	40
BA	37	1	59	69	45	AN	47	0	48	59	32
BA	42	2	76	61	68	AN	27	0	47	56	47
BA	30	3	72	71	46	AN	27	0	49	60	58
BA	28	1	48	55	65	AN	21	0	59	57	51
BA	38	1	49	75	61	AN	22	0	44	65	47
BA	49	2	59	50	44	AN	39	0	45	41	25
BA	50	2	65	56	59	AN	26	0	43	47	24
BA	34	2	69	42	59	AN	45	1	45	39	21
BA	31	2	90	55	61	AN	25	0	42	31	32
BA	49	3	75	52	42	AN	25	0	45	33	53
BA	33	1	61	61	60	AN	50	1	48	64	59
BA	43	0	69	62	42	AN	33	0	53	44	21

1.2 Case Study Problems	17

addresses, in which case the echo reply will go to the forged source. Further, it is possible to ping multiple machines by sending an echo request to a network broadcast address. This causes every machine on that IP broadcast domain to respond to the source address (which, in the attack case, is forged). In such a way, it is possible to cause a very large number of computers across the Internet to send echo replies to a single target host, overwhelming that host's processing or network connection. A probe looking for flaws might precede an attack. One powerful probe software is SAINT – the Security Administrator's Integrated Network Tool, that uses a thorough deterministic protocol to scan various network services for possible problems. Intrusion detection systems collect information of anomalies and other patterns of communication such as compromised user accounts and unusual login behavior. The data set Intrusion in Table 1.6 consists of a hundred communication packages along with some of their features sampled from a set of artificially created data publicly available on webpage of MIT Lincoln Laboratory http://www.ll.mit.edu/mission/communications/ist/corpora/ideval/data/intex.html. Although the value of the data as a source to analyze the attacks is debatable, it does reflect the structure of the problem. The features reflect the packet as well as activities of its source:

1 – Pr, the protocol-type, which is either tcp or icmp or udp (a nominal feature),
2 – BySD, the number of data bytes from source to destination,
3 – SH, the number of connections to the same host as the current one in the past 2 s,
4 – SS, the number of connections to the same service as the current one in the past 2 s,
5 – SE, the rate of connections (per cent in SHCo) that have SYN errors,
6 – RE, the rate of connections (per cent in SHCo) that have REJ errors,
7 – A, the type of attack (ap - apache, sa - saint, sm - smurf, and no attack).

Of the hundred entities in the set, the first 23 are classified as attacking the apache2 server, the 24–69 packets are normal, eleven entities 80–90 are consistent with a SAINT probe, and the last ten, 91–100, appear to be smurf attacks.

These are examples of problems arising in relation to the Intrusion data:

– identify features to judge whether the system functions normally or is it under attack (Correlation);
– is there any relation between the protocol and type of attack (Correlation);
– how to visualize the data reflecting similarity of the patterns (Summarization).

Case 1.2.6: Confusion

Table 1.7 presents results of an experiment on errors in human judgement, specifically, on confusion of human operators between segmented numerals (drawn on Fig. 1.2). In the experiment, a digit flashes for a short time on screen before an individual (stimulus) who is to report then what digit they have seen (response): (i,j)-the

Table 1.6 Intrusion data

Pr	BySD	SH	SS	SE	RE	A	Pr	ByS	SH	SS	Se	RE	A
tcp	62344	16	16	0	0.94	Ap	Tcp	287	14	14	0	0	No
Tcp	60884	17	17	0.06	0.88	Ap	Tcp	308	1	1	0	0	No
Tcp	59424	18	18	0.06	0.89	Ap	Tcp	284	5	5	0	0	No
Tcp	59424	19	19	0.05	0.89	Ap	Udp	105	2	2	0	0	No
Tcp	59424	20	20	0.05	0.9	Ap	Udp	105	2	2	0	0	No
Tcp	75484	21	21	0.05	0.9	Ap	Udp	105	2	2	0	0	No
Tcp	76944	22	22	0.05	0.91	Ap	Udp	105	2	2	0	0	No
Tcp	59424	23	23	0.04	0.91	Ap	Udp	105	2	2	0	0	No
Tcp	57964	24	24	0.04	0.92	Ap	Udp	44	3	8	0	0	no
Tcp	59424	25	25	0.04	0.92	Ap	Udp	44	6	11	0	0	no
Tcp	0	40	40	1	0	Ap	Udp	42	5	8	0	0	no
Tcp	0	41	41	1	0	Ap	Udp	105	2	2	0	0	no
Tcp	0	42	42	1	0	Ap	Udp	105	2	2	0	0	no
Tcp	0	43	43	1	0	Ap	Udp	42	2	3	0	0	no
Tcp	0	44	44	1	0	Ap	Udp	105	1	1	0	0	no
Tcp	0	45	45	1	0	Ap	Udp	105	1	1	0	0	no
Tcp	0	46	46	1	0	Ap	Udp	44	2	4	0	0	no
Tcp	0	47	47	1	0	Ap	Udp	105	1	1	0	0	no
Tcp	0	48	48	1	0	Ap	Udp	105	1	1	0	0	no
Tcp	0	49	49	1	0	Ap	Udp	44	3	14	0	0	no
Tcp	0	40	40	0.62	0.35	Ap	Udp	105	1	1	0	0	no
Tcp	0	41	41	0.63	0.34	Ap	Udp	105	1	1	0	0	no
Tcp	0	42	42	0.64	0.33	Ap	Udp	45	3	6	0	0	no
Tcp	258	5	5	0	0	No	Udp	45	3	6	0	0	no
Tcp	316	13	14	0	0	No	Udp	105	1	1	0	0	no
Tcp	287	7	7	0	0	No	Udp	34	5	9	0	0	no
Tcp	380	3	3	0	0	No	Udp	105	1	1	0	0	no
Tcp	298	2	2	0	0	No	Udp	105	1	1	0	0	no
Tcp	285	10	10	0	0	No	Udp	105	1	1	0	0	no
Tcp	284	20	20	0	0	No	Tcp	0	482	1	0.05	.95	sa
Tcp	314	8	8	0	0	No	Tcp	0	482	1	0.05	.95	sa
Tcp	303	18	18	0	0	No	Tcp	0	482	1	0.05	.95	sa
Tcp	325	28	28	0	0	No	Tcp	0	482	1	0.05	.95	sa
Tcp	232	1	1	0	0	No	Tcp	0	482	1	0.05	.95	sa
Tcp	295	4	4	0	0	No	Tcp	0	482	1	0.05	.95	sa
Tcp	293	13	14	0	0	No	Tcp	0	482	1	0.06	.94	sa
Tcp	305	1	8	0	0	No	Tcp	0	482	1	0.06	.94	sa
Tcp	348	4	4	0	0	No	Tcp	0	482	1	0.06	.94	sa
Tcp	309	6	6	0	0	No	Tcp	0	483	1	0.06	.94	sa
Tcp	293	8	8	0	0	No	Tcp	0	510	1	0.04	.96	sa
Tcp	277	1	8	0	0	no	Icmp	1032	509	509	0	0	sm
Tcp	296	13	14	0	0	no	Icmp	1032	510	510	0	0	sm
Tcp	286	3	6	0	0	no	Icmp	1032	510	510	0	0	sm
Tcp	311	5	5	0	0	no	Icmp	1032	511	511	0	0	sm
Tcp	305	9	15	0	0	no	Icmp	1032	511	511	0	0	sm
Tcp	295	11	25	0	0	no	Icmp	1032	494	494	0	0	sm
Tcp	511	1	4	0	0	no	Icmp	1032	509	509	0	0	sm
Tcp	239	12	14	0	0	no	Icmp	1032	509	509	0	0	sm
Tcp	5	1	1	0	0	no	Icmp	1032	510	510	0	0	sm
Tcp	288	4	4	0	0	no	Icmp	1032	511	511	0	0	sm

1.2 Case Study Problems

Table 1.7 Confusion data: the entries characterize the numbers of those of the participants of a psychological experiment who took the stimulus (row digit) for the response (column digit)

	Response									
St	1	2	3	4	5	6	7	8	9	0
1	877	7	7	22	4	15	60	0	4	4
2	14	782	47	4	36	47	14	29	7	18
3	29	29	681	7	18	0	40	29	152	15
4	149	22	4	732	4	11	30	7	41	0
5	14	26	43	14	669	79	7	7	126	14
6	25	14	7	11	97	633	4	155	11	43
7	269	4	21	21	7	0	667	0	4	7
8	11	28	28	18	18	70	11	577	67	172
9	25	29	111	46	82	11	21	82	550	43
0	18	4	7	11	7	18	25	71	21	818

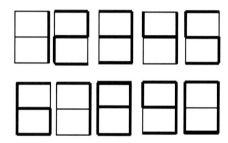

Fig. 1.2 Simplified digit numerals over a rectangle with a line in the middle

entry in Table 1.7 is the proportion of response j to stimulus i (Keren and Baggen 1981). The confusion matrix is understandably not symmetric, whereas its diagonal entries contain by far the larger proportions of observations, which is typical for confusion data as well as switch data.

The problem: are there any patterns of confusion, especially if represented by clusters? If yes, can any numeral shape features be found to describe the confusion clusters more or less exclusively?

Case 1.2.7: Amino Acid Substitution Rates

Table 1.8 is a symmetric table of the so-called amino acid substitution scores that are used mainly as weight coefficients at various schemes for alignment of protein amino acid sequences. Table 1.9 lists amino acid names and codes. A protein amino acid sequence represents the protein prime structure that may change during the process of evolution. The main assumption for studying the evolution is that each two organisms share a common ancestry. The more similar their protein sequences are the more recent was their common ancestor. The likelihood of the event of amino acid i substituted by amino acid j is estimated by using blocks of evolutionarily related protein sequences from various databases. These allow

Table 1.8 Amino acid substitution rates: **BLOSUM62** matrix of substitution scores between amino acids presented using 1-letter code (see Table 1.9 for decoding)

Aa	A	B	C	D	E	F	G	H	I	K	L	M	N	P	Q	R	S	T	V	W	X	Y	Z
A	4	-2	0	-2	-1	-2	0	-2	-1	-1	-1	-1	-2	-1	-1	-1	1	0	0	-3	-1	-2	-1
B	-2	6	-3	6	2	-3	-1	-1	-3	-1	-4	-3	-1	-1	0	-2	0	-1	-3	-4	-1	-3	2
C	0	-3	9	-3	-4	-2	-3	-3	-1	-3	-1	-1	-3	-3	-3	-3	-1	-1	-1	-2	-1	-2	-4
D	-2	6	-3	6	2	-3	-1	-1	-3	-1	-4	-3	1	-1	0	-2	0	-1	-3	-4	-1	-3	2
E	-1	2	-4	2	5	-3	-2	0	-3	1	-3	-2	0	-1	2	0	0	-1	-2	-3	-1	-2	5
F	-2	-3	-2	-3	-3	6	-3	-1	0	-3	0	0	-3	-4	-3	-3	-2	-2	-1	1	-1	3	-3
G	0	-1	-3	-1	-2	-3	6	-2	-4	-2	-4	-3	0	-2	-2	-2	0	-2	-3	-2	-1	-3	-2
H	-2	-1	-3	-1	0	-1	-2	8	-3	-1	-3	-2	1	-2	0	0	-1	-2	-3	-2	-1	2	0
I	-1	-3	-1	-3	-3	0	-4	-3	4	-3	2	1	-3	-3	-3	-3	-2	-1	3	-3	-1	-1	-3
K	-1	-1	-3	-1	1	-3	-2	-1	-3	5	-2	-1	0	-1	1	2	0	-1	-2	-3	-1	-2	1
L	-1	-4	-1	-4	-3	0	-4	-3	2	-2	4	2	-3	-3	-2	-2	-2	-1	1	-2	-1	-1	-3
M	-1	-3	-1	-3	-2	0	-3	-2	1	-1	2	5	-2	-2	0	-1	-1	-1	1	-1	-1	-1	-1
N	-2	1	-3	1	0	-3	0	1	-3	0	-3	-2	6	-2	0	0	1	0	-3	-4	-1	-2	0
P	-1	-1	-3	-1	-1	-4	-2	-2	-3	-1	-3	-2	-2	7	-1	-2	-1	-1	-2	-4	-1	-3	-1
Q	-1	0	-3	0	2	-3	-2	0	-3	1	-2	0	0	-1	5	1	0	-1	-2	-2	-1	-1	2
R	-1	-2	-3	-2	0	-3	-2	0	-3	2	-2	-1	0	-2	1	5	-1	-1	-3	-3	-1	-2	0
S	1	0	-1	0	0	-2	0	-1	-2	0	-2	-1	1	-1	0	-1	4	1	-2	-3	-1	-2	0
T	0	-1	-1	-1	-1	-2	-2	-2	-1	-1	-1	-1	0	-1	-1	-1	1	5	0	-2	-1	-2	-1
V	0	-3	-1	-3	-2	-1	-3	-3	3	-2	1	1	-3	-2	-2	-3	-2	0	4	-3	-1	-1	-2
W	-3	-4	-2	-4	-3	1	-2	-2	-3	-3	-2	-1	-4	-4	-2	-3	-3	-2	-3	11	-1	2	-3
X	-1	-1	-1	-1	-1	-1	-1	-1	-1	-1	-1	-1	-1	-1	-1	-1	-1	-1	-1	-1	-1	-1	-1
Y	-2	-3	-2	-3	-2	3	-3	2	-1	-2	-1	-1	-2	-3	-1	-2	-2	-2	-1	2	-1	7	-2
Z	-1	2	-4	2	5	-3	-2	0	-3	1	-3	-1	0	-1	2	0	0	-1	-2	-3	-1	-2	5

Table 1.9 Amino acids and their encoding as 3-letter and 1-letter symbolics from web-site http://icb.med.cornell.edu/education/courses/introtobio (accessed 8 December 2009)

1-letter	3-letter	Protein residue	Codons
A	Ala	Alanine	GCT, GCC, GCA, GCG
B	Asp, Asn	Asp. acid/Asparagine	GAT, GAC, AAT, AAC
C	Cys	Cysteine	TGT, TGC
D	Asp	Aspartic acid (Aspartate)	GAT, GAC
E	Glu	Glutam. acid/Glutamate	GAA, GAG
F	Phe	Phenylalanine	TTT, TTC
G	Gly	Glycine	GGT, GGC, GGA, GGG
H	His	Histidine	CAT, CAC
I	Ile	Isoleucine	ATT, ATC, ATA
K	Lys	Lysine	AAA, AAG
L	Leu	Leucine	TTG, TTA, CTT, CTC, CTA, CTG
M	Met	Methionine	ATG
N	Asn	Asparagine	AAT, AAC
P	Pro	Proline	CCT, CCC, CCA, CCG
Q	Gln	Glutamine	CAA, CAG
R	Arg	Arginine	CGT, CGC, CGA, CGG, AGA, AGG
S	Ser	Serine	TCT, TCC, TCA, TCG, AGT, AGC
T	Thr	Threonine	ACT, ACC, ACA, ACG
V	Val	Valine	GTT, GTC, GTA, GTG
W	Trp	Tryptophan	TGG
X	Xaa	Any amino acid	Any
Y	Tyr	Tyrosine	TAT, TAC
Z	Glu, Gln	Glutamic acid–Glutamine	GAA, GAG, CAA, CAG
*	STOP	Terminator	TAA, TAG, TGA

estimation of probabilities $p(i)$, $p(j)$ and $p(ij)$ of i, j and mutual substitution of i and j. Given these probabilities, the substitution scores are defined as integers proportional to logarithms of odd-ratios, $log[p(ij)/(p(i)p(j))]$. Elements of matrix in Table 1.8 were derived by Henikoff and Henikoff (1992) using such protein sequences from database BLOCK for which pair-wise alignments involve not more than 62% of identity, which explains the name of the matrix.

This matrix leads to more reasonable results than other scoring matrices; practitioners of protein alignment have selected this matrix as a standard (see Betts and Russell 2003). We consider BLOSUM62 as a similarity matrix and are interested in finding clusters of amino acids that tend to replace each other and looking at physic and chemical properties explaining the groupings.

1.3 An Account of Data Visualization

1.3.1 General

Visualization can be a by-product of the model and/or method, or it can be utilized by itself. The concept of visualization usually relates to the human cognitive abilities, which are not yet well understood. Computationally meaningful studies of structures of visual image streams such as in a movie or video began only recently. A most update account of the developments in information visualization can be found in Mazza (2009), see also Spence (2001).

We are going to be concerned with presenting data as maps or diagrams or digital screen objects in such a way that relations between data entities or features or both are reflected in distances or links, or other visual relations, between their images. Among more or less distinct visualization goals, beyond sheer presentation that appeals to the cognitive domination of visual over other senses, we can distinguish between:

A. Highlighting
B. Integrating different aspects
C. Narrating
D. Manipulating

Of these, manipulating visual images of entities, such as in computer games, seems an interesting area yet to be developed in the framework of data analysis. There can be mentioned, though, operations of mild manipulation readily available at various sites already such as scrolling, representing an overview with possibilities of getting further details of individual fragments by zooming or windowing, and an overview that allows focusing on specific fragments by enlarging them on the same screen (Mazza 2009). The other three will be briefly discussed and illustrated in the remainder of this section.

1.3.2 Highlighting

To visually highlight a feature of an image one may distort the original dimensions. A good example is the London tube scheme by H. Beck (1906) which greatly enlarges relative sizes of the Centre of London part to make them better seen. Such a gross distortion, for a long while being totally rejected by the authorities, is now a standard for metro maps worldwide (see Fig. 1.3).

In fact, this line of thinking has been worked on in geography for centuries, since the mapping of the Earth global surface to a flat sheet is impossible to do exactly. Various proxy criteria have been proposed leading to interesting highlights way beyond conventional geography maps, such as presented on Fig. 1.4 (Fullers' projection) and Fig. 1.5 (August's projection); see website http://en.wikipedia.org/wiki/World_map for more.

More recently this idea was applied by Rao and Card (1994) to table data (see Fig. 1.6); more on this can be found in Card et al. (1999) and Mazza (2009).

It should be noted that the disproportionate highlighting may lead to visual effects bordering with cheating (or being just that). This is especially apparent when relative proportions are visualized through proportions between areas, as in Fig. 1.7. An unintended effect of the picture is that the decline by half in one dimension is

1.3 An Account of Data Visualization

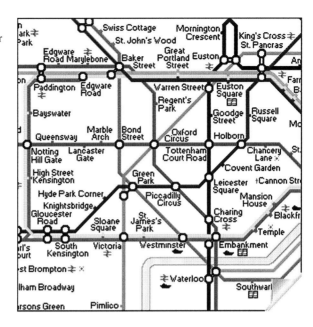

Fig. 1.3 A fragment of London Tube map made after H. Beck; the central part is highlighted by a disproportionate scaling

presented visually by the area of the doctor's body, which is just not half but one fourth of the initial size. This grossly biases the message.

Another typical case of unintentionally cheating is when the relative proportions are visualized using bars that start not at the 0 point but an arbitrary mark, as is the case of Fig. 1.8, on which a newspaper's legitimate satisfaction with its success is visualized using bars that begin at 500,000 mark rather than 0. Another mistake is that the difference between the bars' heights on the picture is much greater than the

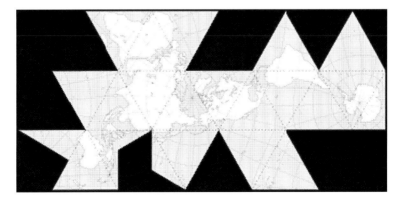

Fig. 1.4 The Fuller Projection, or Dymaxion Map, displays spherical data on a flat surface of a polyhedron using a low-distortion transformation. Landmasses are presented with no interruption

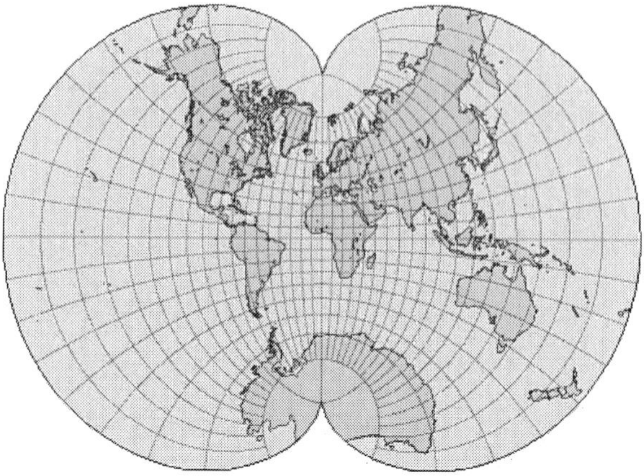

Fig. 1.5 A conformal map: the angle between *any two lines* on the *sphere* is the same between their projected counterparts on the map; in particular, each parallel crosses meridians at *right angles*; and also, the sizes at any point are the same in *all* directions

Fig. 1.6 The Table Lens machine: highlighting a fragment by disproportionally enlarging it (see Card et al. 1999)

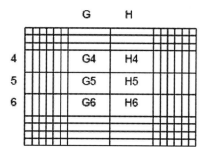

Fig. 1.7 A decline in relative numbers of general practitioner doctors in California was conveniently visualized by Los Angeles Times 5 August 1979 using 1D dimensional scaling whereas the 2D image conveys a quadratic decline – not a half but a quarter of the size, and the like

1.3 An Account of Data Visualization

Fig. 1.8 An unintended distortion: a newspaper's report (July 2005) is visualized with bars that grow from mark 500,000 rather than 0

reported 220,000. Altogether, the rival's circulation bar is more than twice shorter while the real circulation is less by mere 25%.

1.3.3 Integrating Different Aspects

Combining different features of a phenomenon into the same image can make life easier indeed. Fig. 1.9 represents an image that an energy company utilizes for real

Fig. 1.9 An image of Con Edison company's power grid on a PC screen according to website http://www.avs.com/software/soft_b/openviz/conedison.html as accessed in September 2008

time managing, control and repair of its energy network stretching over the island of Manhattan (New York, USA). Operators can view the application on their desktop PCs, monitor the grid and repair problems on the fly by rerouting power or sending a crew out to repair a device on site. This makes "manipulation and utilization of data in ways that were previously not possible," according to the company's website (see reference in the caption).

Bringing features together can be useful for less immediate insights too. A popular story of Dr. John Snow's fight against an outbreak of cholera in Soho, London, 1857, by using visual data mining goes like this. Two weeks into the outbreak, Dr. Snow went over all houses in the vicinity and made as many tics at their locations on his map as many deaths of cholera have occurred there (Fig. 1.10 illustrates a fragment of Dr. Snow's map). The ticks were densest around a water pump, which made Dr. Snow convinced that the pump was the cholera source.

(In fact, Dr. Snow had served in India to become disposed to the idea of the role of water flows in the transmission of the disease). He discussed his findings with the priest of local parish, who removed then the handle of the pump, after which deaths

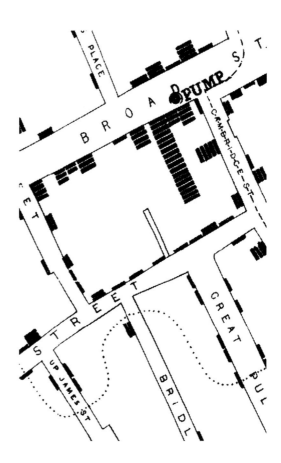

Fig. 1.10 A scheme of a fragment of Dr. Snow's map demonstrating that indeed most deaths (*labeled by bars*) have occurred near the water pump he was dealing with

1.3 An Account of Data Visualization

stopped. This all is true. But there is more to this story. The death did stop - but because too few remained in the district, not because of the removal: the handle was ordered back on the very next day after it had been removed. Moreover, the borough council refused to accept Dr. Snow's "water pump theory" because it contradicted the theory of the time that cholera progressed through stench in the air rather than water flow. More people died in Soho of the next cholera outbreak in a decade. The water pump theory was not accepted until much later, when the science of microbes had become developed. The story is instructive in both the power of visual insight and the fact that data analysis results are not conclusive: a data based conclusion needs a reasonable explanation to get accepted.

The diagram on Fig. 1.11 visualizes relations between features in Company data (Table 1.1) as a decision tree to conceptually characterize their products. For example, the left hand branch distinctly describes Product A by combining "Not retail" and "No e-commerce" edges. (See Section 4.5 for more on decision trees).

One more visual image, Fig. 1.12, depicts relations between confusion patterns of decimal numerals drawn over rectangle's edges (Fig. 1.2) and their descriptions in terms of combinations of edges of the rectangle with which they are drawn. A description may combine both edges present and absent to distinctively characterize a pattern, whereas a profile comprises edges that are present in all elements of its pattern. The confusion patterns are derived from data in Table 1.7 according to clustering of numerals in Section 7.3 and Mirkin 2005.

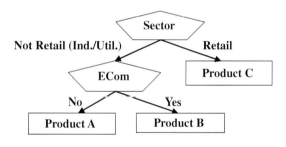

Fig. 1.11 Product decision tree for the Company data in Table 1.1

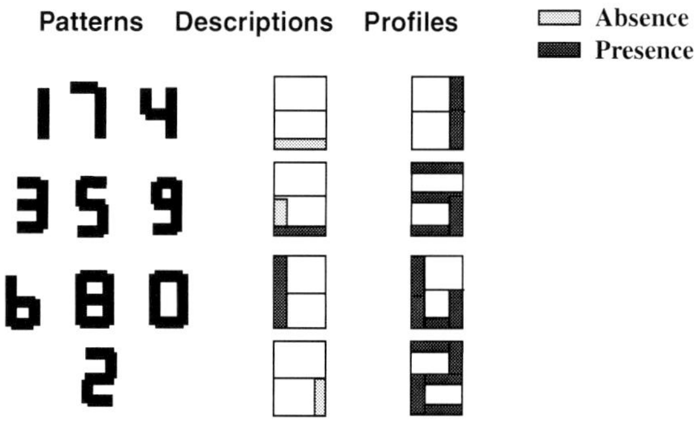

Fig. 1.12 Confusion patterns for numerals visualized from the patterns' data analysis descriptions in terms of edges being present or not. The *right-hand part* presents profiles of the common edges, for comparison

1.3.4 Narrating a Story

In a situation in which data features involve a temporal and/or spatial aspects, integrating them in one image may lead to a visual narrative of a story, with its starting and ending dates, all on the same screen. Such a narrative of a military company from the rich history of Europe (Napoleon's French army invading Russia 1812) is presented in Fig. 1.13. It shows a map of Russia, with Napoleon's army trajectory drawn forth and back, so that the time is enveloping in this static image via the trajectory. The trajectory's width shows the army's strength steadily declining in time on a dramatic scale, in the summer time and in the absence of major fighting.

All the images presented can be considered illustrations of a principle accepted further on. According to this principle, to visualize data, one needs to specify first a "ground" image, such as a map or grid or coordinate plane, which is supposed to be well known to the user. Visualization, as a computational device, can be defined as mapping data to the ground image in such a way that the analyzed properties of the data are reflected in properties of the image. Of the goals considered, integration of data will be of a priority since no temporal aspect is considered in this text.

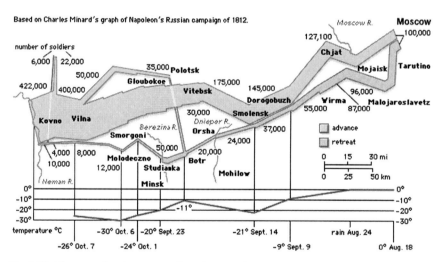

Fig. 1.13 The *upper band* represents the trajectory of Napoleon's army marching to the East and the *lower band* shows it moving back to the West, the band width being proportional to the army's strength (from http://www.emersonkent.com/map_archive/russian_campaign_1812.htm)

1.4 Summary

This chapter introduces four problems in data analysis as related to either summarization or correlation, in either quantitative or categorical way. The former two reflect the structure of theoretical knowledge as comprised, first of all, of concepts and statements of relation among them. Each of these four will be given a specific attention in the text further on. After covering summarization and correlation in 1D and 2D situations (Chapters 2 and 3), we will move on to problems of correlation in Chapter 4, both quantitative, that is, regression (Sections 4.3 and 4.6), and categorical, that is, classifiers (Sections 4.2, 4.4 and 4.5). Chapter 5 is devoted to Principal component analysis and applications, and Chapters 6, 7 and 8 describe clustering: K-means, hierarchical methods and clusters in networks. The latter incorporates some recent developments such as spectral clustering.

Next part of the chapter introduces seven small real-world data sets and related data analysis problems.

The final part of the chapter discusses main goals and some specifics of data visualization. Integrating visualization into the methods discussed further on in a sound way remains a challenge for the future.

References

Abdi, H., Valentin, D., Edelman, B.: Neural Networks, Series: Quantitative Applications in the Social Sciences, 124. Sage Publications, London, ISBN 0-7619-1440-4 (1999)

Berthold, M., Hand D.: Intelligent Data Analysis. Springer, Berlin-Heidelberg (2003)

Betts, M.J., Russell, R.B.: Amino acid properties and consequences of subsitutions. In: Barnes, M.R., Gray, I.C. (eds.) Bioinformatics for Geneticists. Wiley, New York, NY (2003)

Card, S.K., Mackinlay, J.D., Shneiderman B.: Readings in Information Visualization: Using Vision to Think. Morgan Kaufmann Publishers, San Francisco, CA, ISBN 1-55860-533-9 (1999)

Duda, R.O., Hart, P.E., Stork D.G.: Pattern Classification. Wiley-Interscience, New York, NY, ISBN 0-471-05669-3 (2001)

Engelbrecht, A.P.: Computational Intelligence. Wiley, Chichester, ISBN 0-470-84870-7 (2002)

Fisher, R.: The use of multiple measurements in taxonomic problems. Annals Eugen. **7**, 179–188 (1936)

Gama, J.: Knowledge Discovery from Data Streams. Boca Raton, Chapman & Hall/CRC (2010)

Hair, J.F., Black, W.C., Babin, B.J., Anderson, R.E.: Multivariate Data Analysis, 7th edn. Prentice Hall, Upper Saddle River, NJ, ISBN-10: 0-13-813263-1 (2010)

Han, J., Kamber, M., J. Pei: Data Mining: Concepts and Techniques, 2nd edn. Morgan Kaufmann Publishers, San Francisco (2006)

Hartigan, J.A.: Clustering Algorithms. Wiley, New York, NY (1975)

Haykin, S. S.: Neural Networks, 2nd edn. Prentice Hall, Upper Saddle River NJ, ISBN 0132733501 (1999)

Henikoff, S., Henikoff, J.G.: Amino acid substitution matrices from protein blocks. Proc. Natl. Acad. Sci.USA **89**(22), 10915–10919 (1992)

Kendall, M.G., Stewart, A.: Advanced Statistics: Inference and Relationship, 3rd edn. Griffin, London, ISBN: 0852642156 (1973)

Lebart, L., Morineau, A., Piron, M.: Statistique Exploratoire Multidimensionelle. Dunod, Paris, ISBN 2-10-002886-3 (1995)

Lohninger, H.: Teach Me Data Analysis. Springer, Berlin-New York-Tokyo, ISBN 3-540-14743-8 (1999)

Manning, C.D., Raghavan, P., Schütze, H.: Introduction to Information Retrieval. Cambridge University Press, Cambridge, England (2008)

Mazza, R.: Introduction to Information Visualization. Springer, London, ISBN: 978-1-84800-218-0 (2009)

Mirkin, B.: Mathematical Classification and Clustering. Kluwer Academic Press, Boston-Dordrecht (1996)

Mirkin, B.: Clustering for Data Mining: A Data Recovery Approach. Chapman & Hall/CRC, London, ISBN 1-58488-534-3 (2005)
Mitchell, T.M.: Machine Learning. McGraw Hill, New York, NY (2005)
Mitsa, T.: Temporal Data Mining. Chapman & Hall/CRC, Boca Raton (2010)
Murtagh, F.: Multidimensional Clustering Algorithms. Physica-Verlag, Vienna (1985)
Polyak, B.: Introduction to Optimization. Optimization Software, Los Angeles, CA, ISBN: 0911575146 (1987)
Schölkopf, B., Smola, A.J.: Learning with Kernels. The MIT Press, Cambridge, MA (2005)
Spence, R.: Information Visualization. ACM Press, New York, NY, ISBN 0-201-59626-1 (2001)
Tukey, J.W.: Exploratory Data Analysis. Addison-Wesley, Reading, MA (1977)
Vapnik, V.: Estimation of Dependences Based on Empirical Data, 2d edn. Springer Science + Business Media Inc., New York, NY (2006)
Webb, A.: Statistical Pattern Recognition. Wiley, Chichester (2002)
Weiss, S.M., Indurkhya, N., Zhang, T., Damerau, F.J.: Text Mining: Predictive Methods for Analyzing Unstructured Information. Springer Science+Business Media, New York, NY, ISBN 0-387-95433-3 (2005)
Zhang, Z., Zhang, R.: Multimedia Data Mining. Chapman & Hall/CRC, Boka Raton (2009)

Chapter 2
1D Analysis: Summarization and Visualization of a Single Feature

2.1 Quantitative Feature: Distribution and Histogram

1D data is a set of entities represented by one feature, categorical or quantitative. There is no simple criterion to tell a quantitative feature or categorical one. For practical purposes a good criterion is this: a feature is quantitative if averaging it makes sense. Let us first consider the quantitative case.

P2.1.1 Presentation

A most comprehensive, and quite impressive for the eye, way of summarization is the distribution. On the plane, one draws an *x* axis and the feature range boundaries, that is, its minimum and maximum. The range interval is divided then into a number of non-overlapping equal-sized sub-intervals, *bins*. Then the number of entities that fall in each bin is counted, and the counts are reflected in the heights of the bars over the bins, forming a histogram. Histograms of Population resident in Market town dataset and Petal width in Iris dataset are presented on Fig. 2.1.

Q.2.1. Why the bins are not to overlap? **A.** Each entity falls in only one bin if bins do not overlap, and the total of all bin counts equals the total number of entities in this case. If bins do overlap, the principle "one entity – one vote" will be broken.

Q.2.2. Why the bar heights on the left are greater than those on the right in Fig. 2.1? **A.** Because bins on the right are as twice shorter than those on the left; therefore, the numbers of entities falling within them must be smaller.

Q.2.3. Is it true that when there are only two bins, the divider between them must be the midrange point? **A.** Yes, because the bin sizes are equal to each other (see Fig. 2.2).

On Figs. 2.3 and 2.4, two most popular types of histograms are presented. The former corresponds to the so-called power law, sometimes referred to as Pareto distribution. This type is frequent in social systems. According to numerous empirical studies, such features as wealth, group size, productivity and the like are all distributed according to a power law so that very few individuals or entities have

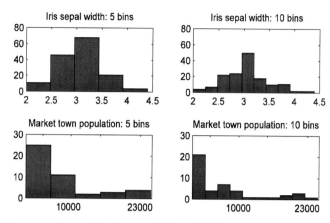

Fig. 2.1 Histograms of quantitative features in Iris and Market town data: the feature represented on x-axis and the counts on y-axis. The histogram shapes depend on the number of bins

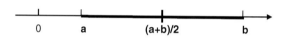

Fig. 2.2 With just two bins on the range, the divider is mid-range

Fig. 2.3 A power type distribution

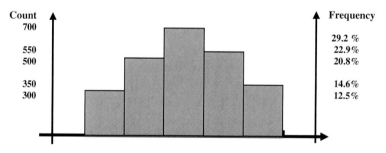

Fig. 2.4 Gaussian type distribution (bell curve)

huge amounts of wealth or members, whereas very many individuals are left with virtually nothing. However, they all are important parts of the same system with the have-nots creating the environment in which the lucky few can strive.

Another type, which is frequent at physical systems, is presented on Fig. 2.4.

2.1 Quantitative Feature: Distribution and Histogram

This type of histograms approximates the so-called normal, or Gaussian, law. Distributions of measurement errors and, in general, features being results of small random effects are thought to be Gaussian, which can be formally proven within a mathematical framework of the probability theory.

Q.2.4. Take a look at the distributions on Fig. 2.1. Can you see which of the two types they are similar to? **A.** The Population's distribution is of power law type, and the Petal width is of Gaussian law type, as one would expect.

Another popular visualization of distributions is known as a pie-chart, in which the bin counts are expressed by the sizes of sectored slices of a round pie (see in the middle of Fig. 2.5).

As one can see, histograms and pie-charts cater for perception of two different aspects of the distribution; the former for the actual envelopment of the distribution along the axis x, whereas the latter caters for the relative sizes of distribution chunks falling into different bins. There are a dozen more formats of visualization of distributions, such as bubble, doughnut and radar charts, easily available in Microsoft Excel spreadsheet.

F2.1.2 Formulation

With N entities numbered from $i = 1, 2, \ldots, N$, data is a set of numbers $x_1, \ldots, x_N$. This set will be usually denoted by $X = \{x_1, \ldots, x_N\}$.

To produce n bins, one needs $n-1$ dividers at points $a + k(b-a)/n$ ($k = 1, 2, \ldots, n-1$). In fact, the same formula works for $k = 0$ and $k = n+1$ leading to the boundaries a as x_0 and b as x_{n+1}, which is useful for the operation of counting the number of entities N_k falling in each of the bins $k = 1, 2, \ldots, n$. Note that bin k has $a + (k-1)(b-a)/n$ and $a + k(b-a)/n$ as, respectively, its left and right boundary. One of them should be excluded from the bin so that the bins are not

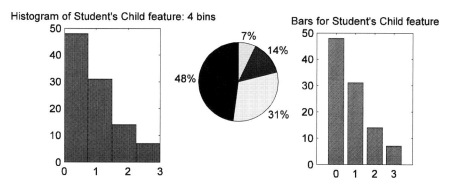

Fig. 2.5 Distribution of the number of children at Student data Child feature visualized as a 4-bin histogram on the *left*, pie-chart in the *middle*, and a bar set on the *right* – this seems the most appropriate of the three at the case

overlapping even on boundaries. These counts, N_k, $k = 1, 2, \ldots, n$ constitute the *distribution* of the feature. A *histogram* is a visual representation of the distribution by drawing a bar of the height N_k over each bin k, $k = 1, 2, \ldots, n$ (see Figs. 2.1, 2.3, 2.4 and 2.5). Note that the distribution is subject to the choice of the number of bins.

The histograms can be thought of as empirical expressions of theoretical probability distributions, the so-called density functions. A density function $p(x)$ expresses the concept of probability, not straightforwardly with $p(x)$ values, but in terms of their integrals, that is, the areas between the $p(x)$ curve and x-axis, over intervals $[a,b]$: such an integral equals the probability that a random variable, distributed according to $p(x)$, falls within $[a,b]$. This implies that the total area between the curve an x-axis must be equal to 1, which is achieved with the corresponding scaling the curve with a constant factor.

The power law density function is $p(x) \approx a/x^\lambda$, where λ reflects the steepness of the frequency's fall (see Fig. 2.6 on the left). Such a law expresses what is called the Matthew's effect referring to the saying "He who has much, will get more; and he who has nothing, will lose even that little that he has," according to Matthew's gospel. The Matthew's effect is expressed, for example, in "the mechanism of preferential attachment": the probability that a new web surfer hits a web-site is proportional to the site's popularity, according to this mechanism.

The normal, or Gaussian, law is $p(x) = C\exp[-(x-a)^2/2\sigma^2]$, where C is a constant, which is sometimes denoted as $N(a, \sigma)$ (see Fig. 2.6 on the right). Distributions of measurement errors and, in general, features being results of small random effects are thought to be Gaussian, which can be formally proven within a mathematical framework of the probability theory. The parameters of this distribution, a and σ, have natural meaning: a expresses the expectation, or mean, and σ^2 – the variance, which naturally translates in data terms in Section 2.2. It should be pointed out that the probability of a value x falling in the interval $a \pm \sigma$ according to the normal distribution is about 88%, and falling in the interval $a \pm 3\sigma$ about 99.7%, virtually unity, so that at modest sample sizes it is highly unlikely that a value x can fall out of this interval, which is referred sometimes as "three sigma rule". The Gaussian distribution can be rescaled to the standard $N(0,1)$ form, with 0 expectation and 1 the variance, by shifting the variable x to the mean, a, and

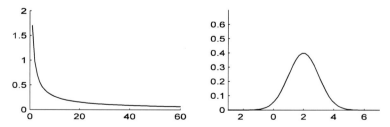

Fig. 2.6 Density functions of the power law with $\lambda = -0.8$, on the *left*, and normal distribution $N(2,1)$, on the *right*

2.1 Quantitative Feature: Distribution and Histogram

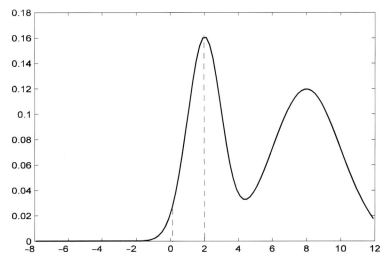

Fig. 2.7 A density function p(x), which is a mixture of two normal distributions, N(2,1) weighted 0.4, and N(8,2) weighted 0.6. The area between the two *dashed lines* is the probability for value x to fall in the interval between 0 and 2 – not too high at this p(x)!

normalizing it afterwards by the square root of σ^2. This transformation, sometimes referred to as z-scoring, is expressed with formula $y = (x-a)/\sigma$, where y is the transformed feature. Sometimes a distribution can be approximated by a mixture of normal distributions (see Fig. 2.7).

One more popular distribution is the uniform distribution, over a range $[l,r]$. Its density is a constant function equal to $p(x) = 1/(r-l)$, so that the probability of an interval (a,b) within the range is just $p = (b-a)/(r-l)$, proportional to the length of the interval.

C2.1.3 Computation

To compute the distributions on Fig. 2.1, one should first load the Iris and Market town data sets with a MatLab command such as

≫ st=load('Data\town.dat');

% the Market data is stored at subfolder "Data" under the name "town.dat"

after which the Population feature can be put in a different variable

≫ pop=st(:,1);

% meaning all the rows of column 1 corresponding to the Population feature.

Then command

≫ h=hist(pop,5);

will produce a 5×1 array h containing counts of entities within each of 5 bins, and command

≫ hist(pop, 5);

will produce a figure of the histogram.

To create a figure with four windows such as on Fig. 2.1, one should use subplot commands, along with corresponding rearrangements of the axes:
≫ subplot(2,2,1);hist(sw,5);axis([2 4.5 0 80]);
≫ subplot(2,2,2);hist(sw,10);axis([2 4.5 0 80]);
% assuming sw denotes Sepal width, column 2 of the Iris data set
≫ subplot(2,2,3);hist(pop,5);axis([2000 24000 0 30]);
≫ subplot(2,2,4);hist(pop,10);axis([2000 24000 0 30]);

The command axis([a b c d]) puts the image coordinate box so that it has the interval [a,b] over x-axis and interval [c,d] over y-axis.

Bar- and pie-charts are produced with `pie` and `bar` commands, respectively.

2.2 Further Summarization: Centers and Spreads

P2.2.1 Centers and Spreads: Presentation

Further summarization of the data leads to presenting a feature with just two numbers, one expressing the distribution's location, its "central" or other important point, and the other representing the distribution's dispersion, the spread. We review some most popular characteristics for both, the center, Table 2.1, and the spread, Table 2.2, see also Lohninger (1999).

Worked example 2.1. Mean

For set X = {1, 1, 5, 3, 4, 1, 2}, mean is c = (1 + 1 + 5 + 3 + 4 + 1 + 2)/7 = 17/7 = 2.42857..., or rounded up to two decimals, c = 2.43.

This is as close an approximation to the numbers as one can get, which is good. A less satisfactory property is that the mean is not stable against outliers. For example, if X in Worked example 2.1 is supplemented with value 23, the mean becomes c = (17+23)/8 = 5, a much greater number. This is why it is a good idea to remove some observations on both extremes of the data range, both the minimum and maximum, before computing the mean, which is utilized in the concept of trimmed mean in statistics.

Worked example 2.2. Median

To compute the median of the set from the previous example, X = {1, 1, 5, 3, 4, 1, 2}, it must be sorted first: 1, 1, 1, 2, 3, 4, 5. The median is defined as the element in the middle, which is 2. This is rather far away from the mean, 2.43, which witnesses that the distribution is biased towards the left end, the smaller entities. With the outlier 23 added, the sorted set becomes 1, 1, 1, 2, 3, 4, 5, 23, thus leading to two

2.2 Further Summarization: Centers and Spreads

Table 2.1 A review of location or central point concepts

#	Name	Explanation	Comments
1	Mean	The feature's arithmetic average	0. Minimizes the summary error squared 1. Estimates the distribution's expected value 2. Sensitive to outliers and distribution's shape
2	Median	The middle of the sorted list of feature values	1. Minimizes the summary absolute error 2. Estimates the distribution's expected value 3. Not-sensitive to outliers 4. Sensitive to distribution's shape
3	Mid-range	Middle of the range	1. Minimizes the maximum absolute error 2. Estimates the distribution's expected value 3. Very sensitive to outliers 4. Not sensitive to distribution's shape
4	P-quantile	A value dividing the entire entity set in proportion P or (1−P) of feature values so that those with higher values constitute P proportion (upper P-quantile) or 1−P proportion (bottom P-quantile)	1. Not-sensitive to outliers 2. Sensitive to distribution's shape
5	Mode	A maximum of the histogram	1. Depends on the bin size 2. Can be multiple

Table 2.2 A review of spread concepts

#	Name	Explanation	Comments
1	Standard deviation	The quadratic average deviation from the mean	1. Minimized by the mean 2. Estimates the square root of the variance
2	Absolute deviation	The average absolute deviation from the median	Minimized by the median
3	Half-range	The maximum deviation from the midrange	Minimized by the mid-range

elements in the middle, 2 and 3. The median in this case is the average of the two, $(2+3)/2 = 2.5$, which is by far lesser change than the mean of the extended set, 5.

The more symmetric a distribution, the closer its mean and median to each other. Sepal width of Iris data set (Table 1.3) has mean = 3.05 and median = 3, quite close values. In contrast, in Market town data (Table 1.4), Population resident's median, 5,258, is predictably much less than the mean, 7,351.4. The mean of a power law distribution is always biased towards the great values achieved by the few outliers;

this is why it is a good idea to use the median as its central value. The median is very stable against outliers: the values on the extremes just do not affect the middle of the sorted set if added uniformly to both sides.

The midrange corresponds to the mean of a flat distribution, in which all bins are equally likely. In contrast to the mean and median, the midrange depends only on the range, not on the distribution. It is obviously highly sensitive to outliers, that is, changes of the maximum and/or minimum values of the sample.

The concept of p-quantile is an extension of the concept of median, which is a 50% quantile.

Worked example 2.3. P-quantile (percentile)

Take $p = 10\%$ and determine the upper 10% quantile of Population resident feature. This should be 5th value in its descending order, that is, 18,966. Why is the 5th value? Because 10% of the total number of entities, 45, is 4.5; therefore, the 5-th value leaves out $p = 10\%$ of the largest towns in the sample. Similarly, the lower 10% quantile of the feature is 5th value in its ascending order, 2,230.

Worked example 2.4. Mode

According to the histograms in the bottom of Fig. 2.1, it is the very first bin which is modal in the Population resident distribution. In the 5-bin setting, it takes one fifth of the feature range, $23,801 - 2,040 = 21,761$, that is, 4,352. In the 10-bin setting, it is one tenth of the feature range, that is, 2,176. In the latter case, the modal bin is interval [2,040, 4,216], and the modal bin is as twice wider, [2,040, 6,392], in the former case.

Each of the characteristics of spread in Table 2.2 parallels, to an extent, a location characteristic under the same number.

These measures intend to give an estimate of the extent of error in the corresponding centrality index. The standard deviation is the average quadratic error of the mean. Its use is related to the least-squares approach that currently prevails in data analysis and can be justified by good properties of the solutions, within the data analysis perspective, and properties of the normal distribution, within the probabilistic perspective. These paradigms are explained later in Section 2.2.2.

The absolute deviation expresses the average absolute deviation from the median. Usually, it is calculated regarding the mean, as the average error in representing the feature values by the mean. However, it is more related to the median, because it is the median that minimizes it.

The half-range expresses the maximum deviation from the mid-range; so they should be used on par, as it is done customarily by the research community involved in building classifying rules.

2.2 Further Summarization: Centers and Spreads

F2.2.2 Centers and Spreads: Formulation

There are two perspectives on data summarization and correlation that very much differ from each other. One, of the classical mathematical statistics, views the data as generated by a probabilistic mechanism and uses the data to recover the mechanism or, at least, some properties of it. The other, of data analysis, does not much care of the mechanism and tries to look for patterns in the data instead.

F2.2.2.1 Data Analysis Perspective

Given a series $X = \{x_1, \ldots, x_N\}$, one defines the centre of X as a minimizing the average distance

$$D(X, a) = [d(x_1, a) + d(x_2, a) + \ldots + d(x_N, a)] / N \qquad (2.1)$$

Depending on the definition of the distance, the optimal a can be expressed as follows.

Consider first the least-squares formulation. According to this approach the distance is measured as the squared difference, $d(x, a) = |x-a|^2$. The minimum distance (2.1) then is reached at a equal to the mean c defined by expression

$$c = \sum_{i=1}^{N} x_i / N \qquad (2.2)$$

and distance $D(X,c)$ itself is equal to the variance s^2 defined by expression

$$s^2 = \sum_{i=1}^{N} (x_i - c)^2 / N \qquad (2.3)$$

At the more traditional distance measure $d(x, a) = |x-a|$ in (2.1), the optimal a (center) is but the median, m, and $D(X,a)$ the absolute deviation from the median,

$$ms = \sum_{i=1}^{N} |x_i - m| / N \qquad (2.4)$$

To be more precise, the optimal a in this problem is median, that is the value $x_{(N+1)/2}$ in the sorted order of X, when N is odd. When N is even, any value between $x_{N/2}$ and $x_{N/2+1}$ in the sorted order of X is a solution, including the median. If $D(X,a)$ is defined not by the sum, but by the maximum of the distances, $D(X, a) = \max(d(x_1, a), d(x_2, a), \ldots, d(x_N, a))$, then the midrange mr is the solution, for $d(x,a)$ specified as both $|x-a|^2$ and $|x-a|$.

These statements explain the parallels between the centers and corresponding spread evaluations reflected in Tables 2.1 and 2.2, with each of the centers minimizing its corresponding measure of spread.

The distance minimization problem can be reformulated in the data recovery perspective. In the data recovery perspective, the observed values are assumed to be but noisy realizations of an unknown value a. This is reflected in the form of an equation expressing x_i through a:

$$x_i = a + e_i, \text{ for all } i = 1, 2, \ldots, N, \tag{2.5}$$

in which e_i are additive errors, or residuals, that are to be minimized.

One cannot minimize all the residuals in (2.5) simultaneously. An integral criterion is needed to embrace them all. A general family of such criteria is known as Minkowski's criterion or L_p norm. It is specified by using a positive number p as

$$L_p = \left(|e_1|^p + |e_2|^p + \ldots + |e_N|^p\right)^{1/p}$$

At a given p, minimizing L_p or, equivalently, its p-th power $L_p{}^p$, would lead to a specific solution. Most popular are values $p = 1$, 2, and ∞ (infinity) leading to:

(1) Least-squares criterion $L_2^2 = e_1^2 + e_2^2 + \ldots + e_N^2$ at $p = 2$.
    Its minimization over unknown a is equivalent to the task of minimizing the average squared distance, thus leading to the mean as the optimal a.
(2) Least-modules criterion $L_1 = |e_1| + |e_2| + \ldots + |e_N|$ at $p = 1$.
    Its minimization over unknown a is equivalent to the task of minimizing the average absolute deviation, thus leading to the median, optimal $a = m$.
(3) Least-maximum criterion $L_\infty = \max(|e_1|, |e_2|, \ldots |e_N|)$ at p tending to ∞.
    Minimization of L_∞ with respect to a is equivalent to the task of minimizing the maximum deviation leading to the midrange, optimal $a = mr$.

The Minkowski's criteria (1)–(3) may look just as trivial reformulations of the distance approximation criterion (2.1). This, however, is not exactly so. The Equation (2.5) adds to the solution one more equation. It allows for a decomposition of the data scatter involving the corresponding data recovery criterion.

This is rather straightforward for the least-squares criterion L_2^2 whose minimal value, at a equal to the mean c (2.1) is $L_2^2 = (x_1-c)^2+(x_2-c)^2+\ldots+(x_N-c)^2$. With little algebra, this becomes $L_2^2 = x_1^2+x_2^2+\ldots+x_N^2 - 2c(x_1+x_2+\ldots+x_N)+Nc^2 = x_1^2 + x_2^2 + \ldots + x_N^2 - Nc^2 = T(X) - Nc^2$. where $T(X)$ is the quadratic data scatter defined as $T(X) = x_1^2 + x_2^2 + \ldots + x_N^2$.

This leads to equation $T(X) = Nc^2 + L_2^2$ decomposing the data scatter in two parts: that explained by the model (2.5), Nc^2, and that unexplained, L_2^2. Since the data scatter is constant, minimizing L_2^2 is equivalent to maximizing Nc^2. The decomposition of the data scatter allows measuring the adequacy of model (2.5) not by just the averaged square criterion, the variance, by the relative value of the explained

2.2 Further Summarization: Centers and Spreads

part $L_2^2/T(X)$. A similar decomposition can be derived for the least modules L_1 (see Mirkin 1996).

Q.2.5. Consider a multiplicative model for the error, $x_i = a(1+e_i)$, assuming that errors are proportional to the values. What center a fits the data with the least-squares criterion? **A.** According to the least squares approach, the fit should minimize the summary errors squared. Every error can be expressed, from the model, as $e_i = x_i/a - 1 = (x_i-a)/a$. Thus the criterion can be expressed as $L_2^2 = e_1^2 + e_2^2 + \ldots + e_N^2 = (x_1/a-1)^2 + (x_2/a-1)^2 + \ldots (x_N/a-1)^2$. Applying the first order optimality condition, let us take the derivative of L_2^2 over a and equate it to zero. The derivative is $L_2^{2'} = -(2/a^3)\Sigma_i(x_i-a)x_i$. Assuming the optimal value of a is not zero, the first order condition can be expressed as $\Sigma_i(x_i-a)x_i = 0$, so that $a = \Sigma_i x_i^2/\Sigma_i x_i = (\Sigma_i x_i^2/N)/(\Sigma_i x_i/N)$. The denominator here is but the mean, c, whereas the numerator can be expressed through the variance s^2 because of equation $s^2 = \Sigma_i x_i^2/N - \Sigma_i x_i/N$ which is not difficult to prove. With little algebraic manipulation, the least-squares fit can be expressed as $a = s^2/c + 1$. The variance to mean ratio s^2/c, equal to $a-1$ according to the model, emerges also in statistics as a good relative estimate of the spread.

It seems rather natural that both, the standard deviation and absolute deviation, are not greater than half the range, which can be proven mathematically (see Section 2.3.2).

Q.2.6. Prove that Minkowski's center is not sensitive with respect to changing the scale factor.

Q.2.7. Prove that Minkovski's center grows whenever power p grows.

Q.2.8. For the Population resident feature in Market town data compute Minkowski center at $p = 0.5, 1, 2, 3, 4, 5$. **A.** See solutions found using the cm.m code developed in Project 1.1 (and confirmed, at $p > 1$, with the steepest descent AG-MC method) in Table 2.3.

Table 2.3 Minkowski's metric centers of the Population resident in Market town dataset for different power values p

p	Minkowski's center	Data scatter unexplained
0.5	2,611	0.7143
1	5,258.0 (median)	0.6173
2	7,351.4 (mean)	0.4097
3	8,894.9	0.2318
4	9,758.8	0.1186
5	10,294.5	0.0584

F2.2.2.2 Probabilistic Statistics Perspective

In classical mathematical statistics, a set of numbers $X = \{x_1, x_2, \ldots, x_N\}$ is usually considered a random sample from a population defined by probabilistic distribution with density $f(x)$, in which each element x_i is sampled independently from the others. This involves an assumption that each observation x_i is modeled by the distribution $f(x_i)$ so that the mean's model is the average of distributions $f(x_i)$. The population analogues to the mean and variance are defined over function $f(x)$ so that the mean, median and the midrange are unbiased estimates of the population mean. Moreover, the variance of the mean is N times less than the population variance, so that the standard deviation tends to decrease by N when N grows.

Let us further assume that the population's probabilistic distribution is Gaussian $N(\mu, \sigma)$ with density function

$$f(u) = Ce^{-(u-\mu)^2/2\sigma^2}, \tag{2.6}$$

where C stands for a constant term equal to $C = (2\pi\sigma^2)^{-1/2}$. Then c in (2.2) is an estimate of μ, and s in (2.3), of σ in (2.6). These parameters amount to the population analogues of the mean and variance defined, for any density function $f(u)$, as $\mu = \int uf(u)du$ and $\sigma^2 = \int (u-\mu)^2 f(u)du$ where the integral is taken over the entire u axis.

Consider now that the set X is a random independent sample from a population with a Gaussian, for the sake of simplicity, probabilistic density function $f(x) = C\exp\{-(x-\mu)^2/2\sigma^2\}$ (2.6) where μ and σ^2 are unknown parameters. The likelihood of randomly obtaining x_i then will be $C\exp\{-(x_i - \mu)^2/2\sigma^2\}$. The likelihood of the entire sample X will be the product of these values, because of the independence assumption. Therefore, the likelihood of the sample is $L(X) = \Pi_{i\in I} C\exp\{-(x_i - \mu)^2/2\sigma^2\} = C^N \exp\{-\Sigma_{i\in I}(x_i - \mu)^2/2\sigma^2\}$. One may even go further and express $L(X)$ as $L(X) = \exp\{N\ln(C) - \Sigma_{i\in I}(x_i - \mu)^2/2\sigma^2\}$ where $\ln$ is the natural logarithm (over base e). A well established approach in mathematical statistics, the principle of maximum likelihood, claims that the values of μ and σ^2 best fitting the data X are those at which the likelihood $L(X)$ or, equivalently its logarithm, $\ln(L(X))$, reaches its maximum. The maximum $\ln(L) = N\ln(C) - \Sigma_{i\in I}(x_i - \mu)^2/2\sigma^2$ is reached at μ minimizing the expression in the exponent, $E = \Sigma_{i\in I}(x_i - \mu)^2$, which is in fact the summary quadratic distance (2.1), that is, the least-squares criterion, which thus can be derived from the assumption that the sample is randomly drawn from a Gaussian population. This, however, does not mean that the least-squares criterion is only meaningful under the normality assumption: the least-squares criterion has a meaning of its own within the data analysis paradigm.

Likewise, the optimal σ^2 minimizes part of $\ln(L)$ depending on it, $g(\sigma^2) = -N\ln(\sigma^2)/2 - \Sigma_{i\in I}(x_i-\mu)^2/2\sigma^2$. It is not difficult to find the optimal σ^2 from the first-order optimality condition for $g(\sigma^2)$. Let us take the derivative of the function over σ^2 and equate it to 0: $dg/d(\sigma^2) = -N/(2\sigma^2) + \Sigma_{i\in I}(x_i-\mu)^2/2(\sigma^2)^2 = 0$. This equation leads to $\sigma^2 = \Sigma_{i\in I}(x_i-\mu)^2/N$, which means that the variance s^2 is the maximum likelihood estimate of the parameter in the Gaussian distribution.

However, when μ is not known beforehand but rather found from the sample according to formula (2.2) for the mean, s^2 in (2.3) is a slightly biased estimate of σ^2 and must be corrected by taking the denominator equal to $N-1$ rather than N which is the case in many statistical packages. The intuition behind the correction is that Equation (2.2) is a relation imposed by us on the N observed values, which effectively decreases the "degree of freedom" in the observations from N to $N-1$.

In situations in which the data entities can be plausibly assumed to have been randomly and independently drawn from a Gaussian distribution, the derivation above justifies the use of the mean and variance as the only theoretically valid estimates of the data center and spread. The Gaussian distribution has been proven to approximate well situations in which there are many small independent random effects adding to each other. However, in many cases the assumption of normality is highly unrealistic, which does not necessarily lead to rejection of the concepts of the mean and variance – they still may be utilized within the general data analysis perspective.

In some real life situations, the assumption that X is an independent random sample from the same distribution seems rather adequate. However, in most real-world databases and multivariate samplings this assumption is far from being realistic.

C2.2.3 Centers and Spreads: Computation

In MatLab, there are commands to compute mean(X) and median(X), which can be done over X being a matrix, not just a vector. The result will be a row of within-column means or medians, respectively. To compute the row of mid-ranges, one can use a combined command mr = (max(X)+min(X))/2. To compute an upper p-quantile of a feature vector x, one should first sort it, in the descending order, with command sx = sort(x, 'descent'), after which the quantile is determined as sx(k) where k = ceil(p*length(x)).

The standard deviation is computed with command std(x), with $N-1$ in the denominator of (2.3), or std(x,1), with N in the denominator.

A stable version of the range that can be used at large N values or when outliers are expected, can be defined by utilizing the concept of quantile. Initially, a value of the proportion p, say 1 or 2% is specified. The upper (lower) p-quantile is a value xp of X such that the proportion of entities with larger (smaller) than xp values is p. The $2p$-quantile range is defined as the interval between these p-quantiles, stretched up according to the proportion of entities taken out, $(xp-px)/(1-2p)$, where xp and px are the upper and lower p-quantiles, respectively. For example, at $p = 0.05\%$ and $N = 100,000$, xp cuts off 50 largest and px, 50 smallest, values of X.

2.3 Binary and Categorical Features

P2.3.1 Presentation

A categorical feature differs from a quantitative one not just because its values are strings, not numbers – they are coded by numbers anyway to be processed. The average of a quantitative feature is always meaningful, whereas the averaging of categories, such as Occupations – BA, IT or AN – in Student data or Sector of Economy – Retail, Utility or Industry – in Company data, makes no sense even after they are coded by numbers. The applicability of the operation of averaging is indeed a defining property of being quantitative. For example, one may claim that a feature like the number of children in Student data (see Fig. 2.5) is not quantitative because its values can only be whole numbers. Still, a statement like "the average number of children per woman is 1.85" does make sense because it can be easily made legitimate by moving to counting by hundreds: there are 185 children per every hundred women.

A feature admitting only two, either "Yes" or "No", values is conventionally considered Boolean in Computer Sciences, thus relating it to Boolean algebra with its "True" and "False" statement evaluations. We do not adhere to this strict logic approach but rather engage the numbers and arithmetic. The values not only can be coded by numerals 1, for "Yes", and 0, for "No", but also arithmetic operations on them can be meaningful too. Two-valued features will be referred to as binary ones.

The mean of a 1/0 coded binary feature is the proportion of its "Yes" values, which is rather meaningful. The other above defined central values bear much less information. The median is 1 only if the proportion of ones is 0.5 or greater; otherwise, it is 0. In a rare event when the number of entities is even and the proportion of ones is exactly one half, the median is one half too. The mode is ether 1 or 0, the same as the median.

For categorical features, there is no need to define bins: the categories themselves play the role of bins. However, their histograms typically are visualized with bars or stems, like on Fig. 2.8 that represents the distribution of categories IT, BA and AN of Occupation feature in Student data.

The distribution of the feature can be expressed in absolute numbers of entities falling in each of the categories, that is, D = (35, 34, 31), or on the relative scale, by using proportions found by dividing frequencies over their total, 35+34+31 = 100, which leads to the relative frequency distribution d = (0.35, 0.34, 0.31).

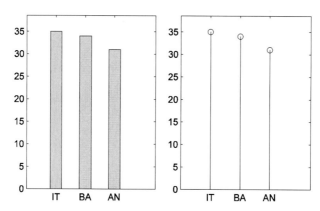

Fig. 2.8 The distribution of categories IT, BA and AN of occupation feature in Student data shown with *bars on the left* and *stems on the right*

2.3 Binary and Categorical Features

Table 2.4 Race distribution of stop-and-search cases in England and Wales in 2005/6

Race	Number of "stop-and-searches"	Relative frequency, (%)
Black (B)	131, 723	15
Asian (A)	70, 250	8
White (W)	676, 180	77
Total	878, 153	100

This distribution is close to the uniform one in which all frequencies are equal to each other. In real life, many distributions are far from that. For example the distribution by race of the 878,153 stop-and-search cases performed by police in England and Wales was widely discussed in the media (see Table 2.4. and BBC's website http://news.bbc.co.uk/1/hi/uk/7069791.stm of 29/10/07.) This is far from uniform indeed: the proportion of W category is thrice greater than of the other two taken together. Yet it was a claim of racial bias because the proportion of W category in the population is even higher than that (for further analysis, see Section 3.4).

Q. 1.9. What is the modal category in the distribution of Table 2.4? In Occupation on Student data? **A.** These are most likely categories, W in Table 2.4 and IT in Student data.

A number of coefficients have been proposed to evaluate how much a distribution differs from the uniform distribution. The most popular are the entropy and Gini index. The latter also is referred to as the categorical variance.

The entropy measures the amount of information in signals being transferred over a communication channel. Intuitively, a rare signal bears more information than a more frequent one. Additionally, the levels of information in independent signals are to be summed up to estimate the total information. These two requirements lead to the choice of logarithm of $1/p$, that is, $-log(p)$, for scoring the level of information in a signal which appears with the probability (frequency) p. The logarithm's base is taken to be 2, because all the digital coding uses the binary number system. The entropy is defined as the averaged level of information in categories of a categorical feature. The unit of entropy has been chosen to be the bit, which is the entropy of a uniformly distributed binary feature, also referred to as a binary digit with two equally likely states. Intuitively, one bit is the level of information given in an answer to a Yes-or-No question in which no prior knowledge of the possible answer is assumed. The maximum entropy of a feature with m categories, $H = log(m)$, is reached when their distribution is uniform. The maximum Gini index, $(m-1)/m$ is reached at the uniform distribution too. Gini index measures the average level of error of the method of proportional classifier. Given a sequence of entities with unknown values of a categorical feature, the proportional classifier assigns entities with values chosen randomly, each with a probability proportional to its frequency. The average error of a category whose frequency is p is equal to $p(1-p) = p-p^2$. If, for example, $p = 20\%$, then the average error is $0.2 - 0.2*0.2 = 16\%$.

Worked example 2.5. Entropy and Gini index of a distribution

Table 2.5 presents all the steps to compute the value of entropy, the summary $-p\log(p)$ value, and Gini index, the summary $p(1-p)$ value where p are probabilities (relative frequencies) of categories.

Entropy is the averaged amount of information in the three categories, $H = -p_1 \log(p_1) - p_2 \log(p_2) - p_3 \log(p_3)$. The entropy in Table 2.5 relative to the maximum is $0.99/1.585 = 0.625$ because at $m = 3$ the maximum entropy is $H = \log(3) = 1.585$. Gini index is defined as the average error of the proportional prediction. The proportional prediction mechanism is defined over a stream of entities of which nothing is known beforehand except for the distribution of categories $\{p_l\}$. This mechanism predicts category l at an entity in p_l proportion of all instances. In our case, $G = p_1(1-p_1) + p_2(1-p_2) + p_3(1-p_3) = 0.378$. The maximum Gini index value, $(m-1)/m$, is reached at the uniform distribution, that is, $G = 2/3$. The relative Gini index, thus, is $0.378/(2/3) = 0.567$, which is not that different from the relative entropy.

Table 2.5 Entropy and Gini index for race distribution in Table 2.4

Distribution		Entropy		Qualitative variance	
Category	Relative frequency p	Information $-\log(p)$	Weighted $-p\log(p)$	Error $1-p$	Variance $p(1-p)$
B	0.15	2.74	0.41	0.85	0.128
A	0.08	3.64	0.29	0.92	0.074
W	0.77	0.38	0.29	0.23	0.177
Total	1.00		0.99		0.378

F2.3.2 Formulation

A categorical feature such as Occupation in Students data or Protocol in Intrusion data, partitions the entity set in such a way that each entity falls in one and only one category. Categorical features of this type are referred to as *nominal* ones.

If a nominal feature has L categories $l = 1, \ldots, L$, its distribution is characterized by amounts $N_1, N_2, \ldots, N_L$ of entities that fall in each of the categories. Because of the partitioning property these numbers sum to the total number of entities, $N_1 + N_2 \ldots, N_L = N$. The relative frequencies, defined as $p_l = N_l/N$ sum to the unity ($l = 1, 2, \ldots, L$).

Since categories of a nominal feature are not ordered, their distributions are better visualized by pie-charts than by histograms.

The concepts of centrality, except for the mode, are not applicable to categorical feature distributions. Spread here is also not quite applicable. However, the variation – or diversity – of the distribution $(p_1, p_2, \ldots, p_L)$ can be measured. There are two rather popular indexes that evaluate dispersion of the distribution, Gini index, or qualitative variance, and entropy.

2.3 Binary and Categorical Features

Gini index G is the average error of the proportional prediction rule. According to the proportional prediction rule, each category l, $l = 1, 2, \ldots, L$ is predicted randomly with the distribution (p_l), so that l is predicted at Np_l cases out of N. The average error of predictions of l in this case is equal to $1 - p_l$, which makes the average arror to be equal to:

$$G = \sum_{l=1}^{L} p_l(1 - p_l) = 1 - \sum_{l=1}^{L} p_l^2$$

Entropy averages the quantity of information in category l as measured by $\log(1/p_l) = -\log(p_l)$ over all l. The entropy is defined as

$$H = -\sum_{l=1}^{L} p_l \log p_l$$

This is not too far away from the Gini index, the qualitative variance, because at small p, $-\log(1-p)$ and $1-p$ coincide, up to a very minor difference, as is well known from calculus (see Fig. 2.9).

A very important class of nominal features consists of features with only two categories – binary features. They may emerge independently as some attributes or divisions. They also can be produced by converting categories of categorical features into binary attributes. For example, IT occupation in Student data can be converted into a question "Is it true that the student's occupation is IT?", that is, a binary feature with answers Yes and No.

These combine properties of both categorical and quantitative features. Indeed, an important difference between categorical and quantitative features is in their admissible coding sets. An admissible numerical recoding of values of a feature changes them consistently, in such a way that the relations between entities according to the feature remain intact. For example, the human heights in centimeters can be recoded in millimeters, by multiplying them by 10, or temperatures at various locations expressed in Fahrenheit can be recoded in Celsius, by subtracting 32 and dividing the result by 1.8. Such a recoding would not change the relations between locations that have been put in effect when Fahrenheit temperatures had been recorded. If, however, we assign arbitrary values to the temperatures, the new set will be inconsistent with the previous one and give a very different information. This is the borderline between quantitative and nominal features: the nominal feature can only compare if the categories are the same or not, thus admitting any

Fig. 2.9 Graphs of functions of the error $f(p) = 1-p$ involved in Gini index (*dashed line*) and the information $f(p) = -\log(p)$

one-to-one recoding as admissible, whereas the quantitative feature can only admit shifts of the origin of the scale and change of the scale factor. This borderline however is not quite hard. The binary features, as nominal ones, admit any numerical recoding. But the recoding, in this case, can always be expressed as a shift of the origin and change of the scale factor. Indeed, for any two numbers, α and β, a conversion of the feature values from 0 to α and from 1 to β can be achieved in a conventional quantitative fashion by using two rescaling parameters: the shift of the origin (α) and scaling factor ($\beta-\alpha$).

Thus, a binary feature can be always considered as coded into a quantitative 1/0 format, 1 for Yes and 0 for No. Thus coded, a binary feature sometimes is referred to as a dummy variable.

To compute the variance of a binary feature with mean $c=p$, sum Np items $(1-p)^2$ and $N(1-p)$ items p^2, which altogether leads to $s^2 = p(1-p) = 1-p^2$. Accordingly, the standard deviation is the square root of the variance, $s = \sqrt{p(1-p)}$. Obviously, this is maximum when $p=0.5$, that is, both binary values are equally likely. The range is always 1. The absolute deviation, in the case when $p < 0.5$ so that median $m=0$, comprises Np items that are 1 and $N(1-p)$ items that are 0, so that $sm=p$. When $p < 0.5$, $m=1$ and the number of unity distances is $N(1-p)$ leading to $sm = 1-p$ That means that, in general, $sm = \min(p, 1-p)$, which is less than or equal to the standard deviation. Indeed, if $p \leq 0.5$, then $p \leq 1-p$ and, thus, $p^2 \leq p(1-p)$, so that $ms \leq s$. Analogously, if $p > 0.5$ then $p > 1-p$ and, thus, $p(1-p) > (1-p)^2$, so that again $sm < s$, which proves the statement.

When a categorical feature is converted into a set of binary features corresponding to its categories, the total variance of the L binary variables is equal to the Gini index, or qualitative variance, of the original feature.

There are some probabilistic underpinnings to binary features. Two models are popular, one by Bernoulli and another by Poisson. Given p, $0 \leq p \leq 1$, Bernoulli model assumes that every x_i is either 1, with probability p, or 0, with probability $1 - p$. Poisson model suggests that, among the N binary numerals, random pN are unities, and $(1-p)N$ zeros. Both models yield the same mathematical expectation, p. However, their variances differ: the Bernoulli distribution's variance is $p(1-p)$, whereas the Poisson distribution's variance is p, which is obviously greater for all positive p, because the factor at Bernoulli standard deviation, $1 - p$, is less than 1 under this condition. Similar models can be considered for nominal features with more than two categories.

There is a rather natural, though somewhat less recognized, relation between quantitative and binary features: the variance of a quantitative feature is always smaller than that of the corresponding binary feature. To explicate this according to Mirkin (2005), assume the interval [0,1] to be the range of data $X = \{x_1,\ldots,x_N\}$. Assume that the mean c divides the interval in such a way that a proportion p of the data is greater than or equal to c, whereas proportion of those smaller than c is $1 - p$. The question then is this: given p, at what distribution of X the variance is maximized. To address the question, assume that X be any given distribution within interval [0,1] with its mean at some interior point c. According to the assumption, there are Np observations between 0 and c. Obviously, the variance can only increase if we move each of these points to the boundary, 0. Similarly, the variance will only increase if we push each of $N(1-p)$ points between c and 1, into the opposite boundary 1. That means that the variance $p(1-p)$ of a binary variable with Np zero and $N(1-p)$ unity values is the maximum, at any p. The following is proven. A binary variable, whose distribution is $(p, 1-p)$, has the maximum variance, and the standard deviation, among all quantitative variables of the same range and p entries below its average.

This implies that no variable over the range [0,1] has its variance greater than the maximum $\frac{1}{4}$ reached by a binary variable at $p=0.5$. The standard deviation of this binary variable is $\frac{1}{2}$, which is just half of the range. Therefore, the standard deviation of any variable cannot be greater than its half-range.

The binary variables also have the maximum absolute deviation among the variables of the same range, which can be proven similarly.

C2.3.3 Computation

If the distribution of a feature is in vector df, then a command like

≫ bar(df, .4);h=axis;axis(1.1*h);

will produce its bar drawing. The parameters here are: 0.4 the width of bars, 1.1 the rescaling to allow some air between the histogram and the border in the drawing frame (see Fig. 2.8).

Computation of the entropy and Gini index for the distribution presented in vector df can be done with commands:
≫ df=df/sum(df); h=−sum(df .*log2(df)); % h is entropy
≫ df=df/sum(df); g=−sum(df .*(1−df)); % h is Gini

Q.2.10 Take nominal features from the Intrusion data set and generate category-based binary features, after which compute their individual means and variances Compare the variances with Gini index for the original features.

2.4 Modeling Uncertainty: Intervals and Fuzzy Sets

2.4.1 Individual Membership Functions

In those cases when the probability distributions are unknown or inapplicable, intervals and fuzzy sets are used to reflect uncertainty in data. When dealing with complex systems, feature values cannot be determined precisely, even for such a relatively stable and homogeneous dimension as the population resident in a country. The so-called "linguistic variables" (Zadeh 1970) express imprecise categories and concepts in terms of appropriate quantitative measures, such as the concept of "normal temperature" of an individual – a body temperature from about 36.0 to 36.9

Celsius or "normal weight" – the Body Mass Index BMI (the ratio of the weight, in kg, to the height, in meters, squared) somewhat between 20 and 25. (Those with BMI > 25 are considered overweight or even obese if BMI > 30; and those with BMI < 20, underweight). In these examples, the natural boundaries of a category are expressed as an interval.

A more flexible description can be achieved using the concept of fuzzy set A expressed by the membership function $\mu_A(x)$ defined, on the example of Fig. 2.10, as:

$$\mu_A(x) = \begin{cases} 0 & if \quad x \leq 18 \quad or \quad x \geq 27 \\ 0.25x - 4.5 & if \quad 18 \leq x \leq 22 \\ 1 & if \quad 22 \leq x \leq 24 \\ -x/3 + 9 & if \quad 24 \leq x \leq 27 \end{cases}$$

This function says that the normal weight does not occur outside of the BMI interval [18, 27]. Moreover, the concept applies in full, with the membership 1, only within BMI interval [22, 24]. There are "grey" areas expressed with the slopes on the left and the right so that, say, a person with BMI = 20 will have the membership value $\mu_A(20) = 0.25*20 - 4.5 = 0.5$ and the membership of that with BMI = 26.1, will be $\mu_A(26.1) = -26.1/3 + 9 = -8.7 + 9 = 0.3$.

In fact, a membership function may have any shape; the only requirement perhaps is that there should exist at least one point or sub-interval at which the function reaches the maximum value 1. A fuzzy set formed with straight lines, such as that on Fig. 2.10, is referred to as a trapezoidal fuzzy set. Such a set can be represented by four points on the axis x :(a, b, c, d) such that $\mu_A(x) = 0$ outside the outer interval *[a,d]* and $\mu_A(x) = 1$ inside the inner interval *[b,c]* (with the straight lines connecting points (a,0) and (b,1) as well as (c,1) and (d,0) (see Fig. 2.10).

Both the precise and interval values can be considered special cases of trapezoidal fuzzy sets. An interval *(a, b)* can be equivalently represented by a trapezoidal fuzzy set *(a, a, b, b)* in which all points of *(a, b)* have their membership value equal to 1, and a point *a* can be represented by trapezoidal fuzzy set *(a, a, a, a)*.

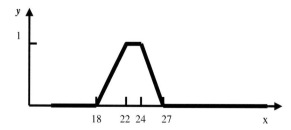

Fig. 2.10 A trapezoidal membership function expressing the concept of normal body mass index; a positive degree of membership is assigned to each point within interval [18, 27] and, moreover, those between 22 and 24 certainly belong to the set

2.4 Modeling Uncertainty: Intervals and Fuzzy Sets

The so-called triangular fuzzy sets are also popular. A triangular fuzzy set A is represented by an ordered triplet (a,b,c) so that $\mu_A(x) = 0$ outside the interval $[a,c]$ and $\mu_A(x) = 1$ only at $x = b$, with values of $\mu_A(x)$ in between are represented by the straight lines between points $(a,0)$ and $(b,1)$ and between $(c,0)$ and $(b,1)$ on the Cartesian plane, see Fig. 2.11.

Fuzzy sets presented on Figs. 2.10 and 2.11 are not equal to each other: only those fuzzy sets A and B are equal at which $\mu_A(x) = \mu_B(x)$ for every x, not just outside of the base interval

A fuzzy set should not be confused with a probabilistic distribution such as a histogram: there may be no probabilistic mechanism nor frequencies behind a membership function, just an expression of the extent at which a concept is applicable. A conventional, crisp set S, can be specified as a fuzzy set whose membership function μ admits only values 0 or 1 and never those between; thus, $\mu(x) = 1$ if $x \in S$ and $\mu(x) = 0$, otherwise.

Q.2.11. Prove that triangular fuzzy sets are but a special case of trapezoidal fuzzy sets. **A.** Indeed a triangular fuzzy set (a,b,c) can be represented by a trapezoidal fuzzy set (a,b,b,c).

There are a number of specific operations with fuzzy sets imitating those with the "crisp" sets, first of all, the set-theoretic complement, union and intersection.

The complement of a fuzzy set A is fuzzy set B such that $\mu_B(x) = 1 - \mu_A(x)$. The union of two fuzzy sets, A and B, is a fuzzy set denoted by $A \cup B$ whose membership function is defined as $\mu_{A \cup B}(x) = \max(\mu_A(x), \mu_B(x))$. Similarly, the intersection of two fuzzy sets, A and B, is a fuzzy set denoted by $A \cap B$ whose membership function is defined as $\mu_{A \cap B}(x) = \min(\mu_A(x), \mu_B(x))$.

It is easy to prove that these operations indeed are equivalent to the corresponding set theoretic operations when performed over crisp membership functions. It should be noted, though, that of all these operations only the union is always correct; the others can bring forward a fuzzy set whose maximum is less than 1.

Q.2.12. Draw the membership function of fuzzy set A on Fig. 2.11.

Q.2.13. What is the union of the fuzzy sets presented in Figs. 2.10 and 2.11.

Q.2.14. What is the intersection of the fuzzy sets presented in Figs. 2.10 and 2.11.

Q.2.15. Draw the membership function of the union of two triangular fuzzy sets represented by triplets (2,4,6), for A, and (3,5,7), for B. What is the membership function of their intersection?

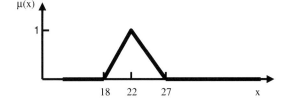

Fig. 2.11 A triangular fuzzy set for the normal weight BMI

Q.2.16. What type of a function is the membership function of the intersection of two triangular fuzzy sets? Of two trapezoidal fuzzy sets? Does it always represent a fuzzy set?

2.4.2 Central Fuzzy Set

The conventional center and spread concepts can be extended to intervals and fuzzy sets. Let us consider an extension of the concept of average to the triangular fuzzy sets using the least-squares data recovery approach.

Given a set of triangular fuzzy sets $A_1, A_2, \ldots, A_N$, the central triangular set A can be defined by such a triplet *(a, b, c)* that approximates the triplets (a_i, b_i, c_i), $i = 1, 2, \ldots, N$. The central triplet can be defined by the condition that the average difference squared,

$$L(a,b,c) = (\Sigma_i(a_i - a)^2 + \Sigma_i(b_i - b)^2 + \Sigma_i(c_i - c)^2)/(3N)$$

is minimized by it. Since the criterion L is additive over the triplet's elements, the optimal solution is analogous to that obtained in the conventional case: the optimal a is the mean of $a_1, a_2, \ldots, a_N$; and the optimal b and c are the means of b_i and c_i, respectively.

Q.2.17. Prove that the average a_i indeed minimizes L. **A.** Let us take the derivative of L over a: $\partial L/\partial a = -2\Sigma_i(a_i - a)/(3N)$. The first-order optimality condition, $\partial L/\partial a = 0$, has the average as its solution described.

Q.2.18. Explore the concepts of central trapezoidal fuzzy set and central interval in an analogous way.

Project 2.1. Computing Minkowski metric's center

Consider a series x_i, $i = 1, 2, \ldots, N$ and given a positive $p > 0$, compute such an a that minimizes the summary Minkowski criterion, p-th power of the distance,

$$Lp = |x_1 - a|^p + |x_2 - a|^p + \ldots |x_N - a|^p \qquad (2.7)$$

When $p \neq 2$, no generally applicable analytic expression can be derived for the minimizer. One way to proceed would be using the mechanisms of hill-climbing, a strategy of iteratively approaching a (local) minimum point by moving step-by-step in the anti-gradient direction which is frequently referred to as the steepest descent direction (see Polyak 1987). Another way is to use a nature-inspired strategy by letting a population of admissible solutions to interatively evolve and keeping track of the "best" points visited (see Engelbrecht 2002).

2.4 Modeling Uncertainty: Intervals and Fuzzy Sets

We take on both approaches to minimization of *Lp*:

(i) Steepest descent iterations, and
(ii) Nature inspired iterations.

(i) Steepest descent computation MC_SD

Before we proceed to computations, let us explore the criterion Lp in (2.7). For the sake of simplicity, assume $p \geq 1$. Consider that the N values in X are sorted in the ascending order so that $x_1 \leq x_2 \ldots \leq x_N$. Then it is easy to prove, first, that the criterion is a convex function shaped like that presented on Fig. 2.12, and, second, the optimal a-value is indeed between the minimum, x_1, and the maximum, x_N.

Assume the opposite: that the minimum is reached outside of the interval, say at $a > x_N$. Then, obviously, $Lp(x_N) < Lp(a)$ because $|x_i - x_N| < |x_i - a|$ for every $i = 1, 2, \ldots, N$, and the same holds for the p-th powers of those. As to the convexity, let us consider any a in the interval between x_1 and x_N. Criterion (2.7) then can be rewritten as:

$$Lp(a) = \sum_{i \in I_+} (a - x_i)^p + \sum_{i \in I_-} (x_i - a)^p \qquad (2.8)$$

where I_+ is set of those indices i for which $a > x_i$, and I_- is set of such i's that $a \leq x_i$. The derivative of $Lp(a)$ in (2.8) is equal to:

$$Lp'(a) = p \left(\sum_{i \in I_+} (a - x_i)^{p-1} - \sum_{i \in I_-} (x_i - a)^{p-1} \right) \qquad (2.9)$$

and the second derivative, to

$$Lp''(a) = p(p-1) \left(\sum_{i \in I_+} (a - x_i)^{p-2} + \sum_{i \in I_-} (x_i - a)^{p-2} \right).$$

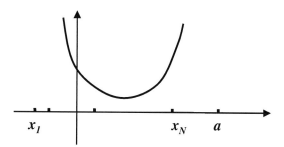

Fig. 2.12 A convex function of a

The latter expression is positive for each a value, provided that $p > 1$, which proves that $Lp(a)$ is convex. This leads to one more property: assume that $Lp(x_{i*})$ is minimum among all the $Lp(x_i)$ values ($i = 1, 2, \ldots, N$), then the minimum of $Lp(a)$ lies within interval $(x_{i'}, x_{i''})$ where $x_{i'}$ is that x_i-value, which is the nearest to x_{i*} among those on the left of it, at which $Lp(x_i) > Lp(x_{i*})$. And, similarly, $x_{i''}$ is that x_i-value, which is the nearest to x_{i*} among those to the right of it, at which $Lp(x_i) > Lp(x_{i*})$.

The properties above justify the following steepest descent algorithm applicable at $p > 1$:

MC_SD

1. Initialize with $a0 = x_{i*}$ and a positive learning rate λ.
2. Compute $a0 - \lambda Lp'(a0)$ according to formula (2.9) and take it as $a1$ if it falls within the interval $(x_{i'}, x_{i''})$. Otherwise, decrease λ a bit and repeat the step.
3. Test whether $a1$ and $a0$ coincide, up to a pre-specified precision threshold. If yes, halt the process and output $a1$ as the optimal value for a. If not, move on.
4. Test whether $Lp(a1) \leq Lp(a0)$. If yes, set $a0 = a1$ and $Lp(a0) \leq Lp(a1)$, and go to (2). If not, decrease λ a bit and go to 2 (without changing $a0$).

(ii) Nature-inspired computation MC_NI

According to the nature-inspired approach, a population of possible solutions rather than a single solution is maintained. In contrast to the classical approaches, the improvements here are a matter of a random evolution of the population from one generation to another, which is organized in such a way that improvements are likely to be acquired. Since this is a 1D search, it is likely that any random moves would approximate the optimal point soon enough. The simple algorithm MC_NI presented below works quite well in experiments:

1. *Determining the area of admissible solutions.* Determine an area A of admissible solutions – a set of points which should contain the optimum point(s).
   This is quite easy in this case: as proven above, the optimum lies between the minimum lb and maximum rb of the series $x_i, i = 1, 2, \ldots, N$. Thus, the area is interval *(lb,rb)*.
2. *Population setting.* Specify the size pe of the population to evolve, say, $pe = 15$, and randomly put points $s_1, s_2, \ldots, s_{pe}$ in the admissible area *(lb,rb)*.
3. *Elite initialization.* Evaluate values of the criterion, frequently referred to as the "fitting function", for each member of the population and store information of the best (elite), that is, the minimum, as the only record s_e to output when needed.
4. *Next generation.* Modify the population by, first, adding random Gaussian noise r:

$$s'_k = s_k + \lambda r$$

and, second, by moving all those of the resulting values that went out of the area of admissible solutions A back to the area.
5. *Elite maintenance.* Evaluate values of the criterion at the new generation, pick the best and worst of them, say s_b and s_w, and compare with the elite s_e. If s_b is better than s_e, change the elite for s_b. Else, that is, if s_b and, more so, s_w are worse than s_e, improve the current population by changing s_w in that for the record s_e.
6. *Stop condition.* If the number of iterations has not reached a pre-specified value, go to (4). Otherwise, output the elite solution.

Experiments show that the gradient based procedure of the steepest descent is faster than the nature-inspired one. But the latter works at any p, whereas the former only at $p > 1$.

Project 2.2. Analysis of a multimodal distribution

Let us take a look at the distributions of OOP and CI marks at the Student data. Assuming that the data file of Table 1.4 is stored as Data\stud.dat, the corresponding MatLab commands can be as follows:

≫ a=load('Data\stud.dat');
≫ oop=a(:,7); % column of OOP mark
≫ coi=a(:,8); % column of CI mark
≫ subplot(1,2,1); hist(oop);
≫ subplot(1,2,2); hist(coi);

With ten bins used in MatLab by default, the histograms are on Fig. 2.13.
The histogram on the left seems to have three humps, that is, three-modal. Typically, a homogeneous sample should have a unimodal distribution, to allow interpretation of the feature as its modal value with random deviations from it. The

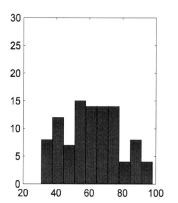 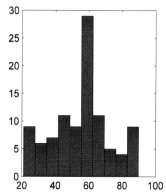

Fig. 2.13 Histograms of the distributions of marks for OOP (*on the left*) and CI (*on the right*) from students data

three modes on the OOP mark histogram require an explanation. For example, one may hypothesize that the modes can be explained by the presence of three different groups of students represented by their occupations so that IT group should have higher marks than BA group whose marks should still be higher than those at AN group.

To test this hypothesis, one needs to compare distributions of OOP marks at each of the occupations. To make the distributions comparable, we need to specify an array with boundaries between 10 bins that can be used for each of the samples. This array, b, can be computed as follows:

≫ r=max(oop)−min(oop);for i=1:11;b(i)=min(oop)+(i−1)*r/10;end;

Now we are ready to produce comparable distributions for each of the occupations with MatLab command histc:

≫ for ii=1:3;li=find(a(:,ii)==1);hp(:,ii)=histc(oop(li),b);end;

This generates a list, li, of student indices corresponding to each of the three occupations presented by the three binary columns. Matrix hp stores the three distributions in its three columns. Obviously, the total distribution of OOP, presented on the left of Fig. 2.13 is the sum of these three columns. To visualize the distributions, one may use "bar" command in MatLab:

≫ bar(hp);

which produces bar histograms for each of the three occupations (see Fig. 2.14). One can see that the histograms differ indeed and concur with the hypothesis, so that IT concentrates in top seven bins and, moreover, it shares the top three bins with no other occupation. The other two occupations overlap more, though AN takes over on the leftmost, worst marks, positions indeed.

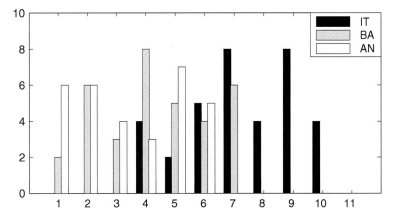

Fig. 2.14 Histograms of OOP marks for each of three occupations, IT, BA and AN, each presented with bars filled in according to the legend

2.4 Modeling Uncertainty: Intervals and Fuzzy Sets

Q.2.19. What would happen if array *b* is not specified once for all but the histogram is drawn by default for each of the sub-samples? **A.** The 10 default bins depend on the data range, which may be different at different sub-samples; if so, the histograms will be incomparable.

There can be other hypotheses as well, such as that the modes come from different age groups. To test that, one should define the age group boundaries first.

Project 2.3. Computational validation of the mean by bootstrapping

The data file short.dat in Appendix A5 is a 50 × 3 array whose columns are samples of three data types described in Table 2.6.

The normal data is in fact a sample from a Gaussian N(10,2), that has 10 as its mean and 2 as its standard deviation. The other two are Two-modal and Power law samples. Their histograms are on the left-hand sides of Figs. 2.15, 2.16, and 2.17. Even with the aggregate data in Table 2.6 one can see that the average of Power law does not make much sense, because its standard deviation is more than three times greater than the average.

Table 2.6 Aggregate characteristics of columns for short.dat array

Data type		Normal	Two-modal	Power law
Mean		10.27	16.92	289.74
Standard deviation	Real value	1.76	4.97	914.50
	Related to $\sqrt{N}$	0.25	0.70	129.33

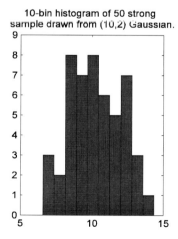
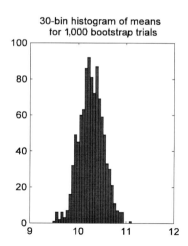

Fig. 2.15 The histograms of a 50 strong sample from a Gaussian distribution (*on the left*) and its mean's bootstrap values (*on the right*): all falling between 9.7 and 10.1

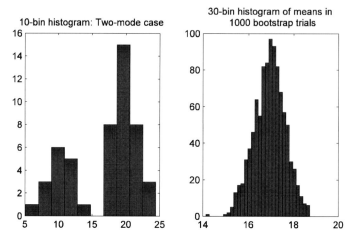

Fig. 2.16 The histograms of a 50 strong sample from a Two-mode distribution (*on the left*) and its mean's bootstrap values (*on the right*)

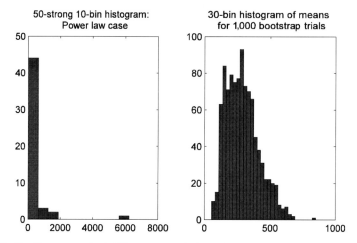

Fig. 2.17 The histograms of a 1,000 strong sample from a Power law distribution (*on the left*) and its mean's bootstrap values (*on the right*): all falling between 260 and 560

Many statisticians would argue the validity of characteristics in Table 2.6 not because of the distribution shapes – which would be a justifiable source of concern for at least two of the three distributions – but because of small sizes of the samples. Is the 50 entities available a good representation of the entire population indeed? To address these concerns, the Mathematical Statistics have worked out principles based on the assumption that the sampled entities come randomly and independently from a – possibly unknown but stationary – probabilistic distribution. The mathematical thinking would allow then, in reasonably well-defined situations, to arrive at a theoretical distribution of an aggregate index such as the mean, so that the distribution may lead to some confidence boundaries for the index.

2.4 Modeling Uncertainty: Intervals and Fuzzy Sets

Typically, one would obtain the boundaries of an interval at which 95% of the population falls, according to the derived distribution. For instance, when the distribution is normal, the 95% confidence interval is defined by its mean plus/minus 1.96 times the standard deviation related to the square root of the number observations, which is 7.07 at $N = 50$. Thus, for the first column data, the theoretically derived 95% confidence interval will be $10 \pm 1.96*2/7.07 = 10 \pm 0.55$, that is, (9.45, 10.55) (if the true parameters of the distribution are known) or $10.27 \pm 1.96*1.76/7.07 = 10.27 \pm 0.49$, that is, (9.78,10.76) (at the observed parameters in Table 2.6). The difference is rather minor, especially if one takes into account that the 95% confidence is a rather arbitrary notion. In probabilistic statistics, the so-called Student's distribution is used to make up for the fact that the sample-estimated standard deviation value is used instead of the exact one, but that distribution differs little from the Gaussian distribution when there are more than several hundred entities.

In many real life applications the shape of the underlying distribution is unknown and, moreover, the distribution is not necessarily stationary. The theoretically defined confidence boundaries are of little value then. This is why a question arises whether any confidence boundaries can be derived computationally by re-sampling the data at hand rather than by imposing some debatable assumptions. There have been developed several approaches to computational validation of sample based results. One of the most popular is bootstrapping which will be used here in its two basic, "pivotal" and "non-pivotal", formats as defined in Carpenter and Bithell (2000) (see also Efron and Tibshirani 1993).

Bootstrapping is based on a pre-specified number, say 1,000, of random trials. *A trial* involves randomly drawn N entities, with replacement, from the entity set. Note that N is the size of the entity set. Since the sampling goes with replacement, some entities may be drawn two or more times so that some others are bound to be left behind. Recalling that $e = 2.7182818\ldots$ is the natural logarithm base, it is not difficult to see that, on average, only approximately $(e-1)/e = 63.2\%$ entities get selected into a trial sample. Indeed, at each random drawing from a set of N, the probability of an entity being not drawn is $1-1/N$, so that the approximate proportion of entities never selected in N draws is $(1-1/N)^N \approx 1/e = 1/2.71828 \approx 36.8\%$ of the total number of entities. For instance, in a bootstrap trial of 15 entities, the following numbers have been drawn: 8, 11, 7, 5, 3, 3, 11, 5, 9, 3, 11, 6, 13, 13, 9 so that seven entities have been left out of the trial while several multiple copies have got in.

A trial set of a thousand randomly drawn entity indices (some of them, as explained, would coincide) is assigned with the corresponding row data values from the original data table so that coinciding entities get identical rows. Then a method under consideration, currently "computing the mean", applies to this trial data to produce the trial result. After a number of trials, the user gets enough results to represent them with a histogram and derive confidence boundaries from that.

The bootstrap distributions for each of the three types of data generation mechanism, after 1,000 trials, are presented in Figs. 2.15, 2.16 and 2.17 on the right hand side.

The pivotal validation method is based on the assumption that the bootstrap distribution of means is Gaussian, so that having estimated its average m_b and standard deviation s_b, the 95% confidence interval is estimated as usual, with formula $m_b \pm 1.96 * s_b = 10.24 \pm 1.96 * 0.24 = 10.24 \pm 0.47$, which is the interval between 9.77 and 10.71 – which is very similar to that obtained under the hypothesis of Gaussian distribution – this is no wonder here because the hypothesis is true.

The non-pivotal method makes no assumption of the distribution of bootstrap means and uses the empirical bootstrap found distribution to cut it at its 2.5% upper and bottom quantiles. To do this, we can sort values of the vector of bootstrap means and find the values at its 26th and 975th components that cut out exactly 2.5% of the set each. This action produces interval between 9.78 and 10.70, which is very close to the previously found boundaries for the 95% confidence interval for the mean value of the first sample.

There is theoretical evidence, presented by E. Bradley (1993), to support the view that the bootstrap can produce somewhat tighter estimate of the marks deviation than the estimate based on the original sample. In our case, we can see in Table 2.7 that indeed, with the means almost unchanged, the standard deviations have been slightly reduced.

Unfortunately, the bootstrap results are not that helpful in analyzing the other two distributions: as can be seen in our example, both of the means, the Two-modal and Power law ones, are assigned rather decent boundaries while, in most applications, the mean of either of these two distributions may be considered meaningless. It is a matter of applying other data analysis methods such as clustering to produce more homogeneous sub-samples whose distributions would be more similar to that of a Gaussian.

The reader is requested to provide pivotal and not-pivotal estimates of 95% confidence interval for the other two samples in short.dat dataset (Two-modal and Power law).

Project 2.4. K-fold cross validation

Another set of validation techniques utilizes randomly splitting the entity set in two parts of pre-specified sizes, the so-called train and test sets, so that the method's results obtained for the train set are compared with the data on the test set. To

Table 2.7 Aggregate characteristics of the results of 1,000 bootstrap trials over short.dat array

Data type		Normal	Two-mode	Power law
Mean		10.27	16.94	287.54
Standard deviation	Original sample	0.25	0.70	129.33
	Bootsrap value	0.25	0.69	124.38
	Relative to mean, %	2.46	4.05	43.26

2.4 Modeling Uncertainty: Intervals and Fuzzy Sets

guarantee that each of the entities gets into a train/test set the same number of times, the so-called cross-validation methods have been developed.

The so-called K-fold cross validation works as follows. Randomly split entity set in K parts $Q(k)$, $k = 1,\ldots,K$, of equal sizes.[1] Typically, K is taken as 2 or 5 or 10. In a loop over k, each part $Q(k)$ is taken as test set while the rest forms the train set. A data analysis method under consideration is run over the train set ("training phase") with its result applied to the test set. The average score of all the test sets constitutes a K-fold cross-validation estimate of the method's quality.

The case when K is equal to the number of entities N is especially popular. It was introduced earlier under the term "jack-knife", but currently term "leave-all one-out" is more popular as better reflecting the idea of the method: N trials are run over the entire set except for just each one entity removed from the training.

Let us apply the 10-fold cross-validation method to the problem of evaluation of the means of the three data sets. First, let us create a partition of our 1,000 strong entity set in 10 non-overlapping classes, a hundred entities each, with randomly assigning entities to the partition classes. This can be done by randomly putting entities one by one in each of the 10 initially empty buckets. Or, one can take a random permutation of the entity indices and divide then the permuted series in 10 chunks, 100 strong each. For each class $Q(k)$ of the 10 classes ($k = 1, 2, \ldots, 10$), we calculate the averages of the variables on the complementary 900 strong entity set, and use these averages for calculating the quadratic deviations from them – not from the averages of class $Q(k)$ – on the class $Q(k)$. In this way, we test the averages found on the complementary training set.

The results are presented in Table 2.8. The values found at the original distribution and with a ten fold cross validation are similar. Does this mean that there is no need in applying the method? Not at all, when more complex data analysis methods are used, the results may differ indeed. Also, whereas the ten quadratic deviations calculated on the ten test sets for the Gaussian and Two-modal data are very similar to each other, those at the Power law data set drastically differ, ranging from 391.60 to 2,471.03.

Table 2.8 Quadratic deviations from the means computed on the entity set as is and by using ten fold cross validation

Data type		Normal	Two-modal	Power law
Standard deviation	On set	1.94	5.27	1744.31
	ten fold cr.-val.	1.94	5.27	1649.98

Q.2.20. What is the bin size in the example of Fig. 2.18? **A.** 2.

---

[1] To do this, one may start from all sets $Q(k)$ being empty and repeatedly run a loop over $k = 1 : K$ in such a way that at each step, a random entity is drawn from the entity set (with no replacement!) and put into the current $Q(k)$; the process halts when no entities remain out of $Q(k)$.

Fig. 2.18 Range [2,12] divided in five bins

Q.2.21. Consider feature x whose range is between 1 and 10. When the range of x is divided in 9 bins (in this case, intervals of the lengths one: [1,2), [2,3), ..., [9,10]), the x frequencies in the corresponding bins are: 10, 20, 10, 20, 30, 20, 40, 20, 30. Please answer these questions:

(i) How many observations of x are available?
(ii) What can be said about the value of the median of x?
(iii) Provide the minimum and maximum estimates of the average of x.
(iv) What can be said of 20% quantiles of x?
(v) What is the distribution of x when the number of bins is 3? What is the qualitative variance (Gini coefficient) for this distribution?

A.

(i) There are 200 observations.
(ii) The median lies between 100-th and 101-th values in a sorted order, that is, in the 6-th bin, that is, between 6 and 7.
(iii) The minimum estimate of the mean is computed with the minimal values in bins:
$(1*10+2*20+3*10+4*20+5*30+6*20+7*40+8*20+9*30)/200 = 5.7$
The maximum estimate is calculated using the same formula with all bin values increased by 1, which should lead to $5.7+1 = 6.7$.
(iv) 20% of 200 is 40. That means that the 20% quantile on the left end of x is 4, while that on the right end must be in the 8-th bin, that is, between 8 and 9.
(v) The three-bin distribution will be 40, 70, 90 or, in the relative frequencies, 0.2, 0.35, 0.45, which leads to the Gini index equal to $1-0.2^2-0.35^2-0.45^2 = 0.635$.

Q.2.22. Occurrence and co-occurrence. Of 100 Christmas shoppers, 50 spent £60 each, 20 spent £100 each, and 30 spent £150 each. What are the (i) average, (ii) median and (iii) modal spending? Tip: How one can take into account in the calculation that there are, effectively, only three different types of customers?

A. Average: First, let us see that the proportions of shoppers who spent £60, £100 and £150 each are, respectively, 0.5, 0.2 and 0.3. The average can be calculated by weighting the expenditures by the proportions so that Average = $60*0.5 + 100*0.2 + 150*0.3 = 30.0 + 20.0 + 45.0 = 95$.

Median: According to definition, the median of 100 numbers is the mid value between 50th and 51st entries in their sorted order, which are 60 and 100 in this case. Thus the median spending is £80.

Mode: The modal value is the most likely one, that is, 60.

2.4 Modeling Uncertainty: Intervals and Fuzzy Sets   63

Q.2.23. Consider two geological formations that are represented by 7 and 5 ore specimens, respectively. The mineral contents in formation A is described by vector a = (7.6, 11.1, 6.8, 9.8, 4.9, 6.1, 15.1), and in formation B, by vector b = (4.7, 6.4, 4.1, 3.7, 3.9). The average content in A is 8.77 and in B, 4.56. Test the hypothesis that the mineral contents in A is richer than in B (with 95% confidence) by using bootstrap. **A.** Because the sets are quite small, the number of trials should be taken rather moderate, not greater than 200. At 200 trials, 95% confidence interval will be found with boundaries at 6-th and 195-th values in the sorted series of bootstrap means. In our computation, this is interval (6.66, 11.09) for A and (3.82, 5.44) for B. Since all of the former interval is greater than all of the latter interval, the hypothesis can be considered confirmed. (There is a flaw in this solution, because of some imprecision in the notion that A is richer than B. If we define that A is richer than B with 95% confidence if a random sample from A is richer than a random sample from B in 95% of the cases, then the 95%-intervals are not enough – they cover only 0.95*0.95 = 90.25% of all possible pairs of bootstrap mean values, which means the hypothesis is proven with 90% confidence. Yet if we take a look at the minimum and maximum bootstrap mean values, we find that the entire range of means is (6.33, 11.94) for A and (3.82, 5.82) for B, which means that the hypothesis is proven now since 5.82 < 6.33 – within the limits of the method.)

Q.2.24. Central triangular fuzzy set. Given three triangular fuzzy sets defined by triples (0, 2, 3), (0, 3, 4), and (3, 4, 8), determine the corresponding central triangular fuzzy set. **A.** The central triangular fuzzy set is defined by the average values such as (0+0+3)/3 = 1, for the first component; so that it is (1, 3, 5).

Q.2.25. Iris feature distributions. Consider histograms of Iris dataset features and demonstrate that two of them are bimodal. **A:** With a MatLab command like
>> for k = 1:4;subplot(2,2,k);hist(iris(:,k),15);end; a figure like Fig. 2.19 will appear. Obviously the third and fourth features are bimodal.

Q.2.26. To run a computational experiment, a student is to randomly generate distributions of relative frequencies for a three-category nominal variable. The student decides first generate random numbers in interval (0,1) and then make them sum to unity by relating them to thir sum. Thus, for example, random numbers 0.7116, 0.1295, 0.6598 are first generated, and then divided by their sum 1.5009 to produce values 0.4741, 0.0863, 0.4396 totaling to 1. Is it a right way to go? **A.** Not exactly. A bias towards equal frequencies will be created. For example, take a look at Fig. 2.20a presenting the distribution of the first element of a pair of frequencies found by the described method: generate a pair of random numbers and then divide them by the sum. This distribution is far from that of a uniformly random value presented on Fig. 2.20b. (Can you explain the difference?) An appropriate way for generating uniformly random frequency triplets would be this.

First, generate just two random numbers, then sort them in ascending order and add 0 and 1 into the series: $r_0 = 0 < r_1 < r_2 < r_3 = 1$. Then define the frequencies as differences of neighboring values in the series, $p_k = r_k - r_{k-1}$ $(k = 1, 2, 3)$. For example, if 0.8775, 0.5658 were first generated, then the frequencies would be

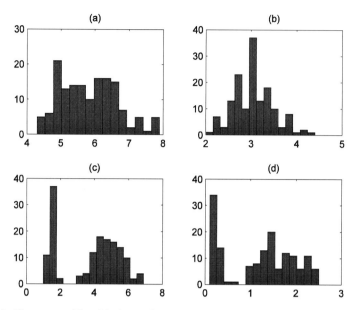

Fig. 2.19 Histograms of four Iris dataset featues; (**c**) and (**d**) are bimodal

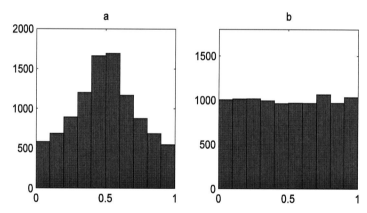

Fig. 2.20 Histograms of a 100,000 strong random sample of (**a**) the first element of a random pair after division by the pair summary value, and (**b**) just a random number

defined as $p_1 = 0.5658$, $p_2 = 0.8775 - 0.5658 = 0.3117$, and $p_3 = 1 - 0.8775 = 0.1225$. This method is easily extendable to any number of categories.

2.5 Summary

This chapter presents summaries of one-dimensional data, first of all, histograms, central points and spread evaluations. Two perspectives are outlined: one is the classical probabilistic and the other of approximation, naturally extending into the data recovery approach to supply a decomposition of the data scatter in the explained and unexplained parts.

A difference between categorical and quantitative features is pointed out: the latter admit averaging whereas the former not. This difference is somewhat blurred at binary features especially the so-called dummies, 1/0 variables representing individual categories – they can be considered quantitative too.

Some attention is given to modeling uncertainty by using intervals and fuzzy sets, but not much. In fact, most of further methods can be extended to these more complex data types.

Several projects are presented to show how questions may arise and get computational answers. Computational intelligence and cross validation approaches are involved.

References

Carpenter, J., Bithell, J.: Bootstrap confidence intervals: when, which, what? A practical guide for medical statisticians. Stat. Med. **19**, 1141–1164 (2000)
Efron, B., Tibshirani, R.: An Introduction to Bootstrap. Chapman & Hall, Boca Raton, FL (1993)
Engelbrecht, A.P.: Computational Intelligence. Wiley, New York, NY (2002)
Lohninger, H.: Teach Me Data Analysis. Springer, Berlin-New York-Tokyo (1999)
Polyak, B.: Introduction to Optimization. Optimization Software, Los Angeles, CA, ISBN: 0911575146 (1987)
Zadeh, L.A.: The concept of a linguistic variable and its application to approximate reasoning I–II. Inf. Sci., 8, 199-249, 301-375 (1975)

Chapter 3
2D Analysis: Correlation and Visualization of Two Features

3.1 General

Analysis of two features on the same entity set can be of interest assuming that the features are related in such a way that certain changes in one of them tend to co-occur with changes in the other. Then the relation – if observed indeed – can be used in various ways, of which two types of application are typically discernible: those oriented at

(i) prediction of values of one variable from those of the other;
(ii) addition of the relation to the knowledge of the domain by interpreting and explaining it in terms of the existing knowledge.

Goal (ii) is a subject in the discipline of knowledge bases as part of the so-called inferential approach, in which all relations are assumed to have been expressed as logical predicates and treated within a formal logic system – this approach will not be described here. We concentrate on another approach, referred to as the inductive one and related to the analysis of what type of information the data can provide with respect to the goals (i) and (ii). Typically, the feature whose values are to be predicted is referred to as the target variable and the other as the input variable. Examples of goal (i) are: prediction of an intrusion attack of a certain type (Intrusion data) or prediction of exam mark (Student data) or prediction of the number of Primary schools in a town whose population is known (Market town data). One may ask: why bother – all numbers are already in the file! Indeed, they are. But in the prediction problem, the data at hand are just a sample from a large population so that it is used as a training ground for devising a decision rule for prediction of the target feature at other, yet unobserved, entities. Typically, the input feature is readily available while the target feature is not. As to the goal (ii), the data usually are just idle empirical facts not necessarily noticeable unless they are generalized into a decision rule.

The mathematical structure and the visual portrayal of the problem differ depending on the type of feature scales involved, which leads us to considering all possible cases (see also Lohninger 1999):

(1) both features are quantitative,
(2) one feature is quantitative, the other categorical, and
(3) both features are categorical.

3.2 Two Quantitative Features Case

P3.2.1 Scatter-Plot, Linear Regression and Correlation Coefficients

In the case when both features are quantitative, the three following concepts are popular: scatter plot, correlation and regression. We consider them in turn by using two features from the Market towns dataset, Population Resident and Number of Primary Schools. The data are taken from Table 1.4 (see below an extract for four towns out of 45):

	Pop (x)	PSchools (y)	(x,y)-point
Tavistock	10,222	5	(10,222,5)
Bodmin	12,553	5	(12,553,5)
Saltash	14,139	4	(14,139,4)
Brixham	15,865	7	(15,865,7)

Scatter plot is a presentation of entities as 2D points in the plane of two pre-specified features. On the left-hand side of Fig. 3.1, a scatter-plot of Market town features Pop (Axis x) and PSchools (Axis y) is presented.

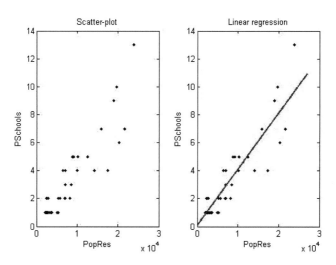

Fig. 3.1 Scatter plot of PopRes versus PSchools in Market town data. The *right hand graph* includes a regression line of PSchools over PopRes

3.2 Two Quantitative Features Case

One can think that these two features are related by a linear equation $y = ax+b$ where a and b are some constant coefficients, referred to as the slope and intercept, respectively, because the number of schools should be related to the number of children which is related to the number of residents. This equation is referred to as the linear regression of y over x. Obviously, most relations are not necessarily that simple because they also depend on other factors such as school sizes, population's age, etc. It would be a miracle if one equation fitted well all 45 towns. The possible inconsistencies in the equation can be modeled as additive errors, or residuals. The slope a and intercept b are taken in such a way that the inconsistencies of the equation on the 45 towns are minimized.

When a linear regression equation is fitted, its validity should be checked. A valid equation can be used for both (i) prediction and (ii) description.

The Galton-Pearson theory of linear regression involves a useful and very popular parameter, the correlation coefficient that shows the extent of linearity in the relation between the two features. Its square, referred to as the determination coefficient, can be used for a quick check of the validity of the regression: it shows the proportion of the variance of y that is taken into account by the regression. The correlation coefficient between the two features, Pop and PSchools, is 0.909. The correlation coefficient, in general, ranges between -1 and 1, and a value close to 1 or -1 indicates a high extent of the linear dependence between the features. In physics or chemistry, a high value of the correlation coefficient is rather usual; in social sciences, rather not – that is, the current features are highly related indeed.

Most other features in Market town data – such as the numbers of Post offices or Doctors – are also highly related to Pop feature, but not the number of Farmers markets. This latter feature appears to be binary here: a town either has a farmers market or not. The low value of the correlation coefficient, just below 0.15, shows that the size of the town does not much matter in this part of the world: a farmers market is as likely in a small town as it is in a larger town.

A low or even zero value of the correlation coefficient does not necessarily mean "no relation at all", but rather just "no *linear* relation". A zero correlation coefficient may hide a different type of functional relation, as shown on Fig. 3.2, which presents three different cases of the zero correlation. Only one of these, that on the left, case is genuine – there is no relation between x and y according to the picture indeed. Each of the other two cases relates to a rather high association between x and y. Specifically, the figure in the middle refers to a quadratic dependence and the figure on the right, to a split between two subsamples of highly linear but inverse relations.

Then the regression equation, estimated according to formulas (3.4–3.6) in Section 3.2.3.2, is this:

$$\text{PSchool} = 0.401^*\text{Pop} + 0.072 \tag{3.1}$$

where Population resident (Pop) is expressed in thousands to make the slope the thousand times greater than it would be if population is expressed in the absolute numbers. The slope expresses how much target changes when the input changes

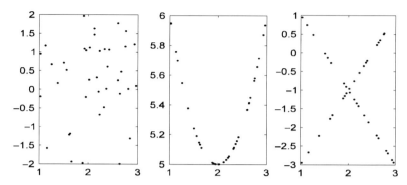

Fig. 3.2 Three scatter-plots corresponding to zero or almost zero correlation coefficient ρ; the case *on the left*: no relation between x and y; the case in the middle: a non-random quadratic relation $y = (x - 2)^2 + 5$; the case *on the right*: two symmetric linear relations, $y = 2x - 5$ and $y = -2x + 3$, each holding at a half of the entities

by 1. Because the target's values are integers, the value of slope can be rephrased as follows: the growth of population in a town by 2.5 thousand would lead, on average, to building one more primary school.

P3.2.2 Validity of the Regression

A regression function built over a data set should be validated. Three types of validity checks can be considered:

(a) The proportion of the variance of target variable taken into account by the regression, the determination coefficient: the greater the determination the better the fit.
(b) The confidence intervals of regression parameters – their ranges can give an idea of how stable the regression is.
(c) The direct testing of the accuracy of prediction both on data used for building the regression and data not used for that.

Worked example 3.1. Determination coefficient

Consider feature PSchools as target versus Pop as input, in Market Data (Fig. 3.1). The correlation coefficient between them is 0.909. The determination coefficient, in the case of linear regression, is its square, that is, $0.909^2 = 0.826$, which shows that the linear dependence on Pop decreases the variance of PSchools by 82.6%, a rather high value.

If the determination coefficient is not that high, still the hypothesis of linear relation may hold – depending on the distribution of residuals, that is, differences

3.2 Two Quantitative Features Case

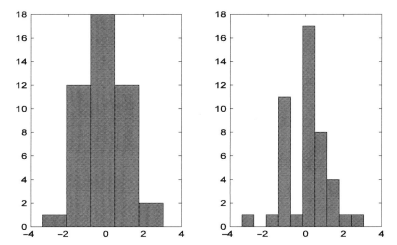

Fig. 3.3 Histograms of the residuals, the differences between values of PSchool as observed and those computed from Pop by using Equation (3.1), with 5 bins (*on the left*) and 10 bins (*on the right*). The dents in the finer histogram can be attributed to the fact that the sample of 45 instances is too small to have 10 bins

between the observed values of PSchool and those computed from Pop according to Equation (3.1). This distribution should be Gaussian or approximately Gaussian, so that the principle of maximum likelihood and formulas derived from it are appropriate. The distribution for the case under consideration is presented on Fig. 3.3. It is similar to a Gaussian distribution indeed, at the 5 bin histogram. The histogram with 10 bins is less so because it is somewhat dented – probably the sample is too small for this level of granularity: on average, only 4–5 entities fall in each of the bins.

A more straightforward validity test can be performed without any statistic theory at all – by purely computational means using the so-called bootstrapping which is a procedure for obtaining a multitude of random estimates of the parameters of interest by using random samples from the dataset as illustrated in Worked example 3.2.

Worked example 3.2. Bootstrap validity testing

Consider the linear regression of PSchools over PopRes in Equation (3.1) in the previous section. How stable are its slope and intercept regarding change of the sample? This can be tested by using bootstrap. One bootstrap trial involves three stages:

1. Randomly choose, with replacement, as many entities as there are in the sample – 45 in this case. Here is the sequence of indices of the entities randomly drawn

with replacement while writing this text: r = {26,17,36,11,29,39,32,25,27,26,29, 4,4,33,10,1,5,45,17,16,13,5,42,43,28,26,35,2,37,44,6,39,33,21,15,11,33,1,44,30, 26,25,5,37,24}. Some indices made it into the sample more than once, most notably 26 – four times, whereas many others did not make it into the sample at all – altogether, 16 entities such as 3,7,8 are absent from the sample. The proportion of the absent indices is 16/45= 0.356, which is rather close to the theoretic estimate 1/e=0.3679 derived in Project 2.3.
2. Take "resampled" versions of *Pop* and *PSchools* as their values on the elements drawn on step 1.
3. Find values of the slope and intercept for the resampled Pop and PSchools and store them.

The MatLab computation steps are similar to those in Project 3.1. After 400 trials the stored slopes and intercepts form distributions presented as 20 bin histograms on Fig. 3.4a, b respectively. After 4,000 trials, the respective histograms are c and d. One can easily see the smoothing effect of the increased number of trials on the histogram shapes – at 4,000 trials they do look Gaussian.

The bootstrapping trials give a diversity needed for estimating the average values of the slope and intercept. Moreover, one can draw confidence boundaries for the values.

How can one obtain, say, 95% confidence boundaries? According to the non-pivotal method, lower and upper 2.5% quantiles are cut out from the distribution in a symmetric way: 95% of the observations fall between the quantiles. For the case of 400 trials, 2.5% equals 10, so that the lower quantile corresponds to 11th and the upper quantile to 390th elements in the sorted set of values. For the case of 4,000 trials, 2.5% equals 100: these quantiles correspond to 101st and 3,900th elements of the sorted sets. They are shown in Table 3.1 at both of the cases, 400 and 4,000 trials. One can see that these provide consistent and rather tight boundaries for the slope: it is between 0.303 and 0.488 in 95% of all trials, according to 4,000-trial data, and

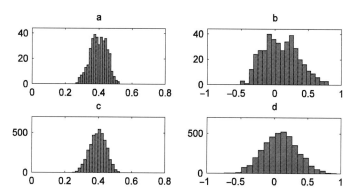

Fig. 3.4 Histograms of the distributions of the slope, *on the left* (**a**) and (**c**), and intercept, *on the right* (**b**) and (**d**), found at 400 (*on top*) and 4,000 (*below*) bootstrapping trials on PopResid, expressed in thousands, and PSchool features in Market town data

3.2 Two Quantitative Features Case

Table 3.1 Parameters of the linear regression of PopResid over PSchool found on the original set, as well as on 400 and 4,000 trials. The latter involves the average values as well as the lower and upper 2.5% quantiles

Regression Parameters	Set	400 trials			4,000 trials		
		Mean	2.5%	97.5%	Mean	2.5%	97.5%
Slope	0.401	0.399	0.296	0.486	0.398	0.303	0.488
Intercept	0.072	0.089	−0.343	0.623	0.092	−0.400	0.594

more or less the same at 400-trial data. The values of intercept are distributed with a greater dispersion and provide for a worsened accuracy. Symmetric 95% confidence intervals for the intercept are [−0.343, 0.623] at 400 trials and [−0.400, 0.594] at 4,000 trials.

Q.3.1. How a pivotal bootstrapping rule can be applied here? This would provide more stable evaluations than empirical distributions. The standard deviations of the slope and intercept are 0.0493 and 0.2606, respectively, at 400 boostrapping trials; they are somewhat smaller, 0.0477 and 0.2529, at the 4,000 trials. Can one derive from this a symmetric 95% confidence interval for the slope or intercept? Tip: in a Gaussian distribution, 95% of all values fall within interval mean ± 1.96*std. This is the so-called pivotal bootstrapping method.

Q.3.2. Can you give an estimate of the level of variance of the differences between PSchool observed and computed values?

A final validity test of the regression equation is probably the toughest one – by the prediction error (see Worked example 3.3).

Worked example 3.3. Prediction error of the regression equation

Compare the observed values of PSchool with those computed through Pop according to Equation (3.1). Table 3.2 presents a few examples taken from both ends of the sorted Pop feature.

Table 3.2 Observed numbers of Primary schools versus those predicted from the Population resident data on some Market towns

PS obs.	PS comp.	Pop	PS obse.	PS comp.	Pop
1	0.89	2,040	2	2.35	5,676
2	0.97	2,230	2	2.90	7,044
2	1.06	2,452	4	4.12	10,092
2	1.19	2,786	7	6.44	15,865
1	1.54	3,660	4	7.05	17,390

On average, the predictions are close, but, in some cases, are less so. One can easily estimate the relative error, which is [(1 − 0.89)/1]*100 = 11% at the first case, [(2 − 0.97)/2]*100 = 51.5% at the second case, etc. The average relative error of Equation (3.1) is equal to 30.7%. Can it be made smaller? On the first glance, no, it cannot, because Equation (3.1) minimizes the error. But, the error minimized by Equation (3.1) is the average quadratic error, not the relative error under consideration. The two errors do differ, and Equation (3.1) is not necessarily optimal with regard to the relative error.

The classical optimization theory has virtually nothing to propose for the minimization of the relative error – this criterion is neither linear, nor quadratic, nor convex. Yet the evolutionary optimization approach can be applied to the task. This approach uses a population of solutions randomly evolving, iteration after iteration, in the search for better solutions as explained in Project 3.2. Applying the algorithm from that project to minimize the criterion of relative error, one can find a different solution, in fact, a set of solutions each leading to the average relative error of 26.4%, a reduction of 4.3 points, one seventh of the relative error of Equation (3.1). The new solution is PSchool= 0.28*Pop + 0.33 expressing a smaller rate of increase in school numbers at the growth of population.

F3.2.3 Linear Regression: Formulation

F3.2.3.1 Fitting Linear Regression

Let us derive parameters of linear regression. Given target feature y and predictor x at N entities (x_1, y_1), (x_2, y_2)..., (x_N, y_N), we are interested at finding a linear equation relating them so that

$$y = ax + b \qquad (3.2)$$

The exact fit can occur only if all pairs (x_i, y_i) belong to the same straight line on (x,y)-plane, which is rather unlikely on real-world data. Therefore, Equation (3.2) will have an error at each pair (x_i, y_i) so that the equation should be rewritten as

$$y_i = ax_i + b + e_i \quad (i = 1, 2, \ldots, N) \qquad (3.2')$$

where e_i are referred to as errors or residuals. The problem is of determining the two parameters, a and b, in such a way that the residuals are least-squares minimized, that is, the average square error

$$L(a, b) = \Sigma_i e_i^2 / N = \Sigma_i (y_i - ax_i - b)^2 / N, \qquad (3.3)$$

reaches its minimum over all possible a and b, given x_i and y_i $(i = 1, 2, \ldots, N)$. This minimization problem is easy to solve with the elementary calculus tools.

3.2 Two Quantitative Features Case

Indeed $L(a, b)$ is a "bottom down" parabolic function of a and b, so that its minimum corresponds to the point at which both partial derivatives of $L(a, b)$ are zero (the first-order optimality condition):

$$\partial L/\partial a = 0 \quad \text{and} \quad \partial L/\partial b = 0.$$

Leaving the task of actually finding the derivatives to the reader as an exercise, let us focus on the unique solution to the first-order optimality equations defined by the following formulas (3.4), for a, and (3.6), for b:

$$a = \rho \, \sigma(y)/\sigma(x) \qquad (3.4)$$

where

$$\rho = \left[\Sigma_i (x_i - m_x)(y_i - m_y)\right] / \left[N\sigma(x)\sigma(y)\right] \qquad (3.5)$$

is the so-called correlation coefficient, m_x, m_y are means of x_i, y_i, respectively and $\sigma^2(x), \sigma^2(y)$ are standard deviations;

$$b = m_y - a m_x \qquad (3.6)$$

By putting these optimal a and b into (3.3), one can express the minimum criterion value as

$$L_m(a, b) = \sigma^2(y)(1 - \rho^2) \qquad (3.7)$$

The Equation (3.2) is referred to as the linear regression of y over x, index ρ in (3.4) and (3.5) as the correlation coefficient, its square ρ^2 in (3.7) as the determination coefficient, and the minimum criterion value L_m in (3.7) is referred to as the unexplained variance.

F3.2.3.2 Correlation Coefficient and Its Properties

The meaning of the coefficients of correlation and determination, in the data recovery framework of data analysis, is provided by Eqs. (3.3), (3.4), (3.5), (3.6) and (3.7). Here are some formulations.

Property 1 Determination coefficient ρ^2 shows the relative decrease of the variance of y after its linear relation to x has been taken into account by the regression (follows from (3.7)).

Property 2 Correlation coefficient ρ ranges between -1 and 1, because ρ^2 is between 0 and 1, as follows from the fact that value L_m in (3.7) cannot be negative because the items in its expression (3.3) are all squares. The closer ρ to either 1 or

−1, the smaller are the residuals in the regression equation. For example, $\rho = 0.9$ implies that y's unexplained variance L_m is $1 - \rho^2 = 19\%$ of the original value.

Property 3 The slope a is proportional to ρ according to (3.4); a is positive or negative depending on the sign of ρ. If $\rho = 0$, the slope is 0: in this case, y and x are referred to as not correlated.

Property 4 The correlation coefficient ρ does not change under shifting and rescaling of x and/or y, which can be seen from Equation (3.5). Its formula (3.5) becomes especially simple if the so-called z-scoring has been applied to standardize both x and y.

To perform z-scoring over a feature, its mean m is subtracted from all the values and the results are divided by the standard deviation σ:

$$x_i' = (x_i - m_x)/\sigma(x) \quad \text{and} \quad y_i' = (y_i - m_y)/\sigma(y), \quad i = 1, 2, \ldots, N$$

Using the z-score standardization, formula (3.5) can be rewritten as

$$\rho = \Sigma_i x_i' y_i'/ = <x', y'>/N \qquad (3.5')$$

where $<x', y'>$ denotes the inner product of vectors $x' = (x_i')$ and $y' = (y_i')$.

The next property refers to one of the fundamental discoveries by K. Pearson, an interpretation of the correlation coefficient in terms of the bivariate Gaussian distribution. A generic formula for the density function of this distribution, in the case in which the features have been pre-processed by using z-score standardization described above, is

$$f(u, \Sigma) = C \exp\left\{-u^T \Sigma^{-1} u/2\right\} \qquad (3.8)$$

where $u = (x, y)$ is a two-dimensional vector of the two variables x and y under consideration and Σ is the so-called correlation matrix

$$\Sigma = \begin{pmatrix} 1 & \rho \\ \rho & 1 \end{pmatrix}$$

In formula (3.8), ρ is a parameter with a very clear geometric meaning. Consider a set of points $u = (x, y)$ on (x, y)–plane making function $f(u, \Sigma)$ in (3.8) equal to a pre-specified constant. Such a set makes the values of $u^T \Sigma u$ constant too. That means that a constant density set of points $u = (x, y)$ must satisfy equation $x^2 - 2\rho xy + y^2 = $ const. This equation is known to define a well-known quadratic curve, the ellipsis. At $\rho = 0$ the equation becomes an equation of a circle, $x^2 + y^2 = const$, and the greater the difference between ρ and 0, the more skewed is the ellipsis, so that at $\rho = \pm 1$ the ellipsis becomes a bisector line $y = \pm x + b$ because the left part of the equation makes a full square, in this case, $x^2 \pm 2xy + y^2 = const$, that is,

$(y \pm x)^2 = const$. The size of the ellipsis is proportional to the constant: the greater the constant the greater the size.

Property 5 The correlation coefficient (3.5) is a sample based estimate of the parameter ρ in the Gaussian density function (3.8) under the conventional assumption that the sample points (y_i, x_i) are drawn from a Gaussian population randomly and independently.

This striking fact is behind a long standing controversy. Some say that the usage of the correlation coefficient is justified only when the sample is taken from a Gaussian distribution, because the coefficient has a clear-cut meaning only in this model. This logic seems somewhat overly restrictive. True, the usage of the coefficient for estimating the density function is justified only when the function is Gaussian. However, when trying to linearly represent one variable through the other, the coefficient has a very different meaning in the approximation context, which has nothing to do with the Gaussian distribution, as expressed above with Eqs. (3.4), (3.5), (3.6) and (3.7).

F3.2.3.3 Linearization of Non-linear Regression

Non-linear dependencies also can be fit by using the same criterion of minimizing the square error. Consider a popular case of exponential regression, that is, representing correlation between target y and predictor x as $y = ae^{bx}$ where a and b are unknown constants and e the base of natural logarithm. Given some a and b, the average square error is calculated as

$$E = \left([y_1 - a\exp(bx_1)]^2 + \ldots + [y_N - a\exp(bx_N)]^2 \right)/N = \Sigma_i [y_i - a\exp(bx_i)]^2/N \tag{3.9}$$

There is no method that would straightforwardly lead to a globally optimal solution of the problem of minimization of E in (3.9) because it is too complex function of the unknown values. This is why conventionally the exponential regression is fit by what should be referred to as its linearization: transforming the original problem to that of linear regression. Indeed, let us take the logarithm of both parts of the equation that we want to fit, $y = ae^{bx}$. The resulting equation is $\ln(y) = \ln(a) + bx$. This equation has the format of linear equation, $z = \alpha x + \beta$, where $z = \ln(y)$, $\alpha = b$ and $\beta = \ln(a)$. This leads to the following idea. Let us take the target be $z = \ln(y)$ with its values $z_i = \ln(y_i)$. By fitting the linear regression equation with data x_i and z_i, one finds optimal α and β, so that the original exponential parameters are found as $a = \exp(\beta)$ and $b = \alpha$. These values do not necessarily minimize (3.9), but the hope is that they are close to the optimum anyway. Unfortunately, this may be very wrong sometimes as the material in Project 3.2 clearly demonstrates.

Q.3.3. Find the derivatives of L over a and b and solve the first-order optimality conditions.

Q.3.4. Derive the optimal value of L in (3.7) for the optimal a and b.

Q.3.5. Prove or find a proof in the literature that any linear equation $y = ax+b$ corresponds to a straight line on Cartesian xy plane for which a is the slope and b intercept.

Q.3.6. Find the inverse matrix Σ^{-1} for $\Sigma = \begin{pmatrix} 1 & \rho \\ \rho & 1 \end{pmatrix}$. **A.** $\Sigma^{-1} = \begin{pmatrix} 1 & -\rho \\ -\rho & 1 \end{pmatrix} \Big/ (1-\rho^2)$.

C3.2.4 Linear Regression: Computation

Regression is a technique for representing the correlation between x and y as a linear function (that is, a straight line on the plot), $y = slope*x + intercept$ where *slope* and *intercept* are constants, the former expressing the change in y when x is added by 1 and the latter the level of y at x=0. The best possible values of slope and intercept (that is, those minimizing the average square difference between real y's and those found as *slope*x + intercept*) are expressed in MatLab, according to formulas (3.4), (3.5) and (3.6), as follows:

≫ rho = corrcoef(x,y);
%2×2 matrix whose off-diagonal entry is correlation coefficient
≫ slope = rho(1,2)*std(y)/std(x);
≫ intercept = mean(y) − slope*mean(x);

Here rho(1, 2) is the Pearson correlation coefficient between *x* and *y* (3.5) that can be determined with MatLab operation "corrcoef" which leads to an estimate of the matrix Σ above.

Project 3.1. 2D analysis, linear regression and bootstrapping

Let us take the Students data table as a 100×8 array a in MatLab, pick any two features of interest and plot entities as points on the Cartesian plane formed by the features. For instance, take Age as x and Computational Intelligence mark as y:

≫ x = a(:,4); % Age is 4-th column of array "a"
≫ y = a(:,8); % CI score is in 8-th column of "a"

Then student 1 (first row) will be presented by point with coordinates x = 28 and y = 90 corresponding to the student's age and CI mark, respectively. To plot them all, use command:

≫ plot(x,y,'k.')
% k refers to black colour, "." dot graphics; 'mp' stands for magenta pentagram;
% see others by using "help plot"

3.2 Two Quantitative Features Case

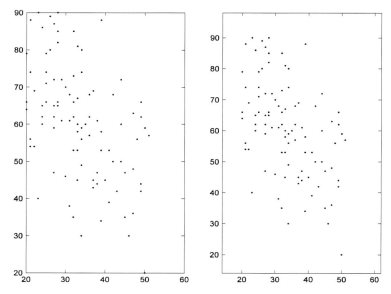

Fig. 3.5 Scatter plot of features "Age" and "CI score"; the display *on the right* is a rescaled version of that *on the left*

Unfortunately, this gives a very tight presentation: some points are on the borders of the drawing. To make the borders stretched out, one needs to change the axis, for example, as follows:

≫ d = axis; axis(1.2*d−10);

This transformation is presented on the right part of Fig. 3.5. To make both plots presented on the same figure, use "subplot" command of MatLab:

≫ subplot(1,2,1)
≫ plot(x,y,'k.');
≫ subplot(1,2,2)
≫ plot(x,y,'k.');
≫ d = axis; axis(1.2*d−10);

Whichever presentation is taken, no regularity can be seen on Fig. 3.5 at all. Let's try then whether anything better can be seen for different occupations. To do this, one needs to handle entity sets for each occupation separately:

≫ o1=find(a(:,1)==1); % set of indices for IT
≫ o2=find(a(:,2)==1); % set of indices for BA
≫ o3=find(a(:,3)==1); % set of indices for AN
≫ x1=x(o1);y1=y(o1); % the features x and y at IT students
≫ x2=x(o2);y2=y(o2); % the features at BA students
≫ x3=x(o3);y3=y(o3); % the features at AN students

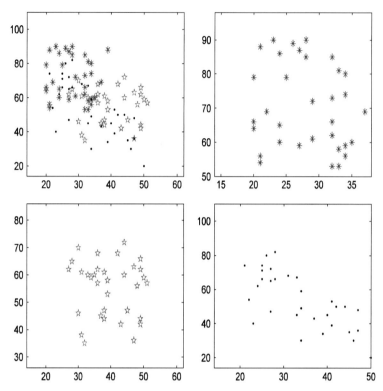

Fig. 3.6 Joint and individual displays of the scatter-plots at the occupation categories (IT *star*, BA *pentagrams*, AN *dots*)

Now we are in a position to put, first, all the three together, and then each of these three separately (again with the command "subplot", but this time with four windows organized in a two-by-two format, see Fig. 3.6).

≫ subplot(2,2,1); plot(x1,y1, '*b',x2,y2,'pm',x3,y3,'.k');% all three
≫ d=axis; axis(1.2*d−10);
≫ subplot(2,2,2); plot(x1,y1, '*b'); % IT plotted with blue stars
≫ d=axis; axis(1.2*d−10);
≫ subplot(2,2,3); plot(x2,y2,'pm'); % BA plotted with magenta pentagrams
≫ d=axis; axis(1.2*d−10);
≫ subplot(2,2,4); plot(x3,y3,'.k'); % AN plotted with black dots
≫ d=axis; axis(1.2*d−10);

Of the three occupation groups, some potential relation can be seen only in the AN group: it is likely that "the greater the age the lower the mark" regularity holds in this group (black dots in the Fig. 3.4's bottom right). To check this, let us utilize the linear regression.

Linear regression equation, $y = slope*x + intercept$ is estimated by using MatLab, according to formulas (3.4), (3.5) and (3.6), as follows:

3.2 Two Quantitative Features Case

≫cc= corrcoef(x3,y3);rho=cc(1,2);% producing rho=−0.7082
≫ slope = rho*std(y3)/std(x3); % this produces slope =−1.33;
≫ intercept = mean(y3) − slope*mean(x3); % this produces intercept = 98.2;

Since we are interested in group AN only, we apply these commands at AN-related values x3 and y3 to produce the linear regression as y3 = 98.2 − 1.33*x3. The slope value suggests that every year added to the age, in general decreases the mark by 1.33, so that aging by 3 years would lead to the loss of 4 mark points. Obviously, care should be taken to draw realistic conclusions.

Altogether, the regression equation explains rho^2 = 0.50 = 50% of the total variance of y3 – not too much, as is usual in social and human sciences.

Let us take a look at the reliability of the regression equation with bootstrapping, the popular computational experiment technique for validating data analysis results that was introduced in Project 2.3 (see also Carpenter and Bithell 2000, Davison and Hinkley 2005).

Bootstrapping is based on a pre-specified number of random trials, for instance, 5,000. Each trial consists of the following steps:

(i) randomly selecting an entity N times, with replacement, so that the same entity can be selected several times whereas some other entities may be never selected in a trial. (As shown above in Project 2.3, on average only 63% entities get selected into the sample.) A sample consists of N entities because this is the number of entities in the set under consideration. In our case, N=31. One can use the following MatLab command:

≫ N=31;ra=ceil(N*rand(N,1));
% rand(N,1) produces a column of N random real numbers, between 0 and 1 each.
% Multiplying this by N stretches them to (0,N) interval; ceil rounds the numbers up to integers.

(ii) the sample ra is assigned with their data values according to the original data table:

≫xt=xx(ra);yt=yy(ra);
% here xx and yy represent the predictor and target, respectively;
% they are x3 and y3, respectively, which can be taken into account with assignments
% xx=x3; and yy=y3.

so that coinciding entities get identical feature values.

(iii) a data analysis method under consideration, currently "linear regression", that basically computes the rho, the slope and the intercept, applies to this data sample to produce the trial result.

To do a number (5,000, in this case) of trials, one should run (i)–(iii) in a loop:

≫ for k=1:5000; ra=ceil(N*rand(N,1));
xt=xx(ra);yt=yy(ra);

```
cc=corrcoef(xt,yt);
rh(k)=cc(1,2);
sl(k)=rh(k)*std(yt)/std(xt); inte(k)=mean(yt)−sl(k)*mean(xt);
end
```
% the results are 5000-strong columns rh (correlations), sl (slopes)
% and inte (intercepts)

Now we can check the mean and standard deviation of the obtained distributions. Commands

≫mean(sl); std(sl)

produce values −1.33 and 0.24. That means that the original value of slope=−1.33 is confirmed with the bootstrapping, but now we have obtained its standard deviation, 0.24, as well. Similarly mean/std values for the intercept and rho are computed. They are, respectively, 98.2 / 9.0 and −0.704 / 0.095.

We can plot the 5,000 values found as 30-bin histograms (see Fig. 3.7):

≫ subplot(1,2,1); hist(sl,30)
≫ subplot(1,2,2); hist(in,30)

Command subplot(1,2,1) creates one row consisting of two windows for plots and puts the follow-up plot into the first window (that on the left). Command subplot(1,2,2) changes the action into the second window which is on the right.

To derive the 95% confidence boundaries for the slope, intercept and correlation coefficient, one may use both pivotal and non-pivotal methods.

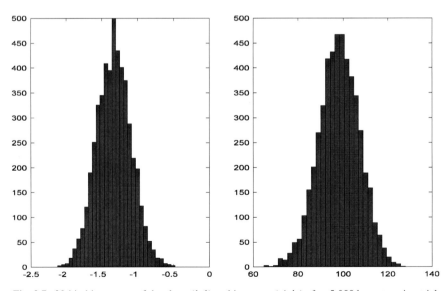

Fig. 3.7 30-bin histograms of the slope (*left*) and intercept (*right*) after 5,000 bootstrapping trials

3.2 Two Quantitative Features Case

The pivotal method uses the hypothesis that the bootstrap sample is indeed a random sample from a Gauusian distribution. Parameters of this distribution for slope are determined with the following commands:

≫ msl=mean(sl); ssl=std(sl);

Since 95% of the Gaussian distribution fall within interval of plus-minus 1.96*std, the 95% confidence boundaries are derived, for the slope, as follows:

≫ lbsl=msl − 1.96*ssl; rbsl=msl + 1.96*ssl

The non-pivotal estimates require no such a hypothesis and are based on the bootstrap distribution as is. One just sorts all the values and takes 2.5% quantiles on both extremes of the range:

≫ ssl=sort(sl); lbn=ssl(126);rbn=ssl(4875);

Indeed, we need to cut out 5% items from the sample, to make a 95% confidence interval. Since 5% of 5,000 is 250, conventionally divided in two halves, this requires cutting off first 125 observations as well as the last 125 observations of the presorted list of the bootstrap values, which brings us to ssl(126) and ssl(4875) as the non-pivotal boundaries for the slope value.

All these estimates are presented in Table 3.3. The pivotal and non-pivotal estimates do not fall too far apart. Either can be taken as parameters of the boundary regressions.

This all can be visualized by, first, defining the three regression lines, the regular one and two corresponding to the lower and upper estimate boundaries, respectively, with

≫ y3reg=slope*x3+intercept;
≫ y3regleft=lbsl*x3+lbintercept;
≫ y3regright=rbsl*x3+rbintercept;

and then plotting the four sets onto the same figure Fig. 3.8:

≫ plot(x3,y3,'*k',x3,y3reg,'k',x3,y3regleft,'r',x3,y3regright,'r')
% x3,y3,'*k' presents student data as black stars; x3,y3reg,'k' presents the
% real regression line in black
% x3,y3regleft,'g' and x3,y3regright,'g' for boundary regressions in green

The lines on Fig. 3.8 show the boundaries of the regression line for 95% of trials.

Table 3.3 Parameters of the bootstrap distributions and pivotal and non-pivotal boundaries

	Mean	St. dev.	Pivotal boundaries		Non-pivotal boundaries	
			Left	Right	Left	Right
Slope	−1.337	0.241	−1.809	−0.865	−1.800	−0.850
Intercept	98.51	9.048	80.776	116.244	80.411	116.041
Corr. coef.	−0.707	0.094	−0.891	−0.523	−0.861	−0.493

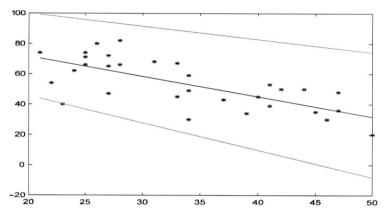

Fig. 3.8 Regression of CI score over Age (*black line*) within occupation category AN with boundaries covering 95% of potential biases due to sample fluctuations

Project 3.2. Non-linear and linearized regression: a nature-inspired algorithm

In many domains the correlation between features is not necessarily linear. For example, in economics, processes related to the inflation over time are modeled by using the exponential function. A similar way of thinking applies to the processes of growth in biology. Variables describing climatic conditions obviously have a cyclic character. The power law in social systems is nonlinear too.

Consider, for example, a power law function $y = ax^b$ where x is predictor and y predicted variables whereas a and b are unknown constant coefficients. Given the values of x_i and y_i on a number of observed entities $i = 1, \ldots, N$, the power law regression problem can be formulated as the problem of minimizing the summary squared or absolute error over all possible pairs of coefficients a and b. There is no method that would straightforwardly lead to a globally optimal solution of the problem because minimizing a sum of many exponents is a complex problem.

This is why conventionally the power law regression is fit by transforming it into a linear regression problem. Indeed, the equation of the power law regression, taken with no errors, is equivalent to the equation of linear regression with $\log(x)$ being predictor and $\log(y)$ target: $\log(y) = b \log(x) + \log(a)$. This gives rise to the very popular strategy of linearization of the problem. First, transform x_i and y_i to $v_i = \log(x_i)$ and $z_i = \log(y_i)$ and fit the linear regression equation for given v_i and z_i; then convert the found coefficients into those of the original exponential function. This strategy seems especially suitable since the logarithm of a variable typically is much smoother so that the linear fit is better under the logarithm transformation.

There is one caveat, however: the fact that found coefficients are optimal in the linear regression problem does not necessarily imply that the converted exponents are necessarily optimal in the original problem. This we are going to explore in this project.

3.2 Two Quantitative Features Case

Nature-inspired optimization is a computational intelligence approach to minimize a non-linear function. Rather than look and polish a single solution to the optimization problem under consideration, this approach utilizes a population of solutions iteratively evolving from generation to generation, according to rules imitating a real-world evolutionary process. The rules typically include: (a) random changes from generation to generation such as "mutations" and "crossovers" in earlier, genetic, algorithms, and (b) policies for selecting and maintaining the best found solutions, the "elite". After a pre-specified number of iterations, the best solution among those observed is reported as the outcome.

To start the evolutionary optimization process, one should first define a restricted area of admissible solutions so that no member of the population may leave the area. This warrants that the population will not explode by moving solutions to the infinity. Under the hypothesis of a power law relation $y = ab^x$, for any two entities i and j, the following equations should hold: $z_i = b*v_i + c$ and $z_j = b*v_j + c$ where $c = \log(a)$, $z_j = \log(y_i)$ and $v_i = \log(x_i)$. From these, b and c can be expressed as follows: $b = (z_i - z_j)/(v_i - v_j)$, $c = (v_i^*z_j - v_j^*z_i)/(v_i - v_j)$, which may lead to different values of b and c at different i and j. Denote bm and bM the minimum and the maximum of $(z_i - z_j)/(v_i - v_j)$, and cm and cM the minimum and maximum of $(v_i^*z_j - v_j^*z_i)/(v_i - v_j)$ over those i and j for which $v_i = v_j \neq 0$. One would expect that the admissible b and c should be within these boundaries, which means that the area of admissible solutions should be defined by the inequalities $(bm, cm) \leq (b, c) \leq (bM, cM)$. Since the optimal values of (b,c) should be around the averages of the ratios above, that is, lie deep inside the area between their maxima and minima, it helps to speed up the computation if one takes only those pairs (i,j) at which the values of v_i, v_j and z_i, z_j are not too close to 0 so that their logarithms are not that far away from 0, and, similarly, the differences between them should be neither that small nor that high. This approach is implemented in MatLab code ddr.m in Appendix A4.

For the step of producing the next generation, let us denote the population's $p \times 2$ array by f, at the current iteration, and by f', at the next iteration. The transition from f to f' is done in three steps. First, take the row of mean values within the columns of f and repeat it p times in a $p \times 2$ array mf. Then make a Gaussian random move:

$$fn = f + \text{randn}(p,2).*mf/20 \quad (3.10)$$

Here randn(p,2) is a $p \times 2$ array of (pseudo) random numbers generated according to Gaussian distribution $N(0,1)$ with 0 expectation and 1 variance. The symbol .* denotes the operation of multiplication of corresponding elements in matrices, so that (aij).*(bij) is a matrix whose (i,j)-th elements are products aij*bij. This random matrix is scaled down by $mf/20$ so that the move accounts for about 5% (one twentieth) of the average f values.

Since the move is to be restricted within the admissibility area, any a-element (first column of fn) which is greater than aM, is to be changed for aM, and any a-element smaller than am is to be changed for am. Similar trimming applies to b-elements. Denote result by fr.

At the next step, take a $p\times 2$ array el whose rows are the same stored elite solution and arrive at the next generation f' by using the following "elite mix":

$$f' = 0.7fr + 0.3el$$

The elite mix moves all population members in the direction of the best solution found so far by 30%, which has been found work well in the examples of our interest.

This procedure is implemented in MatLab code nlr.m that relies on ddr.m at step 1 and a subroutine, delta, for evaluating the fitness (see A4 in Appendix).

Consider now this experiment. Generate predictor x as a 50-long vector of random positive entries between 0 and 10, x=10*rand(1,50), and define $y = 2*x^{1.07}$ with the normal additive noise $2*N(0, 1)$ where 0 is the mean and 1 the variance, which is suppressed when overly negative, according to the Matlab code line

≫for ii=1:50;yy=2*x(ii)^1.07 +2*randn;y(ii)=max(yy,1.01);end;

When using the conventional linearized regression model by linearly mapping log(x) to log(y), to extract b and a (as the exponent of the found c) from this, the program llr.m implementing this approach produces $a = 3.0843$ and $b = 0.8011$ leading to the averaged squared error $y - ax^b$ equal to 4.41, so that the standard error is 2.10, about 20% of the mean y value, 10.1168. It is not only that the error is high, but also a wrong law is identified. The generated function y stretches x out ($b>1$), whereas the found function stretches x in ($b<1$).

Minimization of the averaged squared error $y - ax^b$ of the original model directly by using the code nlr.m, that implements the nature-inspired algorithm, the values are $a = 2.0293$ and $b = 1.0760$ leading to the average squared error of 0.0003 and the standard error of 0.0180. In contrast to the values found at the linearized scheme, the parameters a and b here are very close to those generated.

This obviously considerably outperforms the conventional procedure. Similar results can be found at different values of the noise variance.

Case-study 3.1. Growth of Investment

Let us apply a similar approach to the following example involving variables x and y defined over a period of 20 time moments as presented in Table 3.4.

Table 3.4 Data of investment y at time moments x from 0.10 to 2.00

x	0.10	0.20	0.30	0.40	0.50	0.60	0.70	0.80	0.90	1.00	1.10	1.20	1.30	1.40	1.50
	1.60	1.70	1.80	1.90	2.00										
y		1.30	1.82	2.03	4.29	3.30	3.90	3.84	4.24	4.23	6.50	6.93	7.23	7.91	9.27	9.45
	11.18	12.48	12.51	15.40	15.91										

3.2 Two Quantitative Features Case

Variable x can be thought of as related to the time periods whereas y may represent the value of a fund. In fact, the components of x are numbers from 1 to 20 divided by 10, and y is obtained from them in MatLab according to formula $y=2*\exp(1.04*x)+0.6*\mathrm{randn}$ where randn is the normal (Gaussian) random variable with the mathematical expectation 0 and variance 1.

Let us, first, try a conventional approach of finding the average growth of the fund during all the period.

The average growth of the investment according to these data is conventionally expressed as the root 19, or power 1/19, of the ratio y_{20}/y_{01}, that is, 1.14. This estimates the average growth as 14% per period – which is by far greater than 4% in the data generating model.

Let us now try to make sense of the relation between x and y by applying the conventional linearization strategy to this data.

The strategy of linearization of the exponential equation outlined in Section 3.2.3.3 leads to values 1.1969 and 0.4986 for b and c, respectively, to produce $a = e^c = 1.6465$ and $b = 1.1969$ according to formulas there. As one can see, these differ from the original $a = 2$ and $b = 1.04$ by the order of 15–20%. The value of the squared error here is $E = 13.90$. See Fig. 3.9 representing the data.

Let us now apply the nature inspired approach to the original non-linear least-squares problem.

The program nlrm.m implementing the evolutionary approach described in Project 3.2 found $a = 1.9908$ and $b = 1.0573$. These are within 1–2% of the error from the original values $a = 2$ and $b = 1.04$. The summary squared error here is $E = 7.45$, which is by far smaller than that found with the linearization strategy.

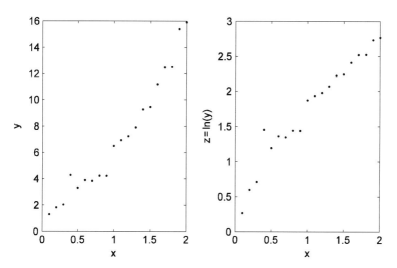

Fig. 3.9 Plot of the original pair (x,y) in which y is a noisy exponential function of x (*on the left*) and plot of the pair (x,z) in which $z = \ln(y)$. The plot *on the right* looks somewhat straighter indeed, though the correlation coefficients are rather similar, 0.970 for the plot *on the left* and 0.973 for the plot *on the right*

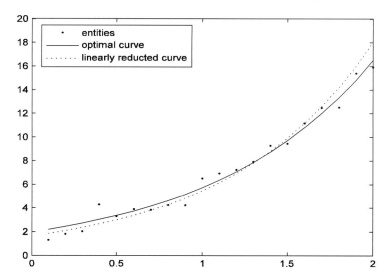

Fig. 3.10 Two fitting exponents are shown, with *stars and dots*, for the data in case study 3.1

The two found solutions can be represented on the scatter-plot graph, see Fig. 3.10. One can see that the linearized version has a much steeper exponent, which becomes visible at later periods.

Q.3.7. Consider a binary feature defined on seven entities so that it is category A on the first three of them, and category B on the next four. Let us draw two dummy 1/0 variables, xA and xB, corresponding to each so that xA = 1 on the first three entities and xA = 0 on the rest, whereas xB = 0 on the first three entities and xB = 1 on the rest. What can be said of the correlation coefficient between xA and xB? **A.** The correlation coefficient between xA and xB is −1 because xA+xB=1 for all entities so that xA = −xB+1.

Q.3.8. Extend the nature-inspired approach to the problem of fitting a linear regression with a nonconventional criterion such as the average relative error defined by formula $1/N \sum_{i=1}^{N} |e_i/y_i|$.

Case-study 3.2. Correlation Between Iris Sepal Length and Width

Take x and y from the Iris set in Table 1.3 as the Sepal's length and width, respectively.

A scatter plot of x and y is presented on the left part of Fig. 3.11. This is a loose cloud of points which looks similar to that on the left part of Fig. 3.2, of no correlation. Indeed the correlation coefficient value here is not only very small,

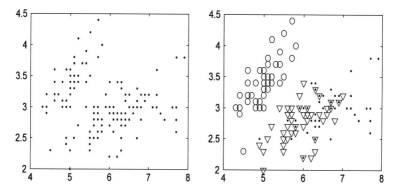

Fig. 3.11 Scatter plot of Sepal length and Sepal width from Iris data set (Table 1.3), as a whole *on the left* and taxon-wise *on the right*. Taxon 1 is presented by *circles*, taxon 2 by *triangles*, and taxon 3 by *dots*

−0.12, but also negative, which is somewhat odd, because intuitively the features should be positively correlated as reflecting the size of the same flower.

To see a particular reason for the low, and negative, correlation, one should take into account that the sample is not homogeneous: the Iris set consists of 50 specimens of each of three different taxa. When the taxa are separated (see Fig. 3.11 on the right), the positive correlation is restored. The correlation coefficients are 0.74, 0.53 and 0.46 for taxon one, two and three, respectively. Here is a nice example of the negative effect of the non-homogeneity of the sample on the data analysis results.

3.3 Mixed Scale Case: Nominal Feature Versus a Quantitative One

P3.3.1 Box-Plot, Tabular Regression and Correlation Ratio

Consider x a categorical feature on the same entities as a quantitative feature y, such as Occupation and Age at Students data set. The within-category distributions of y can be used to investigate the correlation between x and y. The distributions can be visualized by using just ranges as follows: present categories with equal-size bins on x axis, draw two lines parallel to x axis to present the minimum and maximum values of y (in the entire data set), and then present the within category ranges of y as shown on Fig. 3.12.

The correlation between x and y is higher when the within-category spreads are tighter because the tighter the spread within an x-category, the more precise is prediction of y at it. Figure 3.13 illustrates an ideal case of a perfect correlation – all within-category y-values are the same leading to an exact prediction of Age when Occupation is known.

90   3 2D Analysis: Correlation and Visualization of Two Features

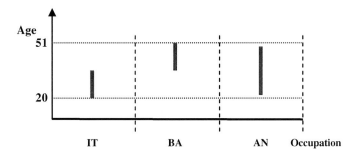

Fig. 3.12 Graphic presentation of within category ranges of Age at Student data

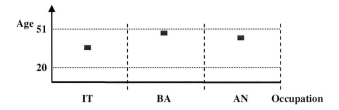

Fig. 3.13 In a situation of ideal correlation, with zero within-category variances, knowledge of the Occupation category would provide an exact prediction of the Age within it

Figure 3.14 presents another extreme, when knowledge of an Occupation category does not lead to a better prediction of Age than when the Occupation is unknown.

A simple statistical model extending that for the mean will be referred to as tabular regression. The tabular regression of quantitative y over categorical x is a table comprising three columns corresponding to:

(1) Category of x
(2) Within category mean of y
(3) Within category standard deviation of y

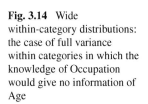
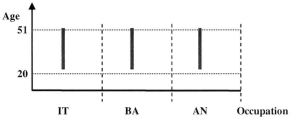

Fig. 3.14 Wide within-category distributions: the case of full variance within categories in which the knowledge of Occupation would give no information of Age

3.3 Mixed Scale Case: Nominal Feature Versus a Quantitative One   91

The number of rows in the tabular regression thus corresponds to the number of *x*-categories; there should be a marginal row as well, with the mean and standard deviation of *y* on the entire entity set.

Worked example 3.4. Tabular regression of Age (quantitative target) over Occupation (categorical predictor) in Students data

Let us draw a tabular regression of Age over Occupation in Table 3.5. The table suggests that if we know the Occupation category, say IT, then we can safely predict the Age as being 28.2 within the margin of plus/minus 5.6 years. With no knowledge of the Occupation category, we could only say that the Age is on average 33.7 plus/minus 8.5, a somewhat less precise estimate.

The table can be visualized in a manner similar to Figs. 3.12, 3.13 and 3.14, this time presenting the within category averages by horizontal lines and the standard deviations by vertical strips (see Fig. 3.15).

One more way of visualization of categorical/quantitative correlation is the so-called box-plot. The within-category spread is expressed here with a quantile (percentile) box rather than with the standard deviation. First, a quantile level should be defined such as, for instance, 40%, which means that we are going to show the within-category range over only 60% of its contents by removing 20% off of both its top and bottom extremes. These are presented with box' heights such as on Fig. 3.16; the full within-category ranges are shown with whiskers.

Table 3.5 Tabular regression of Age over Occupation in Students data

Occupation	Age Mean	Age StD
IT	28.2	5.6
BA	39.3	7.3
AN	33.7	8.7
Total	33.7	8.5

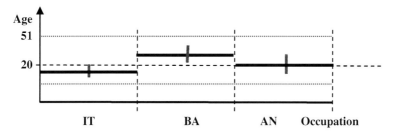

Fig. 3.15 Tabular regression visualized with the within-category averages and standard deviations represented by the position of *solid horizontal lines* and *vertical line sizes*, respectively. The *dashed line's* position represents the overall average (*grand mean*)

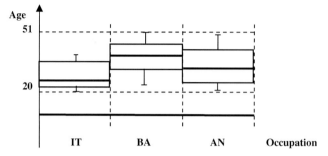

Fig. 3.16 Box-plot of the relationship between Occupation and Age with 20% quantiles; *the box heights* reflect the Age within-category 60% ranges, whiskers show the total ranges. Within-box *horizontal lines* show the within category averages

Worked example 3.5. Box-plot of Age at Occupation categories at Students data

With the quantile level specified at 40%, at the category IT, Age ranges between 20 and 39, but if we sort it and remove 7 entities of maximal Age and 7 entities of minimal Age (there are 35 students in IT so that 7 makes 20% exactly), then the Age range on the remaining 60% is from 22 to 33. Similarly, Age 60% range is from 32 to 47 on BA, and from 25 to 44 on AN (see box heights on Fig. 3.16). The whiskers reflect 100% within category ranges, which are intervals [20, 39], [27, 51] and [21, 50], respectively.

The box-plot proved useful in studies of quantitative features too: one of the features is partitioned into a number of bins that are treated then as categories.

Consider now one more tabular regression, this time of the OOProgramming mark over Occupation (Table 3.6)

A natural question emerges: In which of the tables the correlation is greater, 3.5 or 3.6?

This can be addressed with an integral characteristic of the tabular regression, the correlation ratio. This coefficient scores the extent at which the within group variance is smaller on average than the variance of the feature on the set before the split – a determination coefficient for the tabular regression.

Table 3.6 Tabular regression OOProg/Occupation

Occupation	OOP Mean	OOP StD
IT	76.1	12.9
BA	56.7	12.3
AN	50.7	12.4
Total	61.6	16.5

3.3 Mixed Scale Case: Nominal Feature Versus a Quantitative One

Worked example 3.6. Correlation ratio

Let us address the question above: Is the correlation in Table 3.5 is greater than in Table 3.6?

Correlation ratios for the tables computed by using formulas (3.14) and (3.12) are:

Occupation/Age	28.1%
Occupation/OOProg	42.3%

The drop in variance expressed by the correlation ratio is greater at the second table, that is, the correlation between Occupation and OOProgramming is greater than that between Occupation and Age.

Q.3.9. In Table 3.6, there is a positive relation between the Occupation and the OOP mark, with the largest mark, 76.1, going to IT and the smallest mark, 50.7, to AN. There is no such a relation in Table 3.5 in which AN's Age is in the middle between that at the other two groups. Is it that feature of Table 3.6 that leads to a higher correlation ratio? **A.** No; the order of means is irrelevant at the tabular regression. The correlation ratio is higher at Table 3.6 than at Table 3.5 because of the tighter boundaries on the quantitative feature within the groups in Table 3.6.

F3.3.2 Tabular Regression: Formulation

Given a quantitative feature y, with no further information, its average, $\bar{y} = \sum_{i \in I} y_i / |I|$, would represent a proper summarization of the data. If, however, a set of categories of another variable, x, is additionally present, a more detailed summarization can be provided: the within category averages. Let S_k denote the set of entities falling in k category of x, then the within-category averages are $\bar{y}_k = \sum_{i \in S_k} y_i / |S_k|$.

This can be considered the least-squares solution to the model of tabular regression which extends the data recovery model (2.5) for the mean on page 40 as follows. Find a set of c_k values such that the summary square error $L = \sum_{i \in I} e_i^2$ is minimized, where $e_i = y_i - c_k$ according to equations

$$y_i = c_k + e_i \text{ for all } i \in S_k \quad (3.11)$$

The equations underlie the tabular regression and are referred to sometimes as the piece-wise regression. It is not difficult to prove that the optimal c_k in (3.11) is the within category average $\bar{y}_k$, which implies that the minimum value of L is equal to $L_m = \sum_{k=1}^{K} \sum_{i \in S_k} (y_i - \bar{y}_k)^2$. By dividing and multiplying the interior sum by the number of elements in S_k, $|S_k|$, we can see that in fact $L_m = N\sigma^2_w$ where σ^2_w is the average within category variance defined as

$$\sigma^2_w = \Sigma_k p_k \sigma^2_k \qquad (3.12)$$

where $p_k = |S_k|/N$ is the proportion of category k and σ^2_k the variance of y within S_k.

To further analyze this, consider equation

$$(y_i - \bar{y}_k)^2 = y_i^2 + \bar{y}_k^2 - 2y_i\bar{y}_k$$

and sum it up over all $i \in S_k$. This would lead to the summary right-hand item being similar to that in the middle, thus producing $\sum_{i \in S_k}(y_i - \bar{y}_k)^2 = \sum_{i \in S_k} y_i^2 - |S_k|\bar{y}_k^2$.

Summing these equations over k and moving the right-hand item to the other side of the equation, would lead to the following decomposition:

$$\sum_{i \in I} y_i^2 = \sum_{k=1}^{K} |S_k|\bar{y}_k^2 + \sum_{k=1}^{K}\sum_{i \in S_k}(y_i - \bar{y}_k)^2 \qquad (3.13)$$

Note that the right-hand item in (3.13) is the summary least-squares criterion of model in (3.11) L_m. This allows us to interpret the Equation (3.13) as a decomposition of the scatter of variable y, the item on the left, in two parts on the right: the explained part, in the middle, and the unexplained part L_m.

The explained part sums contributions of individual categories k, $|S_k|\bar{y}_k^2$. The value of the contribution is proportional to both the category frequency and the squared value – the greater the better.

Another expression of decomposition (3.13) can be obtained under the assumption that variable y is centered, so that its mean is 0, by relating it to N:

$$\sigma^2 = \sum_{k=1}^{K} p_k \bar{y}_k^2 + \sum_{k=1}^{K} p_k \sigma^2_k \qquad (3.14)$$

where σ^2 is the variance of y, the item on the right the minimum value L_m/N from (3.12), and the item in the middle, the weighted summary squared distance between the grand mean $\bar{y}=0$ and within-category means $\bar{y}_k$.

Equation (3.14) is very popular in statistics as the decomposition of the variance into the within-group variance, the item on the right, and the between-group variance, the item in the middle, as the base of a popular method for comparison of within-category means which is referred to as ANOVA (ANalysis Of VAriance). In the context of the tabular regression model (3.11) viewed as a data recovery model, the original decomposition (3.13) of the quantitative feature scatter into part

3.3 Mixed Scale Case: Nominal Feature Versus a Quantitative One

explained by the nominal feature and part remaining unexplained is more appropriate, as will be seen later in Sections 4.4 and 6.3. Viewed in this light, decomposition (3.14) shows that the category k contribution to the total variance of y is proportional to its frequency and the squared difference between within-category mean $\bar{y}_k$ and grand mean $\bar{y} = 0$.

The correlation ratio shows the relative drop in the variance of y when it is predicted according to model (3.11) or, in other words, the relative proportion of the explained part of the variance. Correlation ratio is usually denoted by η^2 and can be defined by the following formula:

$$\eta^2 = 1 - \sigma^2_w / \sigma^2 \qquad (3.15)$$

The definition implies the following properties:

- The range of η^2 is between 0 and 1.
- Correlation ratio $\eta^2 = 1$ when all within-category variances σ^2_k are zero (that is, when y is constant within each group S_k).
- Correlation ratio $\eta^2 = 0$ when all σ^2_k are of the order of σ^2_k.

Q.3.10. Consider two quantitative features x and y. Divide the range of x in five equal-sized bins to produce a categorical variable xc. Is there any relation between the correlation coefficient between x and y and the correlation ratio coefficient between xc and y? **A.** None, the former can be greater than the latter in some cases, and smaller in some others.

3.3.3 Nominal Target

In the case when it is the quantitative variable that is predictor while the categorical variable is the target, one can use all the wealth of methods developed for pattern recognition or machine learning. The problem may be stated variously depending on the learning task. A machine learning task typically assumes a training dataset for deriving a rule that can be applied to entities from a testing dataset, under the assumption that structures of the training and testing datasets are similar – see a discussion in Chapter 4. All features under consideration are assumed known on both of the sets, except that the categories are not known on the testing dataset.

A most popular problem to address would be like this: given a value of the quantitative predictor on an entity, tell the category of the target feature on the entity. We present two approaches to this.

3.3.3.1 Nearest Neighbor Classifier

One of the most popular is the so-called Nearest-Neighbor classifier. It is applicable at any data admitting distances or (dis)similarities between entities. The NN classifier works as this: find, in the training dataset, an entity which is the nearest to that under consideration and extrapolate its category to the entity in question. One can take a look at the results of application of the NN classification rule to two feature pairs, one from Intrusion data set, and the other from Student dataset, in the follow up examples. The results are very different – the former, in Table 3.7, is very successful whereas the other, in Table 3.9, not. An explanation to this is the difference in the strength of correlation between the two variables – very strong in one case and rather weak in the other (see Tables 3.8 and 3.10).

The NN classifier can be easily extended to the so-called k-NN classifier; the latter usually supplies the category supported by a majority of the k nearest neighbors of the entity in question. This classifier may also lead to the so-called "reject option" – giving no answer when there is no clear-cut majority.

Table 3.7 Applying NN classifier SH⇒Attack to a random subsample of the Intrusion dataset

Random sample	9	29	37	51	63	70	72	80	86	89
True target category	apa	nor	nor	nor	nor	nor	nor	sai	sai	sai
Predictor's value PV	24	10	1	14	2	3	1	482	482	483
Nearest Neighbor's PV	23	11	1	13	2	3	1	482	482	482
NN predicted category	apa	nor	nor	nor	nor	nor	nor	sai	sai	sai

Table 3.8 Tabular regression of SHCo over Attacks in Intrusion data: comparatively small within-category standard deviations

Attack	Number	Mean	Standard deviation
Apache	23	33.61	12.13
Saint	11	484.64	8.42
Smurf	10	508.40	5.13
Normal	56	5.13	5.59
Total	100	114.75	198.09
Correlation ratio	0.988		

Table 3.9 Applying NN classifier CI⇒Occupation to a random subsample of the Student dataset; wrong category assignments are highlighted in bold

Random sample	4	11	24	42	44	61	87	89	94	100
True target category	IT	IT	IT	BA	BA	BA	AN	AN	AN	AN
Predictor's value PV	72	65	54	65	44	62	72	48	34	45
Nearest Neighbor's PV	72	65	54*	65	44	62**	72	47	35*	45*
NN predicted category	**BA**	**AN**	IT	**AN**	BA	BA	**BA**	**BA**	AN	AN

* – of two other entities having different categories that with the matching one has been selected;
** – of several entities, the most frequent category has been selected.

3.3 Mixed Scale Case: Nominal Feature Versus a Quantitative One

Table 3.10 Tabular regression of CI mark over Occupation in Student data: comparatively high within-category standard deviations

Occupation	Number	Mean	Standard deviation
IT	35	70.57	12.73
BA	34	54.79	10.60
AN	31	53.35	16.29
Total	100	59.87	15.37
Correlation ratio	0.250		

Worked example 3.7. Nearest neighbor classifier

Consider two features from the dataset Intrusion: the type of attack Att, the target, and the number of connections to the same host as the current one in the past two seconds, SH. To make the method work fast, first, sort entities in the ascending order of SH. Take a random 10-element subset (upper row in Table 3.7) along with their Att categories (the second row) and SHCo values (the third row). Now take the entities whose SH values are nearest neighbors of those in the third row: these SH values are in the fourth row, and look at their Att categories (the bottom row). A striking success: all ten are predicted correctly!

Q.3.11. Build a tabular regression of the SHCo over Attack categories and find the correlation ratio. **A.** See Table 3.8.

Q.3.12. Apply NN classifier to predict Occupation from CI Mark over Student dataset. **A.** See Table 3.9.

Q.3.13. Consider a data table for 8 students and 2 features, as follows:

Student	Mark	Occupation
1	50	IT
2	80	IT
3	80	IT
4	60	AN
5	60	AN
6	40	AN
7	40	AN
8	50	AN

(i) Build a regression table for prediction Mark by Occupation.
(ii) Predict the mark for a new student whose occupation is IT.
(iii) Find the correlation ratio for the table.

A. (i) Regression table of Mark over Occupation contains Occupation category frequencies as well as Mark within-category averages and variances is this:

	Mark		
	Frequency	Average	Variance
IT	3	70	14.1
AN	5	50	8.9

(ii) For an IT student the likely mark will be 70±14.1.

(iii) The correlation ratio is determined by the weighted within-category variance, which is $(3*14.1+5*8.9)/8 = (42.3+44.5)/8 = 10.85$, and the total variance, which is calculated on all the data set with the mean = 57.5, and equal to 14.79. Then correlation ratio is $\eta^2 = 1 - 10.85/14.79 = 0.266$. This means that the tabular regression explains only 26.6% of the variance of Mark.

Q.3.14. Build tabular regression of the CI mark over Occupation in Student data and find the correlation ratio. **A.** See Table 3.10.

3.3.3.2 Interval Predicate Classifier

Another, more human friendly, classifier can be built in terms of quantitative feature x intervals. To predict a target feature category k, such a classifier would rely on an interval predicate $x(a(k), b(k))$ which is true if and only if the value of x is between $a(k)$ and $b(k)$. Then an interval predicate rule would be a production $x(a(x), b(k)) \Rightarrow k$. Consider, for example, "Saint" Attack in Intrusion data: there are 11 cases of this type and all, except one, have SHCo values 482 or 483. Thus, the interval predicate rule SHCo(482, 483)⇒ Saint would make only 9% of errors.

How one can infer which of the categories are more likely to be well covered by an interval predicate rule? One of the proposals is to rely on category contributions to the variance of x in (3.13), $p_k(\bar{x} - \bar{x}_k)^2$, in the denotations of this section, where p_k is proportion of entities in category k, $\bar{x}$ is grand mean and $\bar{x}_k$ is within category k mean. The mechanism making sense of this proposal is illustrated on Fig. 3.17 (top): the further away the within-category mean is, the more plausible that the entire category is further away. Yet, in many cases, many entities in a category fall apart from their averages thus leading to errors in the interval based prediction (Fig. 3.17, bottom).

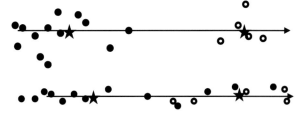

Fig. 3.17 A group of white *circles falls* apart from the rest *on the top*, and much intermixes with the rest *on the bottom*

3.3 Mixed Scale Case: Nominal Feature Versus a Quantitative One

Worked example 3.8. Category contributions for interval predicate productions

Consider the same features Att and SHCo from Intrusion dataset as those considered in Worked example 3.7 and determine the Att category contributions according to formula (3.13), $p_k(\bar{x} - \bar{x}_k)^2$ (see Table 3.11).

With respect to the data in Table 3.11, one can try to build interval predicate based productions for the largest contributing Saint and Smurf categories. We already observed that SH(482, 483)$\Rightarrow$ Saint makes 9% error, which is a false negative. This is caused by a SH value of 510 corresponding to Saint at 90-th row of the Intrusion data table – this does not satisfy the production's subject. Now one can see that rule SH(490, 512) $\Rightarrow$ Smurf would fail only once too, on the same observation – but this time this would be a false positive, satisfying the subject but being not Smurf. The next contributing category, lagging far behind, is "Normal" corresponding to the range of x values from 1 to 28 which overlaps the range (16, 42) of x values corresponding to Apache category. Yet the rule SH(1, 15)$\Rightarrow$ Normal is true for 53 of 56 cases, the three false negative errors making about 5% only. The rule SHCo(16, 42)$\Rightarrow$ Apache has the same three cases as false positives.

Q.3.15. Build a category contribution table like Table 3.11 for CI mark and Occupation features in Students dataset. **A.** See Table 3.12.

A rather successful usage of interval based productions in Worked example 3.8 is due to the tight correlation between SH and Attack. In a less comfortable situation, such as that of pair CI mark – Occupation at Student dataset, the interval based descriptions make no sense at all. Consider the most contributing category IT – its CI mark range is from 53 to 90. If one takes the entire range to make it into rule CI(53, 90)$\Rightarrow$ IT, this would make no false negatives at all. Yet there are 22 entities

Table 3.11 Category contributions according to formula (3.13)

Attack	Proportion	Mean	Contribution
Apache	0.23	33.61	1514.3
Saint	0.11	484.64	15049.8
Smurf	0.10	508.40	15496.0
Normal	0.56	5.13	6729.9
Total	1.00	114.75	38790.0

Table 3.12 Occupation category contributions to CI Mark

Occupation	Proportion	Mean	Squared diff.	Contribution
IT	0.35	70.574	4,980.327	1,743.114
BA	0.34	54.794	3,002.395	1,020.814
AN	0.31	38.774	1,503.438	466.066
Total	1.00	55.350		

of BA category and 15 of AN category whose CI mark falls within (53, 90) interval too, totaling to 37 false positive errors! One can try to somewhat reduce the interval predicate range, to lessen the false positive errors, with the price of admitting some false negatives. Consider, for example, CI(62, 90)$\Rightarrow$ IT rule to admit 12 false negative errors as well as 10 BA and 11 AN false positives, a drop to 33 errors altogether – quite a high error rate! Yet the interval based rules follow human way of thinking, which may lead to overall acceptance of such a rule, possibly amended by another feature interval added.

3.4 Two Nominal Features Case

P3.4.1 Analysis of Contingency Tables: Presentation

P3.4.1.1 Deriving Conceptual Relations from Statistics

To analyze interrelations between two nominal features, they are cross-classified in the so-called contingency table. A contingency table has its rows corresponding to categories of one feature and columns to categories of the other feature, with the entries reflecting the counts of entities falling in the overlap of the corresponding row and column categories.

Worked example 3.9. Contingency table on Market towns data

To cross-classify features Banks and Farmer's Market on Market towns data, we first need to categorize the quantitative feature Banks. Consider, for example, the four-category partition of the range of Banks feature at Market towns set presented in Table 3.13.

These categories are cross-classified with FM "yes" and "no" categories in Table 3.14. Besides the cross-classification counts, the table also contains summary within category counts, the totals, on the margins of the table, the last row and last column – this is why they are referred to as marginal frequencies. The total count balances the sheet in the bottom-right corner.

The same contingency data converted to relative frequencies by relating them to the total number of entities are presented in Table 3.15.

Table 3.13 Definition of Ba categories on the Market town dataset

Category	Definition	Notation
1	Ba $\geq$ 10	10+
2	10 > Ba $\geq$ 4	4+
3	4 > Ba $\geq$ 2	2+
4	Ba = 0 or 1	1−

3.4 Two Nominal Features Case

Table 3.14 Cross classification of the Ba categories with FM categories

FarmMarket	Bank/Building Society categories				Total
	10+	4+	2+	1−	
Yes	2	5	1	1	9
No	4	7	13	12	36
Total	6	12	14	13	45

Table 3.15 BA/FM cross-classification relative frequencies, per cent

FM | Ba	10+	4+	2+	1−	Total
Yes	4.44	11.11	2.22	2.22	20
No	8.89	15.56	28.89	26.67	80
Total	13.33	26.67	31.11	28.89	100

Table 3.16 Protocol/Attack contingency table for Intrusion data

Category	Apache	Saint	Smurf	Norm	Total
Tcp	23	11	0	30	64
Udp	0	0	0	26	26
Icmp	0	0	10	0	10
Total	23	11	10	56	100

Q.3.16. Build a contingency table for features "Protocol-type" and "Attack type" in Intrusion data. **A.** See Table 3.16.

A contingency table can be used for assessment of correlation between two sets of categories. The highest level of correlation is that of a conceptual association. A conceptual association may exist if a row, k, has all its entries, not marginal of course, except just one, say l, equal to 0, which would mean that all of the extent of category k belongs to the column category l. The data, thus, indicate that category k implies category l.

Worked example 3.10. Equivalence and implication from a contingency table

Such are rows "Udp" and "Icmp" in Table 3.16. There is a perfect match in this table: a row category $k=$ "Icmp" and a column category $l =$ "Smurf", that contains the only non-zero count. No other combination (k, l') or (k', l) is possible according to the table. In such a situation, one may claim that, subject to the sampling error, category l may occur if and only if k does, that is, k and l are equivalent.

102

A somewhat weaker, but still very much valuable is the case of "Udp" row in Table 3.16. It appears, Udp protocol implies "Norm" column category – a no-attack situation, though there is no equivalence here because the "Norm" column contains another positive count, in row "Tcp".

Case study 3.3. Trimming Contingency Data: A Bad Option

Unfortunately, there are no zeros in Table 3.14: thus, no conceptual relation between the number of Banks and the presence of a Farmer's market. But some of the entries are really close to 0, which may make us tempted to trim the data a bit. Imagine, for example, that in row "Yes" of Table 3.14, two last entries are 0, not 1s. This would imply that a Farmers Market may occur only in a town with 4 or more Banks. A logical implication, that is, a production rule, "If BA is 4 or more, then a Farmer's market must be present", could be derived then from thus modified table. One may try taking this path and cleaning the data of smaller entries, by removing corresponding entities from the table of course, to not obscure our "vision" of the pattern of correlation. Thus trimmed Table 3.17 is obtained from Table 3.14 by removing just 13 entities from "less popular" entries. This latter table expresses, with no exception, a very simple conceptual statement "A town has a Farmer's market if and only if the number of Banks in it is 4 or greater". However nice the rule may sound, let us not forget the cost of the trimming which is the 13 towns, almost 30% of the sample, that have been removed as those not fitting the stated perspective. Such a data doctoring borders with forgery – one of the reasons for a famous quip attributed to B. Disraeli, a celebrated British politician of XIX century: "There are three gradations of lies: lies, damned lies and statistics." The issue of sample adjustment so far has received no reasonable solution, even with respect to outliers – values falling way beyond the feature range one would expect normally. Anyway, the conclusion of the trimming exercise is that one should try finding ways of expressing conceptual relations without much doctoring the sample.

P3.4.1.2 Capturing Relationships with Quetelet Indexes

Quetelet index provides for a strategy for visualization of correlation patterns in contingency tables without removal of "not-fitting" entities. In 1832, A. Quetelet, a

Table 3.17 A trimmed BA/FM cross classification "cleaned" of 13 towns, to sharpen the view

FMarket	Number of Banks/Build. Societies				Total
	10+	4+	2+	1–	
Yes	2	5	0	0	7
No	0	0	13	12	25
Total	2	5	13	12	32

3.4 Two Nominal Features Case

founding father of statistics, proposed to measure the extent of association between row and column categories in a contingency table by comparing the local count with an average one.

Let us consider correlation between the presence of a Farmer's Market and the category "10 or more Banks" according to Table 3.15. We can see that their joint probability/frequency is the entry in the corresponding row and column: P(Ba = 10+ & FM = Yes) = 2/45 = 4.44% (joint probability/frequency rate). Of the 20% entities that fall in the row "Yes", this makes the proportion of "Ba = 10+" under condition "FM = Yes" equal to P(Ba = 10+ /FM = Yes) = P(Ba = 10+ & FM =Yes)/ P(FM = Yes) = 0.0444/0.20 = 0.222 = 22.2%. Such a ratio expresses the conditional probability/rate.

Is this high or low? Hard to tell without comparing this with the unconditional rate, that is, with the frequency of category "Ba = 10+" in the whole dataset, which is P(Ba = 10+) = 13.33%. Let us compute the (relative) difference between the two, which is referred to as Quetelet index q:

$$q(Ba = 10+/FM = Yes) = [P(Ba = 10 + FM = Yes) - P(Ba = 10+)]/P(Ba = 10+) = [0.2222 - 0.1333]/0.1333 = 0.6667 = 66.7\%$$

That means that condition "FM = Yes" raises the frequency of the Bank category by 66.7%. This logic concurs with our everyday intuition. Consider, for example, the risk of getting a serious illness, say tuberculosis, which may be, say, about 0.1%, one in a thousand, in a given region. Take a condition such as "Bad housing" and count the rate of tuberculosis under this condition, amounting to, say 0.5% – which is very small by itself, yet a five-fold increase over the average tuberculosis rate. This is exactly what Quetelet index measures: $q(l/k) = (0.5 - 0.1)/0.1 = 400\%$ to show that the change of the average rate is 4 times.

Worked example 3.11. Quetelet index in a contingency table

Let us apply the general Quetelet index formula (3.15) to entries in Table 3.14. This leads to Quetelet index values presented in Table 3.18. By highlighting positive values in the table, we obtain the same pattern as on the "purified" data as in Case-study 3.3, but this time in a somewhat more realistic manner, keeping the sample intact. Specifically, one can see that "Yes" FM category provides for a strong increase in the probabilities, whereas "No" category leads to much weaker changes.

Table 3.18 BA/FM Cross classification Quetelet coefficients, % (positive entries highlighted)

FMarket	10+	4+	2+	1–
Yes	**66.67**	**108.33**	−64.29	−61.54
No	−16.67	−27.08	**16.07**	**15.38**

Table 3.19 Quetelet indices for the Protocol/Attack contingency Table 3.16, per cent

Category	Apache	Saint	Surf	Norm
Tcp	**56.25**	**56.25**	−100.00	−16.29
Udp	−100.00	−100.00	−100.00	**78.57**
Icmp	−100.00	−100.00	**900.00**	−100.00

Q.3.17. Compute Quetelet coefficients for Table 3.16. **A.** See Table 3.19 in which positive entries are highlighted in bold.

Case-study 3.4. Has There Been a Bias in S'nS' Policy?

Take on the case of Stop-and-Search policy in England and Wales 2005 represented according to race (B - black, A - asian and W - white), by numbers in Table 2.4 in Section 2.3 – these are overwhelmingly in category W. The criticism of this policy came out of comparison of this distribution with the distribution of the entire population. Such a distribution, according to the latest pre-2005 census 2001, can be easily found on web. By subtracting from that the numbers of Stop-and-Search occurrences, under the assumption that nobody has been subjected to this more than once, Table 3.20 has been drawn. Its last column gives the numbers that were used for the claim of a racial bias: indeed category B members have been subjects of the policy six times more frequently than category W members. A similar picture emerges when Quetelet coefficients are used (see Table 3.21). Category B is subject to Stop-and-Search policy 400% more frequently than on average, whereas category W is 15% less.

Yet some would consider drawing a table like Table 3.20, and of course the derived Table 3.21, as something nonsensical, because it is based on an implicit

Table 3.20 Distribution of Stop-and-Search policy cross-classified with race

	S'n'S	Not S'n'S	Total	S'n'S-to-Total
Black	131, 723	1, 377, 493	1, 509, 216	0.0873
Asian	70, 252	2, 948, 179	3, 018, 431	0.0233
White	676, 178	46, 838, 091	47, 514, 269	0.0142
Total	878, 153	51, 163, 763	52, 041, 916	0.0169

Table 3.21 Relative Quetelet coefficients for cross-classification in Table 3.20, per cent

	S'n'S	Not S'n'S
Black	417.2	−7.2
Asian	37.9	−0.6
White	−15.7	0.3

3.4 Two Nominal Features Case

assumption that the Stop-and-Search policy applies to the population randomly. They would argue that police apply the policy only when they deem it necessary, so that the comparison should involve not all of the total population but only those criminal. Indeed, the distribution of subjects to Stop-and-Search policy by race has been almost identical to that of the imprisoned population of the same year. Therefore, the claim of a racial bias should be declared incorrect.

P3.4.1.3 Chi-Square Contingency Coefficient As a Summary Correlation Index

A somewhat more refined visualization of the contingency table comes from the Quetelet indexes weighted by the probabilities of corresponding entries, as explained in Section 3.4.2. They sum to a most popular concept in the analysis of contingency tables, the celebrated chi-square contingency coefficient. This coefficient was introduced by K. Pearson (1900) to express the deviation of the observed bivariate distribution, represented by the relative frequencies in a contingency table, from the situation of statistical independence between the features.

Worked example 3.12. Visualization of contingency table using weighted Quetelet coefficients

Let us multiply Quetelet coefficients in Table 3.18 by the frequencies of the corresponding entries in Table 3.14. Quetelet coefficients in Table 3.18 are taken relative to unity, not per cent. This leads us to Table 3.22 whose entries sum to the value of Pearson's chi-square coefficient for Table 3.14, 6.86. Note that entries in Table 3.20 can be both positive and negative; those with absolute value greater than $6.86/4 = 1.72$ are highlighted in bold – they show the entries of an extraordinary deviation from the average. Of them, column 4+ supplies the highest positive impact and the highest negative impact.

A pair of categories, one from one nominal feature and the other from another nominal feature, are said to be statistically independent if the probability of their co-occurrence is equal to the product of probabilities of these categories. Take, for example, category "Yes" of FM and "4+" of Banks in Table 3.15: the probability of their co-occurrence is 0.111. On the other hand, the probability of FM = "Yes" is 0.2 and that of Banks = 4+ is 0.267, according to the table. If these two categories were independent they would have co-occurred at the level of $0.2 \times 0.267 = 0.053$,

Table 3.22 BA/FM chi-squared (NQ = 6.86) and its decomposition according to (3.19)

FMarket	10+	4+	2+	1−	Total
Yes	1.33	**5.41**	−.064	−.062	5.48
No	−.067	**−1.90**	**2.09**	**1.85**	1.37
Total	0.67	3.51	1.45	1.23	**6.86**

about twice as less than in reality, which means that the pair highly deviates from the statistical independence. Two features are said to be statistically independent if all pairs of their mutual categories are statistically independent. K. Pearson was concerned with the situation at which two features are independent in the population at large but this may not necessarily be reflected in the sample under consideration because of the randomness of sampling. Thus he proposed to take the squared differences between observed frequencies and those that would occur under the independence assumption and relate them to the "theoretical" probabilities that should be true in the population. The summary index is referred to as the Pearson chi-square coefficient, see (3.18) later. The distribution of the summary chi-square index, under conventional assumptions of independence in sampling, converges to the so-called chi-square distribution, which allows for statistical testing of the hypothesis of independence between the features. This suggests that the coefficient should be used only for testing the hypothesis, but not as a measure of correlation. The claim would be – and often has been – that the index can only distinguish between two cases, statistical independence or not, and thus cannot be used for comparison of the extent of the dependence. Yet practitioners are always tempted to ignore this commandment and do compare the extent of dependence at different pairs of categorical features. Indeed, as formula (3.19) shows there is nothing wrong in using chi-square contingency coefficient as an index of correlation – it is indeed the summary Quetelet index, thus showing the average degree of relationship between two features.

Worked example 3.13. A conventional decomposition of chi-square coefficient

Let us consider a conventional way of visualization of contingency tables, by putting Pearson indexes, the square roots $x(k,l)$ of the chi-square coefficient items in (3.21) as the table's elements. These are in Table 3.23. The table does show a similar pattern of positive and negative associations. However, it is not the entries of the table that sum to the chi-square coefficient but rather the squares of the entries. The fact that the summary values on the margins in Tables 3.22 and 3.23 are the same is not by chance: it exemplifies a mathematical property (see Equation (3.19)).

Table 3.23 Square roots of the items in Pearson chi-squared ($X^2 = 6.86$); the items themselves are in parentheses

FMarket	10+	4+	2+	1–	Total
Yes	**0.73**(0.53)	**1.68**(**2.82**)	−1.08 (1.16)	−0.99 (0.98)	(5.49)
No	−0.36 (0.13)	−0.84 (0.70)	**0.54** (**0.29**)	**0.50** (**0.25**)	(1.37)
Total	(0.67)	(3.52)	(1.45)	(1.23)	(6.86)

3.4 Two Nominal Features Case

Q.3.18. In Table 3.22, all marginal values, the sums of rows and columns, are positive, in spite of the fact that many within-table entries are negative. Is this just due to specifics of the distribution in Table 3.14 or a general property? **A**: A general property: the within-row or within-column sums of the elements, $N_{lk}\, q(l/k)$, must be positive, see (3.19).

Q.3.19. Find a similar decomposition of chi-squared for OOPmarks/Occupation in Student data. Hint: First, categorize quantitative feature OOPmarks somehow: you may use equal bins, or conventional boundary points such as 35, 65 and 75, or any other considerations.

Q.3.20. Can any logical production rules come from the columns of Table 3.16?
A. Yes, both Apache and Saint attacks may occur at the tcp protocol only.

Q.3.21. Among the shoppers in Q.2.21, those who spent £60 each are males only and those who spent £100 each are females only, whereas among the rest 30 individuals half are men and half are women. Build a contingency table for the two features, gender and spending. Find and interpret the value of Quetelet coefficient for females who spent £100 each.
A. The contingency table (of co-occurrence counts):

Spending, £				
Gender	60	100	150	Total
Female	0	20	15	35
Male	50	0	15	65
Total	50	20	30	100

This table of absolute co-occurrence counts coincides with that of proportions expressed per cent because the number of shoppers is 100.

Quetelet coefficient for (Female/£100) entry is

$$Q = 100*20/(20*35) - 1 = 2.86 - 1 = 1.86$$

This means that being female in this category of spending is more likely than the average, by 186%.

F3.4.2 Analysis of Contingency Tables: Formulation

Consider two sets of disjoint categories on an entity set I: $l = 1, \ldots, L$ (for example, occupation of individuals constituting I) and $k = 1, \ldots, K$ (say, family or housing type). Each makes a partition of the entity set I; they are crossed to see if there is any correlation between them. Combine a pair of categories $(k, l) \in K \times L$ and count the number of entities that fall in both. The (k,l) co-occurrence count is denoted by N_{kl}. Obviously, these counts sum to N because the categories are not overlapping and cover the entire dataset. A table housing these counts, N_{kl}, or their relative values, frequencies $p_{kl} = N_{kl}/N$, is referred to as a contingency table or just cross-classification. The totals, that is, within-row sums $N_{k+} = \Sigma_l N_{kl}$ and within-column sums $N_{+l} = \Sigma_k N_{kl}$ (as well as their relative frequency counterparts) are referred to as marginals (because they are located on margins of the contingency table).

The (empirical) probability that category l occurs under condition of k can be expressed as $P(l/k) = p_{kl}/p_{k+} = N_{kl}/N_{k+}$. The probability $P(l)$ of the category l with no condition is just $p_{+l} = N_{+l}/N$. Similar notation is used when l and k are swapped. The relative difference between the two probabilities is referred to as (relative) Quetelet index (Mirkin 2001):

$$q(l/k) = \frac{P(l/k) - P(l)}{P(l)} \qquad (3.16)$$

where $P(l) = N_{+l}/N$, $P(k) = N_{k+}/N$, $P(l/k) = N_{kl}/N_{k+}$ That is, Quetelet index expresses correlation between categories k and l as the relative change in the probability of l when k is taken into account.

With little algebra, one can derive a simpler expression

$$q(l/k) = [N_{kl}/N_{k+} - N_{+l}/N]/(N_{+l}/N) = N_{kl}N/(N_{k+}N_{+l}) - 1 = \frac{p_{kl}}{p_{k+}p_{+l}} - 1 \qquad (3.16')$$

Highlighting high positive and negative values in a Quetelet index table, such as Tables 3.18 and 3.21, visualizes the pattern of correlation between the two sets of categories.

This visualization can be extended to a theoretically sound presentation. Let us define the summary Quetelet correlation index Q as the sum of pair-wise Quetelet indexes weighted by their frequencies/probabilities:

$$Q = \sum_{k=1}^{K}\sum_{l=1}^{L} p_{kl} q(l,k) = \sum_{k=1}^{K}\sum_{l=1}^{L} p_{kl}\left(\frac{p_{kl}}{p_{k+}p_{+l}} - 1\right) = \sum_{k=1}^{K}\sum_{l=1}^{L} \frac{p_{kl}^2}{p_{k+}p_{+l}} - 1 \qquad (3.17)$$

The right-hand expression for Q in (3.17) is very popular in statistical analysis of contingency data. In fact, this is equal to chi-squared correlation coefficient proposed by K. Pearson (1900) in a very different context – as a measure of deviation of the contingency table entries from the statistical independence.

To explain this in more detail, let us first introduce the concept of statistical independence. The sets of k and l categories are said to be statistically independent if $p_{kl} = p_{k+}p_{+l}$ for all k and l. Obviously, such a condition is hard to fulfill in reality. K. Pearson suggested using relative squared errors to measure the deviations of observed frequencies from the statistical independence. Specifically, he introduced the following coefficient usually referred to as Pearson's chi-squared association

3.4 Two Nominal Features Case

coefficient:

$$X^2 = N \sum_{k=1}^{K} \sum_{l=1}^{L} \frac{(p_{kl} - p_{k+}p_{+l})^2}{p_{k+}p_{+l}} = N \left(\sum_{k=1}^{K} \sum_{l=1}^{L} \frac{p_{kl}^2}{p_{k+}p_{+l}} - 1 \right) \quad (3.18)$$

The equation on the right can be proven with little algebra. Consider, for example, this part of the expression on the left in (3.18):

$$\sum_{l=1}^{L} \frac{(p_{kl} - p_{k+}p_{+l})^2}{p_{k+}p_{+l}} = \sum_{l=1}^{L} \frac{p_{kl}^2 - 2p_{kl}p_{k+}p_{+l} + (p_{k+}p_{+l})^2}{p_{k+}p_{+l}}$$

$$= \sum_{l=1}^{L} \frac{p_{kl}^2}{p_{k+}p_{+l}} - 2\sum_{l=1}^{L} p_{kl} + \sum_{l=1}^{L} p_{k+}p_{+l} = \sum_{l=1}^{L} \frac{p_{kl}^2}{p_{k+}p_{+l}} - p_{k+}$$

The expression on the right in the above is derived by using equations $\Sigma_l p_{kl} = p_{k+}$ and $\Sigma_l p_{+l} = 1$. Summing these equations over k will produce (3.18). On the other hand, the expression on the right in (3.18) is obviously equal to $\Sigma_l p_{kl} q(l/k)$ so that

$$\sum_{l=1}^{L} \frac{(p_{kl} - p_{k+}p_{+l})^2}{p_{k+}p_{+l}} = \sum_{l=1}^{L} p_{kl} q(l/k) \quad (3.19)$$

By comparing the right-hand parts of (3.17) and (3.18), it is easy to see that $X^2 = NQ$. The same follows from summing Equation (3.19) over k.

The popularity of X^2 index in statistics and related fields rests on the theorem proven by K. Pearson: if the contingency table is based on a sample of entities independently drawn from a population in which the statistical independence holds (so that all deviations are due to just randomness in the sampling), then the probabilistic distribution of X^2 converges to the chi-squared distribution (when N tends to infinity) introduced by Pearson earlier for similar analyses. The probabilistic chi-squared distribution is defined as the distribution of the sum of squares of random variables distributed according to the standard Gaussian distribution.

This theorem is not always of interest to a computational data analyst, because they analyze data that are not necessarily random or not necessarily independently sampled. However, Pearson's chi-squared coefficient is frequently used just for scoring correlation in contingency tables, and the equation $X^2 = NQ$ gives a credible support to it. According to this equation, X^2 also is not necessarily a measure of deviation from the statistical independence. It also has a different meaning of a measure of interrelation between categories: that of the averaged Quetelet coefficient.

To get more intuition on the underlying correlation concept, let us take a look at the extreme values that X^2 can take and situations at which the extreme values are reached (Mirkin 2001). It appears that at $K \leq L$, that is, the number of columns is not greater than that of rows, X^2 ranges between 0 and $K - 1$. It reaches 0 if there is a statistical independence at all (k,l) entries so that all $q_{kl} = 0$, and it reaches $K - 1$

if each column l contains only one non-zero entry $p_{k(l)l}$, which is thus equal to p_{+l}. The latter can be interpreted as the logical implication $k \to l(k)$.

Representation of chi-squared through Quetelet coefficients,

$$X^2 = \sum_{k=1}^{K} \sum_{l=1}^{L} N p_{kl} q(l/k) \qquad (3.20)$$

amounts to decomposition of X^2 into the sum of $N_{kl}\, q(l/k)$ items and allows for visualization of the items within the contingency table format, such as that presented in Table 3.21.

In fact not only the total sum of these items coincide with that of the original chi-squared items $N(p_{kl} - p_{k+}p_{+l})^2/p_{k+}p_{+l}$, but also the within-column and within-row sums coincide too, as (3.19) clearly demonstrates for the latter case.

However all the original chi-squared items in (3.18) are positive and cannot show whether the contribution of an individual entry is positive or negative. To overcome this shortcoming, another visualization of X^2 is in use. That visualization involves the square roots of the chi-squared items

$$r(k,l) = \frac{p_{kl} - p_{k+}p_{+l}}{\sqrt{p_{k+}p_{+l}}} \qquad (3.21)$$

that are convenient to refer to as Pearson indexes. Obviously, $X^2 = N\Sigma_{k,l}\, r(k,l)^2$. Pearson indexes indeed have the same signs as $q(l/k)$, and in fact are closely related: $q(l/k) = r(k,l)[p_{k+}\, p_{+l}]^{1/2}$. It is less clear what interpretation of its own $r(k,l)$ may have, although they are useful in Correspondence analysis of contingency tables (Section 5.4, see also normalized Laplacian in Section 8.2).

Q.3.22. Take two binary features presented as 1/0 variables and build their contingency table, sometimes referred to as a four-fold table (Table 3.24) when symbols a, b, c, d are used to denote the co-occurrence numbers.

Prove that Quetelet coefficient $q(Yes/Yes)$ expressing the relative difference between $a/(a+c)$ and $(a+b)/N$ is equal to

$$q(Yes/Yes) = \frac{ad - bc}{(a+c)(a+b)},$$

Table 3.24 Four-fold contingency table between binary features

		Feature Y		
		Yes	No	Total
Feature X	Yes	a	b	a+b
	Not	c	d	c+d
Total		a+c	b+d	$N = a+b+c+d$

3.5 Summary

and the summary Quetelet coefficient Q, or Pearson's X^2/N, is equal to

$$Q = \frac{(ad-bc)^2}{(a+c)(b+d)(a+b)(c+d)}.$$

Q.3.23. Prove that the correlation coefficient between two 1/0 binary features can be expressed in terms of the four-fold table as $\rho = \sqrt{Q}$, that is,

$$\rho = \frac{ad-bc}{\sqrt{(a+c)(b+d)(a+b)(c+d)}}.$$

Q.3.24. Given a $K \times L$ contingency table P and a pair of categories, $k \in K$ and $l \in L$, consider an absolute Quetelet index $a(l/k) = P(l/k) - P(l)$ – the change from the frequency of $l \in L$ on the whole entity set I to the frequency of l on entities falling in category $k \in K$. In terms of P, $P(l) = p_{+l}$ and $P(l/k) = p_{kl}/p_{+l}$. Prove that the summary Quetelet index $A = \Sigma_{k,l} \, p_{kl} a(l/k) = \Sigma_{k,l} p_{kl}^2/p_{k+} - \Sigma_l \, p_{+l}^2$ is equal to the following expression, an asymmetric analogue to Pearson chi-squared:

$$A = \sum_{k=1}^{K} \sum_{l=1}^{L} \frac{(p_{kl} - p_{k+}p_{+l})^2}{p_{k+}} \quad (3.22)$$

which also is the numerator of the so called Goodman-Kruskal "tau-b" index (Kendall and Stewart 1973).

A. Indeed, by taking the square of the denominator, expression in (3.22) becomes equal to $\Sigma_{k,l} \, (p_{kl}^2 - 2p_{kl}p_{k+}p_{+l} + p_{k+}^2 p_{+l}^2)/p_{k+}$, which is $\Sigma_{k,l}p_{kl}^2/p_{k+} - 2\Sigma_{k,l}p_{kl}p_{+l} + \Sigma_{k,l} \, p_{k+}p_{+l}^2 = \Sigma_{k,l}p_{kl}^2/p_{k+} - 2\Sigma_{k,l} \, p_{+l}^2 + \Sigma_l p_{+l}^2$ because $\Sigma_k p_{kl} = p_{+l}$ and $\Sigma_k p_{k+} = 1$. This is obviolsly $\Sigma_{k,l}p_{kl}^2/p_{k+} - \Sigma_l \, p_{+l}^2 = \Sigma_{k,l} \, p_{kl}a(l/k) = A$, which proves the statement.

3.5 Summary

The Chapter outlines several important characteristics of summarization and correlation between two features, and displays some of the properties of those. They are:

- linear regression and correlation coefficient for two quantitative variables;
- tabular regression, correlation ratio, decomposition of the quantitative feature scatter, and nearest neighbor classifier for the mixed scale case; and
- contingency table, Quetelet index, statistical independence, and Pearson's chi-squared for two nominal variables.

They all are applicable in the case of multidimensional data as well.

Some of the characteristics are rather unconventional. For example, the concepts of tabular regression and correlation ratio are not terribly popular in data mining. The Quetelet indexes are recognized by neither community, the more so the idea that Pearson chi-squared is a summary correlation measure, not necessarily a criterion of statistical independence.

Some examples of non-linear regression and nature-inspired approaches for fitting that are outlined. Computational bootstrap based validation is considered.

References

Carpenter, J., Bithell, J.: Bootstrap confidence intervals: when, which, what? A practical guide for medical statisticians. Stat. Med. **19**, 1141–1164 (2000)

Davison, A.C., Hinkley, D.V.: Bootstrap Methods and Their Application. Cambridge University Press, Cambridge (7th printing) (2005)

Kendall, M.G., Stewart, A.: Advanced Statistics: Inference and Relationship, 3d edn. Griffin, London, ISBN: 0852642156 (1973)

Lohninger, H.: Teach Me Data Analysis. Springer, Berlin-New York-Tokyo, ISBN 3-540-14743-8 (1999)

Mirkin, B.: Eleven ways to look at the chi-squared coefficient for contingency tables. Am. Stat. **55**(2), 111–120 (2001)

Pearson, K.: On a criterion that a given system of deviations from the probable in the case of a correlated system of variables is such that it can be reasonably supposed to have arisen in random sampling. Phil. Mag. **50**, 157–175 (1900)

Chapter 4
Learning Multivariate Correlations in Data

4.1 General: Decision Rules, Fitting Criteria, and Learning Protocols

To specify a problem of learning correlation in a data table, one has to distinguish between two parts in the feature set: *predictor*, or *input*, features and *target*, or *output*, features. Typically, the number of target features is small, and in generic tasks, there is just one target feature. Target features are usually difficult to measure or impossible to know beforehand. This is why one would want to derive a decision rule relating predictors and targets so that prediction of targets can be made after measuring predictors only. Examples of learning problems include:

(a) chemical compounds: input features are of the molecular structure, whereas target features are activities such as toxicity or healing effects;
(b) types of grain in agriculture: input features are those of the seeds, ground and weather, and target features are of productivity and protein contents,
(c) industrial enterprises: input features refer to technology, investment and labor policies, whereas target features are of sales and profits;
(d) postcode districts in marketing research: input features refer to demographic, social and economic characteristics of the district residents, target features – to their purchasing behavior;
(e) bank loan customers: input features characterize demographic and income, whereas output features are of (potentially) bad debt;
(f) gene expression data: input features relate to levels of expression of DNA materials in the earlier stages of an illness, and output features to those at later stages.

A *decision rule* predicts values of target features from values of input features. A rule is referred to as a classifier if the target is categorical and as a regression if the target is quantitative. A generic categorical target problem is defined by specifying just a subset of entities labeled as belonging to the class of interest – the correlation problem in this case would be of building such a decision rule that would recognize, for each of the entities, whether it belongs to the labeled class or not.

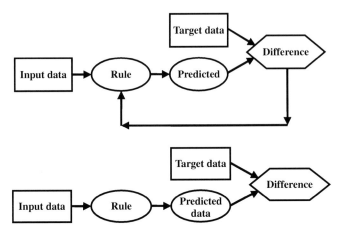

Fig. 4.1 Structure of a training/testing problem: In training, *on the top*, the decision rule is fitted to minimize the difference between the predicted and observed target data. In testing, *the bottom part*, the rule is used to calculate the difference so that no feedback to the rule is utilized

A generic regression problem – the bivariate linear regression – has been considered in Section 3.2; its extension to the multivariate case will be described later in Section 4.3.

A decision rule is learnt over a dataset in which values of the targets are available. These data are frequently referred to as the training data. The idea underlying the process of learning is to look at the difference between predicted and observed target feature values on the training data set and to minimize them over a class of admissible rules. The structure of such a process is presented on the upper part of Fig. 4.1.

The notion that it ought to be a class of admissible rules pre-specified emerges because the training data is finite and, therefore, can be fit exactly by using a sufficient number of parameters. However, this would be valid on the training set only, because the fit would capture all the errors and noise inevitable in data collecting processes. Take a look, for example, at the 2D regression problem on Fig. 4.2 depicting seven points on (x,u)-plane corresponding to observed combinations of input feature x and target feature u.

Fig. 4.2 Possible graphs of interrelation between x and u according to observed data points (*black circles*)

4.1 General: Decision Rules, Fitting Criteria, and Learning Protocols

The seven points on Fig. 4.2 can be exactly fitted by a polynomial of 6th order $u = p(x) = a_0 + a_1 x + a_2 x^2 + a_3 x^3 + a_4 x^4 + a_5 x^6$. Indeed, they would lead to 7 equations $u_i = p(x_i)$ ($i = 1, \ldots, 7$), so that, in a typical case, the 7 coefficients a_k of the polynomial can be exactly determined. Having N points observed would require an N-th degree polynomial to exactly fit them.

However, the polynomial, on which graph all observations lie, has no predictive power both within and beyond the range. The curve may go either course (like those shown) depending on small changes in the data. The power of a theory – and a regression line is a theory in this case – rests on its generalization power, which, in this case, can be cast down as the relation between the number of observations and the number of parameters: the greater the better. When this ratio is relatively small, statisticians would refer to this as an *over-fitted* rule. The overfitting normally produce very poor predictions on newly added observations. The straight line of Fig. 4.2 fits none of the points, but it expresses a simple and very robust tendency and should be preferred because it summarizes the data much deeper: the seven observations are summarized here in just two parameters, slope and intercept, whereas the polynomial line provides no summary: it involves as many parameters as the data entities. This is why, in learning decision rule problems, a class of admissible rules should be selected first. Unfortunately, as of this moment, there is no model based advice, within the data analysis discipline, on how this can be done, except very general ones like "look at the shapes of scatter plots". If there is no domain knowledge to choose a class of decision rules to fit, it is hard to tell what class of decision rules to use.

A most popular advice relates to the so-called *Occam's razor*, which means that the complexity of the data should be balanced by the complexity of the decision rule. A British monk philosopher William Ockham (*c.* 1285–1349) said: "Entities should not be multiplied unnecessarily." This is usually interpreted as saying that all other things being equal, the simplest explanation tends to be the best one. Operationally, this is further translated as the Principle of Maximum Parsimony, which is referred to when there is nothing better available. In the format of the so called "Minimum description length" principle, this approach can be meaningfully applied to problems of estimation of parameters of statistic distributions (see P.D. Grünwald 2007). Somewhat wider, and perhaps more appropriate, explication of the Occam's razor is proposed by Vapnik (2006). In a slightly modified form, to avoid mixing different terminologies, it can be put as follows: "Find an admissible decision rule with the smallest number of free parameters such that explains the observed facts" (Vapnik 2006, p. 448). However, even in this format, the principle gives no guidance about how to choose an adequate functional form. For example, which of two functions, the power function $f(x) = ax^b$ or logarithmic one, $g(x) = b\log(x) + a$ both having just two parameters a and b, should be preferred as a summarization tool for graphs on Fig. 4.3?

Another set of advices, not incompatible with those above, relates to the so-called falsifability principle by K. Popper (1902–1994), which can be expressed as follows: "Explain the facts by using such an admissible decision rule which is easiest to falsify" (Vapnik 2006, p. 451). In principle, to falsify a theory one needs to give

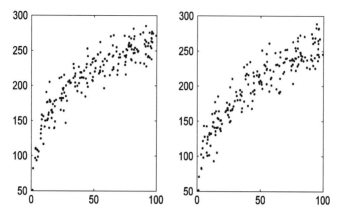

Fig. 4.3 Graph of one of two functions, $f(x) = 65x^{0.3}$ and $g(x) = 50\log(x) + 30$, both with an added normal noise $N(0,15)$, is presented on each plot. Can the reader give an educated guess of which is which? (Answer: *f(x)* is *on the right* and *g(x)* *on the left*.)

an example contradicting to it. Falsifability of a decision rule can be formulated in terms of the so-called *VC-complexity*, a measure of complexity of classes of decision rules: the smaller VC-complexity the greater the falsifability.

Let us explain the concept of VC-complexity for the case of a categorical target, so that a decision rule to be would be a classifier. However many categorical target features are specified, different combinations of target categories can be assigned different labels, so that a classifier is bound to predict a label. A set of classifiers Φ is said to shatter the training sample if for any possible assignment of the labels, a classifier exactly reproducing the labels can be found in Φ. Given a set of admissible classifiers Φ, the VC-complexity of a classifying problem is the maximum number of entities that can be shattered by classifiers from Φ. For example, 2D points have VC complexity 3 in the class of linear decision rules. Indeed, any three points, not lying on a line, can be shattered by a line; yet not all four-point sets can be shattered by lines, as shown on Fig. 4.4, the left and right parts, respectively.

The VC complexity is an important characteristic of a correlation problem especially within the probabilistic machine learning paradigm. Under conventional conditions of the independent random sampling of the data, a reliable classifier "with probability $a\%$ will be $b\%$ accurate, where b depends not only on a, but also on the sample size and VC-complexity" (Vapnik 2006).

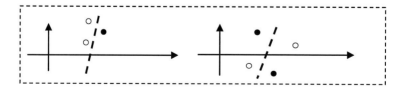

Fig. 4.4 Any two-part split of three points (not on one line) can be made by a linear function, but the presented case on four points cannot be solved by a *line*

4.1 General: Decision Rules, Fitting Criteria, and Learning Protocols

The problem of learning correlation in a data table can be stated, in general terms, as follows. Given N pairs (x_i, u_i), $i = 1, \ldots, N$, in which x_i are predictor/input p-dimensional vectors $x_i = (x_{i1}, \ldots, x_{ip})$ and $u_i = (u_{i1}, \ldots, u_{iq})$ are target/output q-dimensional vectors (usually $q = 1$), build a decision rule

$$\hat{u} = F(x) \qquad (4.1)$$

such that the difference between computed $\hat{u}$ and observed u is minimal over a pre-specified class Φ of admissible rules F.

To specify a correlation learning problem one should specify assumptions regarding a number of constituents including:

(i) Type of target

Two types of target features are considered usually: quantitative and categorical. In the former case, Equation (4.1) is usually referred to as regression; in the latter case, decision rule, and the learning problem is referred to as that of "classification" or "pattern recognition".

(ii) Type of rule

A rule involves a postulated mathematical structure whose parameters are to be learnt from the data. The mathematical structures considered further on are:

- *linear* combination of features
- *neural network* mapping a set of input features into a set of target features
- *decision tree* built over a set of features
- *partition* of the entity set into a number of non-overlapping clusters

(iii) Criterion

Criterion of the quality of fitting depends on the framework in which the learning task is formulated. Most popular criteria are: maximum likelihood (in a probabilistic model of data generation), least-squares (data recovery approach) and relative errors. According to the least-squares criterion, the difference between u and $\hat{u}$ is measured with the average squared error,

$$E = <u - \hat{u}, u - \hat{u}> /N = <u - F(x), u - F(x)> /N \qquad (4.2)$$

which is to be minimized over all admissible F.

(iv) Training protocol

The rule F is learnt from a training dataset. The way the data becomes available can be referred to as the training protocol. Three popular training protocols are: batch, random and on-line. The batch mode is the case when all training set is available and used at once, the other two refer to cases when data entities come one by one so that the learning goes incrementally. In the random protocol, the data are available at once, yet the learning process is organized incrementally by picking up entities randomly one-by-one, possibly many times each. In contrast, in an on-line protocol each entity comes from an external source and can be seen only once.

4.2 Naïve Bayes Approach

4.2.1 Bayes Decision Rule

Consider a situation in which there is only one target, a binary feature labeling two states of the world corresponding to "positive" and "negative" classes of entities. According to Bayes (1702–1761), all relevant knowledge of the world should be shaped by the decision maker in the form of probability distributions. Then, whatever new data may be observed, they may lead to changing the probabilities – hence the difference between prior probabilities and posterior, data-updated, probabilities. Specifically, assume that, $P(1) = p_1$ and $P(2) = p_2$ are prior probabilities of the two states so that p_1 and p_2 are positive and sum to unity. Assume furthermore that there are two probability density functions, $f_1(x_1, x_2, \ldots, x_p)$ and $f_2(x_1, x_2, \ldots, x_p)$, defining the generation of observed entity points $x = (x_1, x_2, \ldots, x_p)$ for each of the classes. That gives us, for any point $x = (x_1, x_2, \ldots, x_p)$ to occur, two probabilities, $P(x/1) = p_1 f_1(x)$ and $P(x/2) = p_2 f_2(x)$, of x being generated from either class. If an $x = (x_1, x_2, \ldots, x_p)$ is actually observed, it leads to a change in probabilities of the classes, from the prior probabilities $P(1) = p_1$ and $P(2) = p_2$ to posterior probabilities $P(1/x)$ and $P(2/x)$, respectively. These can be computed according to the well-known Bayes theorem from the elementary probability theory, so that the posterior probabilities of the classes are

$$P(1|x) = p_1 f_1(x)/f(x) \text{ and } P(2|x) = p_2 f_2(x)/f(x) \tag{4.3}$$

where $f(x) = p_1 f_1(x) + p_2 f_2(x)$.

The decision of which class the entity x belongs to depends on what value, $P(1/x)$ or $P(2/x)$ is greater. The class is assumed to be positive if $P(1/x) > P(2/x)$ or, equivalently,

$$f_1(x)/f_2(x) > p_2/p_1 \tag{4.4}$$

or, negative, if the reverse inequality holds. This rule is referred to as Bayes decision rule. Another expression of the Bayes rule can be drawn by using the difference $B(x) = P(1/x) - P(2/x)$: x is taken to belong to the positive class if $B(x) > 0$, and the negative class if $B(x) < 0$. Equation $B(x) = 0$ defines the so-called separating surface between the two classes.

The proportion of errors admitted by Bayes rule is $1 - P(1/x)$ when 1 is predicted and $1 - P(2/x)$ when 2 is predicted. These are the minimum error rates achievable when both within-class distributions $f_1(x)$ and $f_2(x)$ and priors p_1 and p_2 are known.

Unfortunately, the distributions $f_1(x)$ and $f_2(x)$ are typically not known. Then some simplifying assumptions are to be made so that the distributions could be estimated from the observed data. Among most popular assumptions are: (i) Gaussian probability and (ii) Local independence. Let us consider them in turn:

4.2 Naïve Bayes Approach

(i) Gaussian probability

The class probability distributions $f_1(x)$ and $f_2(x)$ are assumed to be Gaussian, so that each can be expressed as

$$f_k(x) = \exp[-(x-\mu_k)^T \Sigma_k^{-1}(x-\mu_k)/2]/[(2\pi)^p|\Sigma_k|]^{1/2}$$

where μ_k is the central point, Σ_k the $p \times p$ covariance matrix and $|\Sigma_k|$ its determinant ($k = 1, 2$).

The Gaussian distribution is very popular. There are at least two reasons for that. First, it is treatable theoretically and, in fact, may lead to the least squares criterion within the probabilistic approach. Second, some real-world stochastic processes, especially in physics, can be thought of as having the Gaussian distribution. Typical shapes of a 2D Gaussian density function are illustrated on Fig. 4.5: that with zero correlation on the left and 0.8 correlation on the right.

In the case at which the within-class covariance matrices are equal to each other, the Bayes decision function $B(x)$ is linear so that the separating surface $B(x) = 0$ is a hyperplane as explained later in Section 4.4.

(ii) Local independence (Naïve Bayes)

The assumption of local independence states that all variables are independent within each class so that the within-cluster distribution is a product of one-dimensional distributions:

$$f_k(x_1, x_2, \ldots, x_p) = f_{k1}(x_1)f_{k2}(x_2)\ldots f_{kp}(x_p) \quad (4.5)$$

This postulate much simplifies the matters because usually it is not difficult to produce rather reliable estimates of the one-dimensional density functions $f_{kv}(x_v)$ from the training data. Especially simple such a task is when features $x_1, x_2, \ldots, x_p$

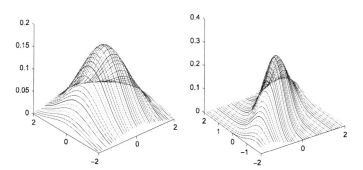

Fig. 4.5 Gaussian bivariate density functions over the origin as the expectation point – with zero correlation *on the left* and 0.8 correlation *on the right*

are binary themselves. In this case Bayes rule is referred to as naïve Bayes rule because in most cases the assumption of independence (4.5) is obviously wrong. Take, for example, the cases of text categorization or genomic analyses – constituents of a text or a protein serving as the features are necessarily interrelated according to the syntactic and semantic structures, in the former, and biochemical reactions, in the latter. Yet naive decision rules based on the wrong assumptions and distributions appear surprisingly good (see discussion in Manning et al. 2008).

Combining the assumptions of local independence and Gaussian distributions in the case of binary variables, one can arrive at equations expressing the conditional probabilities through exponents of linear functions of the variables (as described in Mitchell 2010) so that:

$$P(1/x) = \frac{1}{1 + \exp(c_0 + c_1 x_1 + \ldots + c_p x_p)},$$

$$P(2/x) = \frac{\exp(c_0 + c_1 x_1 + \ldots + c_p x_p)}{1 + \exp(c_0 + c_1 x_1 + \ldots + c_p x_p)},$$

These equations express what is referred to as logistic regression. Logistic regression is a popular decision rule that can be applied to any data on its own right as a model for the conditional probability, and not necessarily derived from the restrictive independence and normality assumptions.

4.2.2 Naïve Bayes Classifier

Consider a learning problem related to data in Table 4.1: there is a set of entities, which are newspaper articles, divided into a number of categories – there

Table 4.1 An illustrative database of 12 newspaper articles along with 10 keywords. The articles are labeled according to their main subjects – F for feminism, E for entertainment, and H for household

Article	Keyword									
	Drink	Equal	Fuel	Play	Popular	Price	Relief	Talent	Tax	Woman
F1	1	2	0	1	2	0	0	0	0	2
F2	0	0	0	1	0	1	0	2	0	2
F3	0	2	0	0	0	0	0	1	0	2
F4	2	1	0	0	0	2	0	2	0	1
E1	2	0	1	2	2	0	0	1	0	0
E2	0	1	0	3	2	1	2	0	0	0
E3	1	0	2	0	1	1	0	3	1	1
E4	0	1	0	1	1	0	1	1	0	0
H1	0	0	2	0	1	2	0	0	2	0
H2	1	0	2	2	0	2	2	0	0	0
H3	0	0	1	1	2	1	1	0	2	0
H4	0	0	1	0	0	2	2	0	2	0

4.2 Naïve Bayes Approach

are three categories in Table 4.1 according to the three subjects: Feminism (F), Entertainment (E) and Household (H). Each article is characterized by its set of keywords presented in the corresponding line. The entries are either 0 – no occurrence of the keyword, or 1 – one occurrence, or 2 – two or more occurrences of the keyword.

The problem is to form a rule according to which any article, including those outside of the collection in Table 4.1, can be assigned to one of these categories using its profile – the data on occurrences of the keywords in the corresponding line of Table 4.1.

Consider the Naïve Bayes decision rule. It assigns each category k with its conditional probability $P(k/x)$ depending on the profile x of an article in question which is similar to Equations in (4.3):

$$P(k/x) = p_k f_k(x) / f(x) \tag{4.6}$$

where $f(x) = \sum_l p_l f_l(x)$. According to the Bayes rule, the category k, at which $P(k/x)$ is maximum, is selected. Obviously, the denominator does not depend on k and can be removed: that category k is selected, at which $p_k f_k(x)$ is maximum.

According to the Naïve Bayes approach, $f_k(x)$ is assumed to be the product of the probabilities of occurrences of the keywords in category k. How one can estimate such a probability? This is not that simple as it sounds.

For example, what is the probability of term "drink" in H category? Probably, it can be taken as 1/4 – since the term is present in only one of four members of H. But what's about term "play" in H – it occurs thrice but in two documents only; thus its probability cannot be taken 3/4; yet 2/4 does not seem right either. A popular convention accepts the "bag-of-words" model for the categories. According to this model, all occurrences of all terms in a category are summed up, to produce 31 for category H in Table 4.1. Then each term's probability in category k would be its summary occurrence in k divided by the bag's total. This would lead to a fairly small probability of the "drink" in H, just 1/31. This bias is not that important, however, because what matters indeed in the Naïve Bayes rule is the feature relative contributions, not the absolute ones. And the relative contributions are all right with "drink", "fuel" and "play" contributing 1/31, 6/31 and 3/31, respectively, to H. Moreover, taking the total account of all keyword occurrences in a category serves well for balancing the differences between categories according to their sizes.

Yet there is one more issue to take care of: zero entries in the training data. Term "equal" does not appear at all in H leading thus to its zero probability in the category. This means that any article with an occurrence of "equal" cannot be classed into H category, however heavy evidence from other keywords may be. One would make a point of course that term "equal" has not been observed in H just because the sample of four articles in Table 4.1 is too small, which is a strong argument indeed. To make up for these, another, a "uniform prior" assumption is widely accepted. According to this assumption each term is present once at any category before the count is started. For the case of Table 4.1, this adds 1 to each numerator and 10 to each denominator, which means that the probability of "drink", "equal", "fuel" and "play" in category H will be $(1+1)/(31+10) = 2/41$, $(0+1)/(31+10) = 1/41$, $(6+1)/(31+10) = 7/41$ and $(3+1)/(31+10) = 4/41$, respectively.

To summarize, the "bag-of-words" model represents a category as a bag containing all occurrences of all keywords in the documents of the category plus one occurrence of each keyword, to be added to every count in the data table.

Table 4.2 contains the prior probabilities of categories, that are taken to be just proportions of categories in the collection, 4 of each in the collection of 12, as well as within-category probabilities of terms (the presence of binary features) computed as described above. Logarithms of these are given too.

Now we can apply Naïve Bayes classifier to any entity presented in the format of Table 4.1 including those in Table 4.1 itself (the training set). Because the probabilities in Table 4.2 are expressed in thousands, we may use sums of their logarithms rather than the probability products; this seems an intuitively appealing operation. Indeed, after such a transformation the score of a category is just the inner product of the row representing the tested entity and the feature scores corresponding to the category. Table 4.3 presents the logarithm scores of article E1 for each of the categories.

It should be mentioned that the Naïve Bayes computations here, as applied to the text categorization problem, follow the so-called multinomial model in which only terms present in the entities are considered – as many times as they occur. Another popular model is the so-called Bernoulli model, in which terms are assumed to be generated independently as binomial variables. The Bernoulli model based computations differ from these on two counts: first, the features are binary indeed so that only binary information, yes or no, of term occurrence is taken, and, second, for each term the event of its absence, along with its probability, is counted too (for more detail, see Manning et al. 2008, Mitchell 2010).

Table 4.2 Prior probabilities for Naïve Bayes rule for the data in Table 4.1 according to the bag-of-word conventions. There are three lines for each of the categories representing, from top to bottom, the term counts from Table 4.1, their probabilities multiplied by 1,000 and rounded to an integer, and the natural logarithms of the probabilities

Category	Prior probability Its logarithm	Total count	Term counts Term probabilities (in thousands) Logarithms of the probabilities								
F	1/3	27	3	5	0	2	2	3	0	5	0	7
			108	162	27	81	81	108	27	162	27	216
	−1.099		4.6	5.1	3.3	4.4	4.4	4.7	3.3	5.1	3.3	5.4
E	1/3	32	3	2	3	6	6	2	3	5	1	1
			95	71	95	167	167	71	95	143	48	48
	−1.099		4.6	4.3	4.6	5.1	5.1	4.3	4.6	5	3.9	3.9
H	1/3	31	1	0	6	3	3	7	5	0	6	0
			49	24	171	98	98	195	146	24	171	24
	−1.099		3.9	3.2	5.1	4.6	4.6	5.3	5	3.2	5.1	3.2

4.2 Naïve Bayes Approach

Table 4.3 Computation of category scores for entity E1 (first line) from Table 4.1 according to the logarithms of within-class feature probabilities. There are two lines for each of the categories: that on top replicates the logarithms from Table 4.2 and that on the bottom computes the inner product

Entity E1		2	0	1	2	2	0	0	1	0	0	
Category	$Log(p_k)$	Feature weights (probability logarithms) Inner product										Score
F	−1.099	4.6	5.1	3.3	4.4	4.4	4.7	3.3	5.1	3.3	5.4	
		2∗4.6	+0+	1∗3.3	+2∗4.4	+2∗4.4	+0+	0+	1∗5.1	+0	+0	35.2
E	−1.099	4.6	4.3	4.6	5.1	5.1	4.3	4.6	5.0	3.9	3.9	
		2∗4.6	+0+	1∗4.6	+2∗5.1	+2∗5.1	+0+	0+	1∗5.0	+0	+0	**39.2**
H	−1.099	3.9	3.2	5.1	4.6	4.6	5.3	5.0	3.2	5.1	3.2	
		2∗3.9	+0	+1∗5.1	+2∗4.6	+2∗4.6	+0+	0+	1∗3.2	+0	+0	34.5

Q.4.1. Apply Naïve Bayes classifier in Table 4.2 to article X = (2 2 0 0 0 0 2 2 0 0) which involves items "drink", "equal", "relief" and "talent" frequently. **A.** The category scores are: s(F/X) = 35.2, s(E/X) = 35.6, and S(H/X) = 29.4 pointing to Entertainment or, somewhat less likely, Feminism.

Q.4.2. Compute Naïve Bayes category scores for all entities in Table 4.1 and prove that the classifier correctly attributes them to their categories. **A.** See Table 4.4.

Table 4.4 Naïve Bayes category scores for the items in Table 4.1

Articles	Category scores		
	F	E	H
F1	**37.7006**	35.0696	29.3069
F2	**28.9097**	25.9362	21.5322
F3	**24.9197**	20.1271	14.8723
F4	**38.2760**	34.6072	30.0000
E1	34.2349	**37.9964**	33.3322
E2	37.2440	**42.1315**	40.2435
E3	43.1957	**44.5672**	40.8398
E4	21.1663	**22.9203**	19.4367
H1	25.8505	29.3940	**34.5895**
H2	34.9290	40.4527	**42.7490**
H3	29.9582	35.3573	**38.3227**
H4	24.7518	28.8344	**34.8408**

4.2.3 Metrics of Accuracy

4.2.3.1 Accuracy and Related Measures: Presentation

Consider a generic problem of learning a binary target feature, so that all entities belong to either class 1 or class 2. A decision rule, applied to an entity, generates a "prediction" which of these two classes the entity belongs to. The classifier may return some decisions correct and some erroneous. Let us pick one of the classes as that of our interest, say 1, then there can be two types of errors: false positives (FP) – the classifier says that an entity belongs to class 1 while it does not, and false negatives (FN) – the classifier says that an entity does not belong to class 1 while it does.

Let it be, for example, a lung screening device for testing against a lung cancer. Whilst established in a hospital cancer ward, on a selected sample of 200 patients sent by local surgeries for investigation, it may produce results that are presented in Table 4.5. Its rows correspond to the diagnosis by the screening device and the columns to the results of further, more elaborate and definitive, tests. This is a cross-classification contingency table, and it is frequently referred to as a confusion table.

There are 94 true positives TP and 98 true negatives TN in the table so that the total accuracy of the device can be rated as $(94 + 98)/200 = 0.96 = 96\%$. Respectively, the numbers of false positives FP = 7, and false negatives FN = 1 sum to 8 leading to 4% error rate. Yet there are significant differences between these two showing that the device is in fact better than the totals show. Indeed, the 7 FP are not that important, because patients with the suspected cancer will be investigated further in depth anyway so that their No-status will be restored, with the cost of further testing. In contrast, 1 FN may go out of the medical system and get their cancer untreated with the potential loss of life because of the error. This is an example of different costs associated with FP and FN errors. The device made just one serious error: of 95 true cancer cases, one error. The TP rate, the proportion of correctly identified true cases, frequently referred to as recall or sensitivity, $94/95 = 98.9\%$, is impressive indeed. On the other hand, the precision, that is, the proportion of the 94 TP cases related to all cancer predicted cases, 101, is somewhat smaller, just 93% to reflect that FP rate is 7%. The difference between precision and sensitivity is somewhat averaged in the value of accuracy rate, 96% in this case, so that the accuracy rate works reasonably well here as a single characteristic of the quality of the testing device.

Yet in a situation in which there is a great disparity in the sizes of Yes and No classes, the accuracy rate fails to reflect the results properly. Consider, for example, results of the same device at a random sample of 200 individuals who have not been sent for the screening by doctors but rather volunteered to be screened from public at large (Table 4.6).

Table 4.5 Confusion table of patients' lung screening test results

		True lung cancer		
		Yes	No	Total
Device's	Yes	94	7	101
diagnosis	Not	1	98	99
Total		95	105	200

4.2 Naïve Bayes Approach

Table 4.6 Contingency table of volunteers' lung screening test results

		True lung cancer		Total
		Yes	No	
Device's	Yes	2	2	4
diagnosis	Not	1	195	196
Total		3	197	200

The accuracy rate at Table 4.6 is even greater than that at Table 4.5, $(2 + 195)/200 = 98.5\%$. Yet both sensitivity, $2/3 = 66.7\%$, and precision, $2/4 = 50\%$, are quite mediocre. The high accuracy rate is caused by the very high specificity, the proportion of correctly identified No cases, $195/197 = 98.9\%$, and by the fact that there are very few Yes cases.

As to a single measure adequately reflecting sensitivity and precision, the one most popular is their harmonic mean, the F-measure, which is equal to $F = 2/(1/(2/3) + 1/(2/4)) = 2/(3/2 + 4/2) = 4/7 = 57.1\%$.

Case study 4.1. Prevalence and Quetelet coefficients

If one looks at the record of the screening device according to Table 4.6 and compares that with the prevalence of the cancer at the sample, 3 cases of 200 – the difference is impressive indeed. This difference is exactly what is caught up in the concept of Quetelet coefficient $q(l/k)$ (see Section 3.3.3.2) at row $k = 1$ and column $l = 1$. This takes the relative difference between the conditional probability $P(1/1) = 2/4$ and the average probability $P(l = 1) = 3/200$ which is referred to sometimes as the prevalence: $q(1/1) = (2/4 - 3/200)/(3/200) = 2*200/(3*4) - 1 - 32.33 - 3233\%$, quite a change. This high value probably explains the difference in sensitivity and specificity between Tables 4.6 and 4.5.

Indeed, a similar Quetelet coefficient at Table 4.5 is $q(1/1) = 94*200/(101*95) - 1 = 0.96 = 96\%$, a less than a 100% increase, which may convey the idea that Table 4.5 is much more balanced than Table 4.6. The accuracy measure works well at balanced tables and it does not at those that are not.

4.2.3.2 Accuracy and Related Measures: Formulation

In general, the situation can be described by a confusion, or contingency, table between two sets of categories related to the class being predicted (1 or not) and the true class (1 or not), see Table 4.7. Of course, if one changes the class of interest, the errors will remain errors, but their labels will change: false positives regarding class 1 are false negatives when the focus is on class 2, and vice versa.

Table 4.7 A statistical representation of the match between the true class and predicted class. The entries are counts of the numbers of co-occurrences

		True class		Total
		1	Not	
Predicted class	1	True positives	False positives	TP+FP
	Not	False negatives	True negatives	FN+TN
Total		TP+FN	FP+TN	N

Among popular indexes scoring the error or accuracy rates are the following:

FP rate = FP/(FP + TN) – the proportion of false positives among those not in 1; 1-FP rate is referred to sometimes as specificity – it shows the proportion of correct predictions among other, not class 1, entities.

TP rate = TP/(TP + FN) – the proportion of true positives in class 1; in information retrieval, this frequently is referred to as recall or sensitivity.

Precision = TP/(TP + FP) – the proportion of true positives in the predicted class 1.

These reflect each of the possible errors separately. There are indexes that try to combine all the errors, too. Among them the most popular are:

Accuracy = (TP + TN)/N – the total proportion of accurate predictions. Obviously, 1-Accuracy is the total proportion of errors.

F-measure = 2/(1/Precision + 1/Recall) – the harmonic average of Recall and Precision.

The latter measure is getting more popularity than the former because the Accuracy counts both types of errors equally, which may be at odds with the common sense in those frequent situations at which errors of one type are "more expensive" than the others. Recall, for example, the case of medical diagnostics in Tables 4.5–4.6: a tumor wrongly diagnosed as malignant would cost much less than the other way around when a deadly tumor is diagnosed as benign. F-measure, to some extent, is more conservative because it, first, combines rates rather than counts, and, second, utilizes the harmonic mean which tends to be close to the minimum of the two, as can be seen from the statements in Q.4.3 and Q.4.4.

Q.4.3. Consider two positive reals, a and b, and assume, say that $a < b$. Prove that the harmonic mean, $h = 2/(1/a + 1/b)$ stays within the interval between a and $2a$ however large the difference b–a is. **A.** Take b be $b = ka$ at some $k > 1$. Then

4.2 Naïve Bayes Approach

$h = 2/(1/a + 1/(ka)) = 2ka/(1 + k)$. The coefficient at a, $2k/(1 + k)$, is less than 2, which proves the statement.

Q.4.4. Consider two positive real values, a and b, and prove that their mean, $m = (a + b)/2$, and harmonic mean, $h = 2/(1/a + 1/b)$, satisfy equation $mh = ab$. A. Take the product $mh = [(a+b)/2][2/(1/a+1/b)]$ and perform elementary algebraic operations.

More elaborate representation of errors of the two types can be achieved with the so-called receiver operating characteristics (ROC) graphs analysis (see, for example, Fawcett 2006). ROC graphs are especially suitable in the cases of classifiers that have a continuous output such as Bayes classifiers. ROC graph is a 2D Cartesian plane plotting TP rate against FP rate so that the latter is shown on x-axis and the former, on y-axis (see Fig. 4.6)

To be specific, let us take a Bayes classifier's rule in (4.4) and change the ratio p_2/p_1 for an arbitrary threshold $d>0$. Take now $d = d_1$ for a specific d_1, so that the rule now predicts class 1 if $f1(x)/f2(x) > d_1$. Count the proportions of true and false positives, $tp1$ and $fp1$, at this threshold and put the point ($fp1$, $tp1$) onto a ROC graph. Then change d to d_2 and count the rates, $tp2$ and $fp2$, at this threshold. If, say, $d_2 > d_1$, then the TP rate can only decrease, because the number of positive predictions can only decrease. The FP rate, in a regular case, should increase at $d_2 > d_1$ so that point (fp2, ft2) would go to the right and above the former point on ROC plot. In this way, by step-by-step changing the threshold d, one can obtain a ROC curve such as curves "a" and "b" on the plot of Fig. 4.6. Such a curve can be utilized as a characteristic of the classifier under consideration that can be used, for instance, for selection of suitable levels of TP and FP rates. In the case shown on Fig. 4.6, one can safely claim that classifier "a" is superior to that of "b", because at each FP rate level, TP rate of "a" is greater than that of "b".

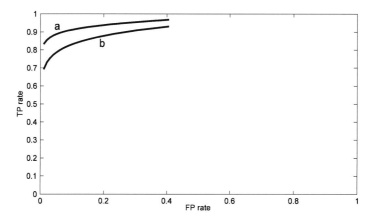

Fig. 4.6 ROC curves for two classifiers; that of a is superior to that of b

4.3 Linear Regression

P4.3.1 Linear Regression: Presentation

Let us extend the notion of linear regression from the bivariate case considered in Section 3.1 to multivariate case, when several features can be used as predictors for a target feature.

Case study 4.2. Linear regression for Market town data

Consider feature Post expressing the number of post offices in Market towns (Table 1.4 on pp. 14) and try to relate it to other features in the table. It obviously relates to the population. For example, towns with population of 15,000 and greater are those and only those where the number of post offices is 5 or greater. This correlation, however, is not as good as to give us more guidance in predicting Post from the Population. For example, at the seven towns whose population is from 8,000 to 10,000 any number of post offices from 1 to 4 may occur, according to the table. This could be attributed to effects of services such as a bank or hospital present at the towns. Let us specify a set of features in Table 1.4 that can be thought of as affecting the feature Post, to include in addition to Population some other features – PS-Primary schools, Do – General Practitioners, Hos- Hospitals, Ba- Banks, Sst – Superstores, and Pet– Petrol Stations; seven features altogether, taken as the set of input variables (predictors).

What we want is to establish a linear relation between the set and target feature Post. A linear relation is an equation representing Post as a weighted sum of input features plus a constant intercept; the weights can be any reals, not necessarily positive. If the relation is supported by the data, it can be used for various purposes such as analysis, prediction and planning.

In the example of seven Market town features used for linearly relating them to Post Office feature, the least-squares optimal weight coefficients are presented in Table 4.8. Each weight coefficient shows how much the target variable would change on average if the corresponding feature is increased by a unity, while the others do not change. One can see that increasing population by a thousand would give a similar effect as adding a primary school, about 0.2, which may seem absurd in the example as Post Office variable can have only integer values. Moreover, the linear function format should not trick the decision maker into thinking that increasing different input features can be done independently: the features are obviously not independent so that increase of, say, the population will lead to respectively adding new schools for the additional children. Still, the weights show relative effects of the features – according to Table 4.8, adding a doctor's surgery in a town would lead to maximally possible increase in post offices. The maximum value is assigned to the intercept in this case. What this may mean? Is it the number of post offices in an empty town with no population, hospitals or petrol stations? Certainly not. The intercept expresses that part of the target variable which is relatively independent

4.3 Linear Regression

Table 4.8 Weight coefficients of input features at Post Office as target variable for Market towns data

Feature	Weight
Pop_Res	0.0002
PSchools	0.1982
Doctors	0.2623
Hospitals	−0.2659
Banks	0.0770
Superstores	0.0028
Petrol	−0.3894
Intercept	0.5784

of the features taken into account. It should be also pointed out that the weight values are relative not to just feature concepts but specific scales in which features measured. Change of a feature scale, say ten fold, would result in a corresponding, inverse, change of its weight (due to the linearity of the regression equation). This is why in statistics, the relative weights are considered for the scales expressed in units of the standard deviation. To find them, one should multiply the weight for the current scale by the feature's standard deviation (see Table 4.9).

The standardized weights are well justified when input features are mutually uncorrelated – indeed, they show the pair-wise correlation with the target feature. Yet in a situation of correlated features, like this, they seem to have much less definite interpretation, except for showing the changes of the target in units of the standard deviations, although some claim that they also reflect feature's correlation with the target or even importance for predicting the target. An argument against their usage as a correlation measure is that, in fact, a regression coefficient multiplied by the standard deviation loses its "purity" as a measure of correlation to the target at constant levels of the other features because the standard deviation does not pertain to constant features. An argument against their usage as measures of importance for prediction is that the standardized coefficient has nothing to do with the change of the coefficient of determination when the corresponding feature is removed from the equation of regression.

Table 4.9 Different indexes to express the idea of importance of a feature in the Post regression problem

Feature	Weights in natural scales, w	Standard deviations, s	Weights in standardized scales, w∗s
Pop_Res	0.0002	6, 193.2	1.3889
PSchools	0.1982	2.7344	0.5419
Doctors	0.2623	1.3019	0.3414
Hospitals	−0.2659	0.5800	−0.1542
Banks	0.0770	4.3840	0.3376
Superstores	0.0028	1.7242	0.0048
Petrol	−0.3894	1.637	−0.6375

J. Bring (1994) proposes to kill two birds with one stone: to clean up the standard deviations from the non-constancy of the other features, which are claimed to reflect the changes in the coefficients of determination. Specifically, take the variance of a feature and take off the proportion of it unexplained by the linear regression of it through the other features. The square root of the result represents the partial standard deviation, which is proportional to the so-called "t-value", and, in the original squared form, to the change of the coefficient of determination inflicted by the removal of the feature from the list of the explanatory variables (Bring 1994). Unfortunately, this is not that simple, as the next case study 4.3 shows.

Case study 4.3. Using feature weights standardized

Table 4.10 presents the feature weights standardized with both the original and partial standard deviations as well as the absolute reductions of the original coefficient of determination 0.8295 after removal of the corresponding variables. There is a general agreement between the absolute values of the first column and those in the third column, but the second column has little in common with either of them. A general analysis of a simpler problem of relation between the regression coefficients and correlation coefficients between the target and input features can be found in Waller and Jones (2010).

Amazingly, the convenient standardization involves negative weights, specifically at features Petrol and Hospitals. This can be an artifact of the method, related to the effect of "replication" of features. One can think of Hospitals being a double for Doctors, and Petrol, for Superstores. Thus, before jumping to conclusions, one should check whether the minus disappears if the "replicas" are removed from the set of features. As Table 4.11 shows, not in this case: the negative weights remain, though they slightly change, as well as other weights. This illustrates that the interpretation of linear regression coefficients as weights should be cautious and restrained.

Table 4.10 Standardized weight coefficients of input features at Post Office as target variable for Market towns

Feature	Weights in standard deviation scales	Weights with partial standard deviations	Determination coefficient reduction
Pop_Res	1.3889	1,602.00	0.0247
PSchools	0.5419	1.02	0.0077
Doctors	0.3414	0.64	0.0055
Hospitals	−0.1542	0.41	0.0023
Banks	0.3376	2.27	0.0059
Superstores	0.0048	1.07	0
Petrol	−0.6375	0.96	0.0251

4.3 Linear Regression

Table 4.11 Weight coefficients for reduced set of features at Post Office as target variable for Market towns data

Feature	Weight
POP_RES	0.0003
PSchools	0.1823
Hospitals	−0.3167
Banks	0.0818
Petrol	−0.4072
Intercept	0.5898

In our example, coefficient of determination $\rho^2 = 0.83$, that is, the seven features explain 83% of the variance of Post Office feature, and the multiple correlation is $\rho = 0.91$. Curiously, the reduced set of five features (see Table 4.11) contributes almost the same, 82.4% of the variance of the target variable. This may make one wonder whether just one Population feature could suffice for doing the regression. This can be tested with the 2D method described in Section 3.2 or with the nD method of this section.

According to the formulation of the multivariate linear regression method further in F3.3, the estimated parameters must be feature weight coefficients – no room for an intercept in the formula. To accommodate the intercept, a fictitious feature whose all values are unities is introduced. That is, an input data matrix X with two columns is to be used: one for the Population feature, the other for the fictitious variable of all ones. According to (4.10), this leads to the slope 0.0003 and intercept 0.4015, though with somewhat reduced coefficient of determination, which is $\rho^2 = 0.78$ in this case. From the prediction point of view this may be all right, but the reduced feature set looses on interpretation.

F4.3.2 Linear Regression: Formulation

The problem of linear regression can be formulated as a particular case of the correlation learning problem with just one quantitative target variable u and linear admissible rules so that

$$u = w_1 x_1 + w_2 x_2 + \ldots + w_p x_p + w_0$$

where $w_0, w_1, \ldots, w_p$ are unknown weights, parameters of the model.

For any entity $i = 1, 2, \ldots, N$, the rule-computed value of u

$$\hat{u}_i = w_1 x_{i1} + w_2 x_{i2} + \ldots + w_p x_{ip} + w_0$$

differs from the observed one by $d_i = |\hat{u} - u_i|$, which may be zero – when the prediction is exact. To find $w_1, w_2, \ldots, w_p, w_0$, one can minimize the summary square error

$$D^2 = \Sigma_i d_i^2 = \Sigma_i (u_i - w_1 * x_{i1} - w_2 * x_{i2} - \ldots - w_p * x_{ip} - w_0)^2 \quad (4.7)$$

over all possible parameter vectors $w = (w_0, w_1, \ldots, w_p)$.

To make the problem treatable in terms of linear operations, a fictitious feature x_0 is introduced such that all its values are 1: $x_{i0} = 1$ for all $i = 1, 2, \ldots, N$. Then criterion D^2 can be expressed as $D^2 = \Sigma_i (u_i - <w_i, x_i>)^2$ using the inner products $<w, x_i>$ where $w = (w_0, w_1, \ldots, w_p)$ and $x_i = (x_{i0}, x_{i1}, \ldots, x_{ip})$ are $(p+1)$-dimensional vectors of which all x_i are known whereas w is not. From now on, the unity feature x_0 is assumed to be part of data matrix X in all correlation learning problems in this text.

The criterion D^2 in (4.7) is but the squared Euclidean distance between the N-dimensional target feature column $u = (u_i)$ and vector $\hat{u} = Xw$ whose components are $\hat{u}_i = <w, x_i>$. Here X is $N \times (p+1)$ matrix whose rows are x_i (augmented with the component $x_{i0} = 1$, thus being $(p+1)$-dimensional) so that Xw is the matrix product of X and w. Vectors defined as Xw for all possible w's form $(p+1)$-dimensional vector space, referred to as X-span.

Thus the problem of minimization of (4.7) can be reformulated as follows: given target vector u, find its projection $\hat{u}$ in the X-span space. The global solution to this problem is well-known: it is provided by a matrix P_X applied to u:

$$\hat{u} = P_X u$$

where P_X is the so-called orthogonal projection operator, an $N \times N$ matrix, defined as:

$$P_X = X(X^T X)^{-1} X^T$$

so that

$$\hat{u} = X(X^T X)^{-1} X^T u \text{ and } w = (X^T X)^{-1} X^T u \quad (4.8)$$

Matrix P_X projects every N-dimensional vector u to its nearest match in the $(p+1)$-dimensional X-span space. The inverse $(X^T X)^{-1}$ does not exist if the rank of X, as it may happen, is less than the number of columns in X, $p+1$, that is, if matrix $X^T X$ is singular or, equivalently, the dimension of X-span is less than $p+1$. In this case, the so-called pseudo-inverse matrix $(X^T X)^+$ can be used as well. This is not a big deal computationally: for example, in MatLab one just puts pinv($X^T X$) instead of inv($X^T X$).

The quality of approximation is evaluated by the minimum value D^2 in (4.7) averaged over the number of entities and related to the variance of the target variable. Its complement to 1, the coefficient of determination, is defined by the equation

$$\rho^2 = 1 - D^2/(N\sigma^2(u)) \quad (4.9)$$

The coefficient of determination shows the proportion of the variance of u explained by the linear regression. Its square root, ρ, is referred to as the coefficient of multiple correlation between u and $X = \{x_0, x_1, x_2, \ldots, x_p\}$.

4.4 Linear Discrimination and SVM

P4.4.1 Linear Discrimination and SVM: Presentation

Discrimination is an approach to address the problem of drawing a rule to distinguish between two classes of entity points in the feature space, a "yes" class and "no" class, such as for instance a set of banking customers in which a, typically very small, subset of fraudsters constitutes the "yes" class and that of the others the "no" class. On Fig. 4.7, entities of "yes" class are presented by circles and of "no" class by squares.

The problem is to find a function $u = f(x)$ that would separate the two classes in such a way that $f(x)$ is positive for all entities in the "yes" class and negative for all the entities in the "no" class. When the discriminant function $f(x)$ is assumed to be linear, the problem is of linear discrimination. It differs from that of the linear regression in that aspect that the target values here are binary, either "yes" or "no", so that this is a classification rather than regression, problem.

The classes on Fig. 4.7 can be discriminated by a straight – dashed – line indeed. The dotted vector w, orthogonal to the "dashed line" hyperplane, represents a set of coefficients at the linear classifier represented by the dashed line. Vector w also shows the direction at which function $f(x) = <w, x> - b$ grows. Specifically, $f(x)$ is 0 on the separating hyperplane, and it is positive above and negative beneath that. With no loss of generality, w can be assumed to have its length equal to unity. Then, for any x, the inner product $<w, x>$ expresses the length of vector x along the direction of w.

To find an appropriate w, even in the case when "yes" and "no" classes are linearly separable, various criteria can be utilized. A most straightforward classifier is defined as follows: put 1 for "yes" and –1 for "no" and apply the least-squares criterion of linear regression. This produces a theoretically sound solution approximating the best possible – Bayesian – solution in a conventional statistics model. Yet, in spite of its good theoretical properties, least-squares solution may be not

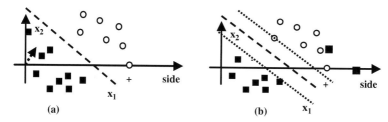

Fig. 4.7 A geometric illustration of a separating hyper-plane between classes of *circles and squares*. The *dotted vector w* on (**a**) is orthogonal to the hyper-plane: its elements are hyper-plane coefficients, so that it is represented by equation $<w, x> - b = 0$. Vector w also points at the direction: at all points above the *dashed line*, the *circles* included, function $f(x) = <w, x> - b$ is positive. The *dotted lines* on (**b**) show the margin, and the *squares and circle* on them are support vectors

necessarily the best at some data configurations. In fact, it may even fail to separate the positives from negatives when they are linearly separable. Consider the following example.

Worked example 4.1. A failure of Fisher discrimination criterion

Let there be 14 2D points presented in Table 4.12 (first line) and displayed in Fig. 4.8a. Points 1,2,3,4,6 belong to the positive class (dots on Fig. 4.8a), and the others to the negative class (stars on Fig. 4.8a). Another set, obtained by adding to each of the components a random number, according to the normal distribution with zero mean and 0.2 the standard deviation; is presented in the bottom line of Table 4.12 and Fig. 4.8b. The class assignment for the disturbed points remains the same.

The optimal vectors w according to formula (4.8) are presented in Table 4.13 as well as that for the separating, dotted, line in Fig. 4.8d.

Note that the least-squares solution depends on the values assigned to classes, leading potentially to an infinite number of possible solutions under different numerical codes for "yes" and "no". A popular discriminant criterion of minimizing the ratio of a "within-class error" over "out-of-class error", proposed by R. Fisher in his founding work of 1936, in fact, can be expressed with the least-squares criterion as well. Just change the target as follows: assign N/N_1, rather than +1, to "yes" class and $-N/N_2$ to "no" class, rather than −1 (see Duda et al. 2001, pp. 242). This means that Fisher's criterion may also lead to a failure in a linear separable situation.

Q.4.5. Why only 10, not 14, points are drawn on Fig. 4.8b? **A.** Because each of the points 11–14 doubles a point 7–10.

Q.4.6. What would change if the last four points are removed so that only points 1–10 remain? **A.** The least-squares solution will be separating again.

By far the most popular set of techniques, Support Vector Machine (SVM), utilize a different criterion – that of maximum margin. The margin of a point x, with

Table 4.12 x–y coordinates of 14 points as given originally and perturbed with a noise generated from the Gaussian distribution N(0, 0.2)

Entity #		1	2	3	4	5	6	7	8	9	10	11	12	13	14
Original data	x	3.00	3.00	3.50	3.50	4.00	1.50	2.00	2.00	2.00	1.50	2.00	2.00	2.00	1.50
	y	0.00	1.00	1.00	0.00	1.00	4.00	4.00	5.00	4.50	5.00	4.00	5.00	4.50	5.00
Perturbed data	x	2.93	2.83	3.60	3.80	3.89	1.33	1.95	2.13	1.83	1.26	1.98	1.99	2.10	1.38
	y	−0.03	0.91	0.98	0.31	0.88	3.73	4.09	4.82	4.51	4.87	4.11	5.11	4.46	4.59

4.4 Linear Discrimination and SVM 135

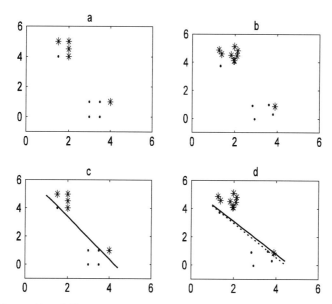

Fig. 4.8 Figures (**a**) and (**b**) represent the original and perturbed data. The least *squares optimal separating line* is added in Figures (**c**) and (**d**) shown by *solid*. Entity 5 falls into the "*dot*" class according to the *solid line* in (**d**); *a separating line* is shown *dotted* there

respect to a hyperplane, is the distance from x to the hyperplane along its perpendicular vector w (Fig. 4.7a), which is measured by the absolute value of inner product $<w, x>$. The margin of a class is defined by the minimum value of the margins of its members. Thus the criterion requires, like L_∞, finding such a hyperplane that maximizes the minimum of class margins, that is, crosses the middle of the line between the nearest entities of two classes. Those entities that fall on the margins, shown by dotted lines on Fig. 4.7b, are referred to as support vectors; this explains the method's name.

It should be noted that the classes are not necessarily linearly separable; moreover in most cases they are not. Therefore, the SVM technique is accompanied with a non-linear transformation of the data into a high-dimensional space which is more likely to make the classes linear-separable, as illustrated on Fig. 4.9. Such

Table 4.13 Coefficients of straight lines on Fig. 4.8

	Coefficients at		
	X	Y	Intercept
LSE at original data	−1.2422	−0.8270	5.2857
LSE at perturbed data	−0.8124	−0.7020	3.8023
Dotted at perturbed data	−0.8497	−0.7020	3.7846

Fig. 4.9 Illustrative example of 2D entities belonging to two classes, *circles* and *squares*. The *separating line* in the space of Gaussian kernel is shown by the *dashed oval*. The support entities are shown by *black*

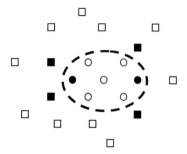

a non-linear transformation is provided by the so-called kernel function. The kernel function imitates the inner product in the high-dimensional space and is represented by a between-entity similarity function such as that defined by formula (4.12) on p. 140.

The intuition behind the SVM approach is this: if the population data – those not present in the training sample – concentrate around training data, then having a wide margin would keep classes separated even after other data points are added (see Fig. 4.7). One more consideration comes from the Minimum Description Length principle: the wider the margin, the more robust the separating hyperplane is and the less information of it needs to be stored. A criticism of the SVM approach is that the support vector machine hyperplane is based on the borderline objects – support vectors – only, whereas the least-squares hyperplanes take into account all the entities so that the further away an entity is the more it may affect the solution, because of the quadratic nature of the least-squares criterion. Some may argue that both borderline and far away entities can be rather randomly represented in the sample under investigation so that neither should be taken into account in distinguishing between classes: it is some "core" entities of the patterns that should be separated – however, there has been no such an approach taken in the literature so far.

Worked example 4.2. SVM for Iris dataset

Consider Iris dataset standardized by subtracting, from each feature column, its midrange and dividing the result by the half-range.

Take Gaussian kernel in (4.15) to find a support vector machine surface separating Iris class 3 from the rest. The resulting solution embraces 21 supporting entities (see Table 4.14), along with their "alpha" prices reaching into hundreds and even, on two occasions, to the maximum boundary 500 embedded in the algorithm.

There is only one error with this solution, entity 78 wrongly recognized as belonging to taxon 3. The errors increase when we apply a cross-validation techniques, though. For example, "leave-all-one-out" cross-validation leads to nine errors: entities 63, 71, 78, 82 and 83 wrongly classified as belonging to taxon 3 (false positives), while entities 127, 133, 135 and 139 are classified as being out of taxon 3 (false negatives).

4.4 Linear Discrimination and SVM

Table 4.14 List of support entities in the problem of separation of taxon 3 (entities 101–150) in Iris data set from the rest (thanks to V. Sulimova for computation)

N	Entity	Alpha	N	Entity	Alpha
1	18	0.203	12	105	2.492
2	28	0.178	13	106	15.185
3	37	0.202	14	115	52.096
4	39	0.672	15	118	15.724
5	58	13.630	16	119	449.201
6	63	209.614	17	127	163.651
7	71	7.137	18	133	500
8	78	500	19	135	5.221
9	81	18.192	20	139	16.111
10	82	296.039	21	150	26.498
11	83	200.312			

F4.4.2 Linear Discrimination and SVM: Formulation

F4.4.2.1 Linear Discrimination

The problem of linear discrimination can be stated as follows. Let a set of N entities in the feature space, $x_i = (x_{i0}, x_{i1}, x_{i2}, \ldots, x_{ip})$ $i = 1, 2, \ldots, N$ is partitioned in two classes, sometime referred to as patterns, a "yes" class and a "no" class, such as for instance a set of banking customers in which a, typically very small, subset of fraudsters constitutes the "yes" class and that of the others the "no" class. The problem is to find a function $u = f(x_0, x_1, x_2, \ldots, x_p)$ that would discriminate the two classes in such a way that u is positive for all entities in the "yes" class and negative for all entities in the "no" class. When the discriminant function is assumed to be linear so that $u = w_1 x_1 + w_2 x_2 + \ldots + w_p x_p + w_0$ at constant w_0, $x_1, \ldots, w_p$, the problem is of linear discrimination.

To make it quantitative, define $u_i = 1$ if i belongs to the "yes" class and $u_i = -1$ if i belongs to the "no" class. The intercept w_0 is referred to, in the context of the discrimination/classification problem, as bias.

A linear classifier is defined by a vector w so that if $\hat{u}_i = <w, x_i> > 0$, predict $\overset{\circ}{u}_i = 1$; if $\hat{u}_i = <<w, x_i> > < 0$, then predict $\overset{\circ}{u}_i = -1$; that is, $\overset{\circ}{u}_i = sign(<w, x_i>)$. (Here the sign function is utilized as defined by the condition that $sign(a) = 1$ when $a > 0$, $= -1$ when $a < 0$, and $= 0$ when $a = 0$.)

To find an appropriate w, even in the case when "yes" and "no" classes are linearly separable, various criteria can be utilized. A most straightforward classifier is defined by the least-squares criterion of minimizing (4.7). This produces

$$w = (X^T X)^{-1} X^T u \quad (4.10)$$

Note that formula (4.10) leads to an infinite number of possible solutions because of the arbitrariness in assigning different u-labels to different classes. A slightly different criterion of minimizing the ratio of the "within-class error" over "out-of-class error" was proposed by R. Fisher (1936). Fisher's criterion, in fact, can be expressed with the least-squares criterion if the output vector u is changed for u_f as follows: put N/N_1 for the components of the first class, instead of +1, and put $-N/N_2$ for the entities of the second class, instead of -1. Then the optimal w (4.10) at $u = u_f$ minimizes the Fisher's discriminant criterion (see Duda, Hart, Stork, 2001, p. 242).

Solution (4.10) has two properties related to the Bayes decision rule. It appears the squared summary difference between the least-square error linear decision rule function $<w, x>$ and Bayes function $B(x)$ approaches minimum when N grows infinitely (Duda et al. p. 243). Moreover, the least-squares linear decision rule is the Bayes function $B(x)$ if the class probability distributions $f1(x)$ and $f2(x)$ are Gaussian with coinciding covariance matrices, so that they can be expressed with formula:

$$f_i(x) = \exp[-(x-\mu_i)^T \Sigma^{-1}(x-\mu_i)/2]/[(2\pi)^p |\Sigma|]^{1/2}$$

where μ_i is the central point and $\sum$ the *pxp* covariance matrix of the Gaussian distribution. In fact, in this case the optimal $w = \Sigma^{-1}(\mu_1 - \mu_2)$ (see Duda, Hart, Stork, p. 40).

F4.4.2.2 Support Vector Machine (SVM) Criterion

Another criterion would put the separating hyperplane just in the middle of an interval drawn through closest points of the different patterns. This criterion produces what is referred to as the support vector machine since it heavily relies on the points involved in the drawing of the separating hyperplane (as shown on the right of Fig. 4.7). These points are referred to as support vectors. A natural formulation would be like this: find a hyperplane $H :< w, x > = b$ with a normed w to maximize the minimum of absolute values of distances $| < w, x_i > -b|$ from points x_i belonging to each of the classes. This, however, is rather difficult to associate with a conventional formulation of an optimization problem because of the following irregularities:

(i) an absolute value to maximize,
(ii) the minimum over points from each of the classes, and
(iii) w being of the length 1, that is, normed.

However, these all can be successfully tackled. The issue (i) is easy to handle, because there are only two classes, on the different sides of H. Specifically, the distance is $< w, x_i > -b$ for "yes" class and $- < w, x_i > +b$ for "no" class – this removes the absolute values. The issue (ii) can be taken care of by uniformly using inequality constraints

4.4 Linear Discrimination and SVM

$< w, x_i > -b \geq \lambda$ for x_i in "yes" class and
$- < w, x_i > +b \geq \lambda$ for x_i in "no" class

and maximizing the margin λ with respect to these constraints. The issue (iii) can be addressed by dividing the constraints by λ so that the norm of the weight vector becomes $1/\lambda$, thus inversely proportional to the margin λ. Moreover, one can change the criterion now because the norm of the ratio w/λ is minimized when λ is maximized. Denote the "yes" class by $u_i = 1$ and "no" class by $u_i = -1$. Then the problem of deriving a hyperplane with a maximum margin can be reformulated, without the irregularities, as follows: find b and w such that the norm of w or its square, $< w, w >$, is minimum with respect to constraints

$$u_i(< w, x_i > -b) \geq 1 \quad (i = 1, 2, \ldots, N)$$

This is a problem of quadratic programming with linear constraints, which is easier to analyze in the format of its dual optimization problem. The dual problem can be formulated by using the so-called Lagrangian, a common concept in optimization, that is, the original criterion penalized by the constraints weighted by the so-called Lagrangian multipliers that are but penalty rates. Denote the penalty rate for the violation of i-th constraint by α_i. Then the Lagrangian can be expressed as

$$L(w, b, \alpha) = < w, w > /2 - \Sigma_i \alpha_i (u_i (< w, x_i > -b) - 1),$$

where $< w, w >$ has been divided by 2 with no loss of generality, just for the sake of convenience. The optimum solution minimizes L over w and b, and maximizes L over non-negative α. The first order optimality conditions require that all partial derivatives of L are zero at the optimum, which leads to equations $\Sigma_i \alpha_i u_i = 0$ and $w = \Sigma_i \alpha_i u_i x_i$. Multiplying the latter expression by itself leads to equation $< w, w > = \Sigma_{ij} \alpha_i \alpha_j u_i u_j < x_i, x_j >$. The second item in Lagrangian L becomes equal to $\Sigma_i \alpha_i u_i < w, x_i > - \Sigma_i \alpha_i u_i b - \Sigma_i \alpha_i = < w, w > -0 - \Sigma_i \alpha_i$. This leads us to the following, dual, problem of optimization regarding the Lagrangian multipliers, which is equivalent to the original problem: Maximize criterion

$$\Sigma_i \alpha_i - \Sigma_{ij} \alpha_i \alpha_j u_i u_j < x_i, x_j > /2 \qquad (4.11)$$

subject to $\Sigma_i \alpha_i u_i = 0$ and $\alpha_i \geq 0$.

Support vectors are defined as those x_i for which penalty rates are positive, $\alpha_i > 0$, in the optimal solution – only they contribute to the optimal vector $w = \Sigma_i \alpha_i u_i x_i$; the others have zero coefficients and disappear.

It should be noted that the margin constraints can be violated, which is not difficult to take into account – by using non-negative values η_i expressing the sizes of violations:

$$u_i(<w, x_i> -b) \geq 1 - \eta_i \quad (i = 1, 2, \ldots, N)$$

in such a way that they are minimized in a combined criterion $<w, w>/2 + C\Sigma_i \eta_i$ where C is a large "reconciling" coefficient that is a user-defined parameter. The dual problem for the combined criterion remains almost the same as above, in spite of the fact that an additional set of dual variables, β_i, needs to be introduced as corresponding to the constraints $\eta_i \geq 0$. Indeed, the Lagrangian for the new problem can be expressed as

$$L(w, b, \alpha, \beta) = <w, w>/2 - \Sigma_i \alpha_i (u_i(<w, x_i> -b) - 1) - \Sigma_i \eta_i (\alpha_i + \beta_i - C)$$

which differs from the previous expression by just the right-side item. This implies that the same first-order optimality equations hold, $\Sigma_i \alpha_i u_i = 0$ and $w = \Sigma_i \alpha_i u_i x_i$, plus additionally $\alpha_i + \beta_i = C$. These latter equations imply that $C \geq \alpha_i \geq 0$ because β_i are non-negative.

Since the additional dual variables are expressed through the original ones, $\beta_i = C - \alpha_i$, the dual problem can be shown to remain unchanged and it can be solved by using quadratic programming algorithms (see Vapnik 2001 and Schölkopf and Smola 2005). Recently, approaches have appeared for solving the original problem as well (see Groenen et al. 2008).

F4.4.2.3 Kernels

Situations at which patterns are linearly separable are very rare; in real data, classes are typically well intermingled with each other. To attack these typical situations with linear approaches, the following trick can be applied. The data are nonlinearly transformed into a much higher dimensional space in which, because of both nonlinearity and higher dimension, the classes may be linearly separable. The transformation can be performed only virtually because of specifics of the dual problem: dual criterion (4.11) depends not on individual entities but rather just inner products between them. This property obviously translates to the transformed space, that is, to transformed entities. The inner products in the transformed space can be computed with the so-called kernel functions $K(x, y)$ so that in criterion (4.11) inner products $<x_i, x_j>$ are substituted by the kernel values $K(x_i, x_j)$. Moreover, by substituting the expression $w = \Sigma_i \alpha_i u_i x_i$ into the original discrimination function $f(x) = <w, x> -b$ we obtain its different expression $f(x) = \Sigma_i \alpha_i u_i <x, x_i> -b$, also involving inner products only, which can be used as a kernel-based decision rule in the transformed space: x belongs to "yes" class if $\Sigma_i \alpha_i u_i K(x, x_i) - b > 0$.

It is convenient to define a kernel function over vectors $x = (x_v)$ and $y = (y_v)$ through the squared Euclidean distance $d^2(x, y) = (x_1 - y_1)^2 + \ldots + (x_V - y_V)^2$ because matrix $(K(x_i, x_j))$ in this case is positive definite – a defining property of matrices of inner products. Arguably, the most popular is the Gaussian kernel defined by:

$$K(x, y) = \exp(-d^2(x, y)) \qquad (4.12)$$

4.5 Decision Trees

Q.4.7. Consider a full set B_n of 2^n binary 1/0 vectors of length n like those presented by columns below for $n = 3$:

1	0	0	0	0	1	1	1	1
2	0	0	1	1	0	0	1	1
3	0	1	0	1	0	1	0	1

These columns can be considered as integers coded in the binary number system; moreover, they are ordered from 0 to 7. Prove that this set shutters any subset of n (or less) points.
A. Indeed, let S be a set of elements $i_1, i_2, \ldots, i_n$ in B_n that are one-to-one labeled by numbers from 1 to n. Consider any partition of S in two classes, S_1 and S_2. Assign 0 to each element of S_1 and 1 to each element of S_2. The partition follows that vector of B_n that corresponds to the assignment.

Q.4.8. Consider set B_n defined above. Prove that its rank is n, that is, there are n columns in matrix B_n that form a base of the space of n-dimensional vectors.
A. Take, for example, n columns e_p that contain unity at p-th position whereas other $n-1$ elements are zero ($p = 1, 2, \ldots n$). These obviously are mutually orthogonal and any vector $x = (x_1, \ldots, x_n)$ can be expressed as a linear combination $x = \Sigma_p x_p e_p$, which proves that vectors e_p form a base of the n-dimensional space.

Q.4.9. What is VC-dimension of the linear discrimination problem at an arbitrary dimension $p \geq 2$? **A.** $p+1$, because each subset of p points can be separated from the others by a hyperplane, but there can be such $(p + 1)$-point configurations that cannot be shattered using liner separators.

4.5 Decision Trees

P4.5.1 General: Presentation

Decision tree is a structure used for learning and predicting quantitative or nominal target features. In the former case it is referred to as a regression tree, in the latter, classification tree. This structure can be considered a multivariate extension of contingency tables in such a way that only meaningful combinations of feature categories are involved.

As illustrated on Fig. 4.10, a decision tree recursively partitions the entity set into smaller clusters by splitting a parental cluster over a single feature. The root of a decision tree corresponds to the entire entity set. Each node corresponds to a subset of entities, cluster, and its children are the cluster's parts defined by values of a single predictor feature x. Note that the trees on Fig. 4.10 are binary: each interior node is split in two parts. This is a most convenient format, currently used in most popular programs. Only binary trees are considered in this section.

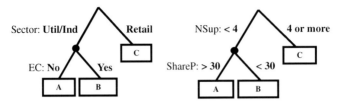

Fig. 4.10 Decision trees for three product based classes of Companies, A, B, and C, made using categorical features, *on the left*, and quantitative features, *on the right*

Decision trees are built from top to bottom in such a way that every split is made to maximize the homogeneity of the resulting subsets with respect to a desired target feature. The splitting stops either when the homogeneity is enough for a reliable prediction of the target feature values or when the set of entities is too small to consider its splits reliable. A function scoring the extent of homogeneity to decide of the stopping is, basically, a measure of correlation between the partition of the entity set being built and the target feature.

When the process of building a tree is completed, each terminal node is assigned with a value of the target that is determined to be characteristic for that node, and thus should be predicted at the conditions leading to the node. For example, both trees on Fig. 4.10 are precise – each terminal class corresponds to one and only one product, which is the target feature, so that each of the trees gives a precise conceptual description of all products by conjunctions of the corresponding branch values. For example, product A can be described as that which is not in Utility sector, nor E commerce utilized in the production process (left-side tree) or as that in which less than 4 suppliers are involved and the share price is greater than 30. Both descriptions are fitting here since both give no errors at all.

Decision trees are very popular because they are simple to understand, use, and interpret. However, one should properly use them, because the decision rules produced with them can be overly simplistic and frequently imprecise. Their effectiveness much depends on the features and samples selected for the analysis. As always in learning correlation, a simpler tree is preferred to a complex one because of the over-fitting problem: a complex tree is more likely reflect noise in the data rather than the true tendencies.

In the next section, we discuss popular homogeneity scoring functions and then proceed to the process of classification tree building, in yet another section.

F4.5.2 General: Formulation

To build a binary decision tree, one needs the following information:

(a) set of input features X,
(b) an output feature u,
(c) a scoring function $W(S, u)$ that scores admissible partitions S against the output feature,

4.5 Decision Trees

(d) rule for obtaining admissible partitions,
(e) stopping criterion
(f) rule for pruning long or unreliable branches, and
(g) rule for the assignment of u-values to terminal nodes.

Let us comment on each of these items:

(a) The input features are, typically, quantitative or nominal. Quantitative features are handled rather easily by testing all possible splits of their ranges. More problematic are categorical features especially those with many categories because the number of possible binary splits can be very large. However, this issue does not emerge at all if categorical features are preprocessed into the quantitative format of binary dummy variables corresponding to individual categories (which is advocated in this text too, see more detail in Section 5.1). Indeed, each of the dummy variables admits only one split – that separating the corresponding category from the rest, which reduces the number of possible splits to consider to the number of categories – an approach advocated by Loh and Shih (1997). A number of such splits can be done in sequence to warrant that any combination of categories is admissible in this approach too.

Since this approach involves one feature at a time only, missing values are not of an issue, because all the relevant information such as means and frequencies can be reasonably well estimated from those values that are available – this is a stark contrast with the other multivariate techniques.

(b) In principle, the decision tree format does not prevent from using multiple target features – just single-target criteria should be summed up when there are several targets (Mirkin 1985). However, all current internationally available programs involve only single target feature. Depending on the scale of the target feature, the learning task differs as well as terminology. Specifically, if the target feature is quantitative, a decision tree is referred to as a regression tree, and if the target feature is categorical, a decision tree is referred to as a classification tree. Yet classification trees may differ on the learning task: (a) learning a partition, if the target is nominal, and (b) learning a category. This section focuses only on the task of learning a classification tree with a partitional target.

(c) Given a decision tree, its terminal nodes (leaves) form a partition S, which is considered then against the target feature u with a scoring function measuring the overall correlation $W(S, u)$. This suggests a context of the analysis of correlation between two features, see Sections 3.3 and 3.4. If the target u is quantitative, then a tabular regression of u over S should be analyzed and scored. Unfortunately, in the data mining literature, this natural approach is not appreciated; thus, the most natural scoring function, the correlation ratio, is not popular. In contrast, at a categorical target, two most popular scoring functions, Gini index and Pearson chi-squared, fit perfectly in the framework of contingency tables and Quetelet indexes described in Section 3.4.1.2, as will be shown in this section further on. Moreover, it will be mathematically proven that these two can be considered as implementations of the same approach of maximizing the contribution to the data scatter of the target categories – the only difference being the way the dummy variables representing the categories are normalized: (i) no normalization to make it Gini index or (ii) normalization by Poissonian standard deviations so that less frequent categories get more important, to make it Pearson chi-squared. This sheds a fresh light on the criteria and suggests the user a way for choosing between the two.

(d) Admissible partitions conventionally are obtained by splitting the entity set corresponding to one of the current terminal nodes over one of the features. To make it less arbitrary, most modern programs do only binary splits. That means that any node may be split only in two parts: (i) that corresponding to a category and the rest, for a categorical feature or (ii) given an a, those "less than a" and those "greater than a", for a quantitative feature. This text attends to this approach as well. All possible splits are tested and that producing the largest value of the criterion is actually made, after which the process is reiterated.

(e) Stopping rule typically assumes a degree of homogeneity of sets of entities, that is, clusters, corresponding to terminal nodes and, of course, their sizes: too small clusters are not stable and should be excluded.

(f) Pruning: In some programs, the size of a cluster is unconstrained so that in the process of splitting nodes over features, some split parts may become very small and, thus, unreliable as terminal nodes. This makes it useful to prune the tree after it is computed, usually by merging the small subset nodes into greater agglomerations. This is typically done not according to the splitting criterion $W(S,u)$ but according to more local considerations such as testing whether proportions of the target categories in a cluster are similar to those used at the assignment of u values to terminal nodes or by removing nodes with small chi-squared values (see, for a review, Esposito et al. 1997).

(g) Assigning a terminal node with a u category conventionally is done by just averaging its values over the node entities if u is quantitative or according to the maximum probability of an u category. Then the quality of quantitative prediction is accessed, as usual, by computing the differences between observed and predicted values of u, and their variance of course. In the nominal target case, this leads to an obvious estimate of the probability of the error: unity minus the maximum probability; these then are averaged over the terminal nodes of the decision tree. To make the error's estimate more robust, cross-validation techniques are used. Consider, say, a ten fold cross validation. The entity set is randomly divided into ten equal-sized subsets. Each of them is used as a testing ground for a decision tree built over the rest: these errors are averaged and given as the error's estimate to the tree built over the entire entity set. These techniques are beyond the scope of the current text.

It should be mentioned that the assignment of a category to a terminal cluster in the tree can be of an issue in some situations: (i) if no obvious winning category occurs in the cluster, (ii) if the category of interest is quite rare, that is, when u's distribution is highly skewed. In this latter case using Quetelet coefficients relating the node proportions with those in the entire set may help by revealing some great improvements in the proportions, thus leading to interesting tendencies discovered (Mirkin 1985).

4.5.3 Measuring Correlation for Classification Trees

P4.5.3.1 Three Approaches to Scoring the Split-to-Target Correlation: Presentation

The process of building a classification tree is, basically, a process of splitting clusters into smaller parts driven by a measure of correlation between the partition S being built and the target feature u. Since our focus here is the case of nominal u's only, the target feature is represented by a partition T which is known to us on the training set.

How to define a function $w(S, T)$ to score correlation between the target partition T and partition S being built? Three possible approaches are:

1. A popular idea is to use a measure of uncertainty, or impurity, of a partition and score the goodness of split S by the reduction of uncertainty achieved when the split is made. If it is Gini index, or nominal variance, which is taken as the measure of uncertainty, the reduction of uncertainty is the popular impurity function utilized in a popular decision tree building program CART (Breiman et al. 1984). If it is entropy, which is taken as the measure of uncertainty, the reduction of uncertainty is the popular Information gain function utilized in another popular decision tree building program C4.5 (Quinlan 1993).
2. Another idea would be to use a popular correlation measure defined over the contingency table between partitions S and T such as Pearson chi-squared. Indeed Pearson chi-squared is used for building decision trees in one more popular program, SPSS (Green and Salkind 2003), as a criterion of statistical independence criterion, though, rather than a measure of association. Yet because Pearson chi-squared is equal to the summary relative Quetelet index (see Section 3.4.1.2), it is a measure of association, and it is in this capacity that Pearson chi-squared is used in this text. Moreover, both the impurity function and Information gain mentioned above also are correlation measures defined over the contingency table as shown in the formulation part of this section. Indeed, the Information gain is just the mutual information between S and T, a symmetric function, and the impurity function, the summary absolute Qutelet index.
3. One more idea comes from the discipline of analysis of variance in statistics (see Section 3.3): the correlation can be measured by the proportion of the target feature variance taken into account by the partition S. How come? The variance is a property of a quantitative feature, and we are talking of a target partition here. The trick is that each class of the target partition is represented by the corresponding dummy feature, which is equal to 1 at entities belonging to the class and 0 at the rest. Each of them can be treated as quantitative, as explained in Section 2.3,

so that the summary explained proportion would make a measure of correlation between S and T. What is nice in this approach, that it is uniform across different types of feature scales: both categorical and quantitative features can be treated the same, which is not the case with other approaches. Although this approach has been advocated by the author for a couple of decades (see, for example, Mirkin 2005), no computational program has come out of it so far. There is a good news though: both the impurity function and Pearson chi-squared can be expressed as the summary explained proportion of the target variance, under different normalizations of the dummy variables course. To get the impurity function (Gini index), no normalization is needed at all, and Pearson chi-squared emerges if each of the dummies is normalized by the square root of its frequency. That means that Pearson chi-squared is underlied by the idea that more frequent classes are less contributing. This might suggest the user to choose Pearson chi-squared if they attend to this idea, or, in contrast, the impurity function if they think that the frequencies of target categories are irrelevant to their case.

There have been developed a number of myths about classification tree building programs and correlation scoring functions involved in them. The following comments are purported to shed light on some of them.

Comment 4.1. There is an opinion lurking in some comments on the web that of two popular programs, CART (Breiman et al. 1984) and CHAID (Green and Salkind 2007), the former is more oriented at prediction whereas the latter, at description. The reason for this perhaps can be traced to the fact that CART involves the impurity function that is defined as the reduction in uncertainty whereas CHAID involves Pearson chi-squared as a measure of the deviation from statistical independence. Yet this opinion is completely undermined by the fact that the measures have very similar predictive powers shaped as the summary Quetelet indexes, the only difference being that one of them involves the relative Quetelet indexes, and the other abosolute ones (see Statements 4.5.2.1 (b) and 4.5.2.2 (b)).

Comment 4.2. The difference between impurity function and Pearson chi-squared amounts to just different scaling options for the dummy variables representing classes of the target partition T (see items (c) in Statements 4.5.2.1 and 4.5.2.2). The smaller T classes get rescaled to larger values, thus contributing more, when using Pearson chi-squared.

Comment 4.3. Pearson chi-squared introduced to measure the deviation of a bivariate distribution from the statistical independence appears also to signify a purely geometric concept, the contribution to the data scatter (see (a) and (c) in Statement 4.5.2.2 on p. 150). This leads to a different advice regarding the zeros in a contingency table. According to classical statistics, the presence of zeros in a contingency table contradicts the hypothesis of statistical independence so that the data are to be trimmed to avoid zeros. However, in the context of data scatter decompositions, the chi-squared is just a contribution with no statistical independence involved so that the presence of zeros is of no issue in this context: thus, no data trimming is needed.

F4.5.3.2 Scoring Functions for Classification Trees: Formulation

Conventional Definitions and Quetelet Coefficients

Consider an entity set I with a pre-specified partition $T = \{T_l\}$ – which can be set according to categories l of a nominal feature – that is to be learnt by producing a classification tree. At each step of the tree building process, a subset $J \subseteq I$ is to be split into a partition $S = \{S_k\}$ in such a way that S is as close as possible to $T(J)$ which is the overalp of T and J. The question is: how the similarity between S and $T(J)$ is to be measured? When $S = T(J)$, there is no confusion between the two. Otherwise, it is the contingency table (see Section 3.3) between S and $T(J)$, $P = (p_{kl})$ where p_{kl} is the proportion of J- entities in $S_k \cap T_l$, that expresses the confusion, which is why it is frequently referred to as a confusion table in this context.

One idea for assessing the extent of similarity is to use a correlation measure over the contingency table such as averaged Quetelet coefficients, Q and A, or chi-squared X^2, as discussed in Section 3.4.1.2.

Seemingly another idea is to score the extent of reduction of uncertainty over $T(J)$ obtained when S becomes available. This idea works like this: take a measure of uncertainty of a feature, in this case partition $T(J)$, $\upsilon(T(J))$, and evaluate it at each of S-classes, $\upsilon(T(S_k))$, $k = 1, \ldots, K$. Then the average uncertainty on these classes will be $\sum_{k=1}^{K} p_{k+}\upsilon(T(S_k))$, where p_{k+} are proportions of entities in classes S_k, so that the reduction of uncertainty is equal to

$$\upsilon(T(J)/S) = \upsilon(T(J)) - \sum_{k=1}^{K} p_{k+}\upsilon(T(S_k)) \qquad (4.13)$$

Of course a function like (4.13) can be considered a measure of correlation over the contingency table P as well, but a nice feature of this approach is that it can be extended from nominal features to quantitative ones – just with an uncertainty index over quantitative T-features (see Q 4.11).

Two very popular measures defined according to (4.13) are so-called impurity function (Breiman at al. 1984) and information gain (Quinlan 1993).

The impurity function builds on Gini coefficient as a measure of variance (see Section 1.3). Let us recall that Gini index for partition T is $G(T) = 1 - \sum_{l=1}^{L} p_l^2$ where p_l is the proportion of entities in T_l. If J is partitioned in clusters $S_k, k = 1, \ldots, K$ and S form a contingency table of relative frequencies $P = (p_{kl})$. Then the reduction (4.13) of the value of Gini coefficient due to partition S is equal to $\Delta(T(J), S) = G(T(J)) - \sum_k p_k G(T(S_k))$. This index $\Delta(T(J), S)$ is referred to as impurity of S over partition T. The greater the impurity the better the split S.

It is not difficult to prove that $\Delta(l, S)$ relates to Quetelet indexes from Section 3.3. Indeed, $\Delta(T, S) = A(T, S)$ where $A(T, S)$ is the summary absolute Quetelet index defined by Equation (3.22) in Q.3.24. Indeed, $\Delta(T, S) = G(T) - \sum_k p_k G(T(S_k)) = 1 - \sum_l p_{+l}^2 - \sum_k (p_{k+} - \sum_l p_{kl}^2/p_{k+})$, where p_{+l} is the proportion of l-th category (class) in set J. This implies indeed that $\Delta(T, S) = \sum_l p_{kl}^2/p_{k+} - \sum_l p_{+l}^2$, which proves the statement.

The information gain function builds on entropy as a measure of uncertainty (see Section 2.3). Let us recall that entropy of partition T is $H(T) = -\sum_{l=1}^{L} p_l \log(p_l)$ where p_l is the proportion of entities in T_l. If J is partitioned in clusters S_k, $k = 1, \ldots, K$, partitions T and S form a contingency table of relative frequencies $P = (p_{kl})$. Then the reduction (4.13) of the value of entropy due to partition S is equal to $I(T(J), S) = H(T(J)) - \sum_k p_k H(T(S_k))$. This index $I(T(J), S)$ is referred to as the information gain due to S. In fact, it is equal to a popular characteristic of the cross-classification of T and S, the mutual information defined as $I(T, S) = H(T) + H(S) - H(ST)$ where $H(ST)$ is entropy of the bivariate distribution represented by contingency table P. (The J argument is omitted here as irrelevant to the statement.)

Please note that the mutual information is symmetric with regard to S and T, in contrast to the impurity function. To prove the statement let us just put forward the definition of the information gain and use the property of logarithm that $\log(a/b) = \log(a) - \log(b)$:

$$I(T, S) = H(T) - \sum_k p_k H(T(S_k)) = H(T) + \sum_k p_{k+} \sum_l p_{kl} \log(p_{kl}/p_{k+}) =$$
$$= H(T) - \sum_k p_{k+} \log(p_{k+}) + \sum_{k,l} p_{kl} \log(p_{kl}) = H(T) + H(S) - H(ST),$$

which completes the proof.

The reduction of uncertainty measures are absolute differences that much depend on the measurement scale and, also, on values of $\upsilon(T)$ and $\upsilon(S)$. This is why it can be of advantage to use relative versions of the reduction of uncertainty measures normalized by $\upsilon(T)$ or $\upsilon(S)$ or both. For example, popular program C4.5 (Quinlan 1993) uses the information gain normalized by $H(S)$ and referred to as the information gain ratio.

Confusion Measures as Contributions to the Data Scatter

Once again we consider a nominal feature over an entity set I of cardinality N (in fact, I and N can be changed for any other symbols – these are just notations in this section), represented by partition $T = \{T_l\}$, and a clustering partition $S = \{S_k\}$ designed from available features to approximate T. This time, though, we are not going to use their contingency table $P = (p_{kl})$, to see the co-occurrence frequencies p_{kl} emerging from a different perspective.

Specifically, assign each target class (category) T_l with a binary variable x_l, a dummy, which is just a 1/0 N-dimensional vector whose elements $x_{il} = 1$ if $i \in T_l$ and $x_{il} = 0$, otherwise ($l = 1, \ldots, L$). Use these dummies as quantitative features to build a tabular regression of each over the partition S. Consider, first, the average of x_l within cluster S_k: the number of unities among x_{il} with $i \in S_k$ is obviously N_{kl}, the size of the intersection $S_k \cap T_l$ because $x_{il} = 1$ only if $i \in S_k$. That means that the within S_k average of x_l is equal to $c_{kl} = N_{kl}/N_{k+}$ where N_{k+} stands for the size of S_k, or, p_{kl}/p_{k+} in terms of the relative contingency table P.

4.5 Decision Trees

Let us now standardize each feature x_l by a scale shift a_l and rescaling factor $1/b_l$, according to the conventional formula $y_l = (x_l - a_l)/b_l$. This will correspondingly change the averages, so that within-cluster averages of standardized features y_l are equal to $c_{kl} = (p_{kl}/p_{k+} - a_l)/b_l$. In mathematical statistics, the issue of standardization is just a routine transforming the probabilistic density function to a standardized format. Things are different in data analysis, since no density function is assigned to data usually. The scale shift is considered as positioning the data against a backdrop of the "norm", whereas the act of rescaling is to balance feature "weights" (see Section 5.1 for discussion). Therefore, choosing the feature means as the "norms" should be reasonable. The mean of feature x_l is obviously the proportion of unities in it, which is p_{+l} in notations related to contingency table P. In fact, the remainder of this section can be considered as another reason for using $a_l = p_{+l}$. The choice of rescaling factors is somewhat less certain, though using all $b_l = 1$ should seem reasonable too because all the dummies are just 1/0 variables measured in the same scale. Incidentally, 1 is the range of x_l. Some other values related to x_l's dispersion could be used as well. With the scale shift value specified, the within cluster average can be expressed as

$$c_{kl} = \frac{p_{kl} - p_{k+}p_{+l}}{p_{k+}b_l}.$$

Let us refer to formula (3.13) on p. 94 in which the feature scatter is presented as the sum of two parts, one explained by partition S and the other unexplained. Using symbolic of this section, the explained part of x_l's scatter can be expressed as $B_l = \sum_{k=1}^{K} N_{k+}c_{kl}^2$. This is the sum of contributions of individual clusters. By using (3.17) each of the individual contributions is equal to

$$B_{kl} = N_{k+}\frac{(p_{kv} - p_{k+}p_{+l})^2}{p_{k+}^2 b_l^2} = N\frac{(p_{kv} - p_{k+}p_{+l})^2}{p_{k+}b_l^2} \qquad (4.14)$$

Accordingly, the total contribution of partition S to the total scatter of the set of standardized dummies representing partition T is equal to

$$B(T/S) = \sum_{k=1}^{K}\sum_{l=1}^{L} B_{kl} = N\sum_{k=1}^{K}\sum_{l=1}^{L} \frac{(p_{kv} - p_{k+}p_{+l})^2}{p_{k+}b_l^2} \qquad (4.15)$$

The total contribution (4.15) reminds us of both the averaged relative Quetelet coefficient (3.19) and the averaged absolute Quetelet coefficient (3.22). The latter, up to the constant N of course, emerges at all rescaling factors $b_l = 1$. The former emerges when rescaling factors $b_l = \sqrt{p_l}$. The square root of the frequency has an appropriate meaning – this is a good estimate of the standard deviation in Poisson model of the variable: according to this model, N_{+l} unities are thrown randomly into the fragment of memory assigned for the storage of vector x_l. In fact, at this scaling system, $B(T/S) = X^2$, Pearson chi-squared!

Let us summarize the proven facts (Mirkin 1996).

Statement 4.5.2.1. The impurity function can be equivalently expressed as

(a) The reduction of Gini uncertainty index of partition T when partition S is taken into account;
(b) The averaged absolute Quetelet index $a(l/k) = p_{kl}/p_{k+} - p_{+l}$ of the same effect;
(c) The total contribution of partition S to the summary data scatter of the set of dummy 1/0 features corresponding to classes of T and standardized by subtracting the mean with no rescaling.

Statement 4.5.2.2. The Pearson chi-square function can be equivalently expressed as

(a) A measure of statistical independence between partitions T and S;
(b) The averaged relative Quetelet index $q(l/k) = (p_{kl}/p_{k+} - p_{+l})/p_{+l}$ between partitions T and S;
(c) The total contribution of partition S to the summary data scatter of the set of dummy 1/0 features x_l corresponding to classes T_l and standardized by subtracting the mean and dividing the result by $b_l = \sqrt{p_l}$.

Statement 4.5.2.3. The information gain can be equivalently expressed as

(a) The reduction of entropy of partition T when partition S is taken into account;
(b) The mutual information $H(T) + H(S) - H(TS)$ between T and S;

C4.5.3.3 Computing Scoring Functions with MatLab: Computation

```
function a=gini(p)         function a=chi(p)          function a=ing(p)
  tot=sum(sum(p));           tot=sum(sum(p));           p=p+1;
% total                    % total                    % to avoid zeros
  pr=p/tot;                  pr=p/tot;                  tot=sum(sum(p));
  rp=sum(pr');               rp=sum(pr');               pr=p/tot;
% row sums                   cp=sum(pr);                rp=sum(pr');
  cp=sum(pr);                ir=find(rp>0);             cp=sum(pr);
%column sums               % nonzero rows               pl=log2(pr);
  ps=pr.*pr;                 ic=find(cp>0);             pp=pr.*pl;
  rps=sum(ps');            %nonzero columns             rpp=sum(pp');
  ir=find(rp>0);             ps=pr.*pr;                 a1=sum(rpp.*rp);
  tr=rps(ir)./rp(ir);        ip=rp'*cp;                 tp=cp.*log2(cp);
  a1=sum(tr);                psi=ps(ir,ic);             a2=sum(tp);
  a2=sum(cp.*cp);            ipi=ip(ir,ic);             a=a1-a2;
  a=a1-a2;                   tp= psi./ipi;              return
  return                     a1=sum(sum(tp));
                             a=a1-1;
                             return
```

4.5 Decision Trees

Three functions discussed above, Gini index, Pearson chi-squared, and Information gain can be coded as presented in columns of the box above where input p is a contingency matrix. Due to a holistic nature of MatLab computation, it is possible to organize the computation without looping through the matrix elements. The subroutines gini, chi and ing in the box can be considered pseudocodes of the functions for coding in any other language as well.

Q.4.10. Consider the variance as an uncertainty measure for a quantitative feature y. Define the uncertainty reduction measure according to formula (4.13), with T changed for y of course, and prove that it is equal to the numerator of the correlation measure – the part of variance of y explained by its tabular regression over S. **A.** The summary contribution of S to the data scatter is equal to $B = \sum_{k=1}^{K} c_k^2 |S_k| = \sum_{i \in I} y_i^2 - \sum_{k=1}^{K} \sigma_k^2 |S_k|$ where σ_k^2 is the within-cluster variance of y (see (3.13) in Section 3.3). Then $B = N(\sigma^2 - \sum_{k=1}^{K} p_k \sigma_k^2)$ where σ^2 is the variance of the standardized feature y (note that the mean of y is 0!) and p_k the proportion of entities in cluster S_k. The last equation clearly shows that the explained part of v is $B = N\sigma^2 \eta^2$. If y has been z-score standardized so that $\sigma^2 = 1$, B equals the correlation ratio.

Q.4.11. What is the formula of summary contribution B of partition S to the set of dummy features representing partition T when they have been normalized by dividing by their Bernoullian standard deviations $b_l = \sqrt{p_{+l}(1 - p_{+l})}$?

Q.4.12. Consider a partition $S = \{S_k\}$ ($k = 1, 2, \ldots, K$) on J and a set of categorical features $v \in V$, each with a set of categories $L(v)$. The category utility function (Fisher 1987) scores partition S against the feature set according to formula:

$$u(S) = \frac{1}{K} \sum_{k=1}^{K} p_k [\sum_{v \in V} \sum_{l \in L(v)} p(v = l/S_k)^2 - \sum_{v \in V} \sum_{l \in L(v)} p(v = l)^2]$$

The term in the square brackets is the increase in the expected number of attribute values that can be predicted given a class, S_k, over the expected number of attribute values that could be predicted without using the class. The assumed prediction strategy follows a probability-matching approach. According to this approach, entities arrive one-by-one in a random order, and the category l is predicted for them with the frequency reflecting its probability, $P(l/k)$ if the class S_k is known, or $p_k = N_k/N$ if information of the class S_k is not provided. Factors p_k weigh classes S_k according to their sizes, and the division by K takes into account the differences in the numbers of clusters: the smaller the better. Prove that the category utility function $u(S)$ is the sum of impurity functions $\Delta(l, s)$ over all features $l \in I$ related to the number of clusters, that is, $u(S) = \sum_{l \in L} \Delta(l, S)/K$.

4.5.4 Building Classification Trees

Building of a classification tree is a recursive process: starting from the entire data set, partition a cluster into a number of parts according to one of the features. To make the partitions less arbitrary, only binary splits are involved in most of the update programs. That means that any node may be split only in two parts: (i) that corresponding to a category and the rest, for a categorical feature, or (ii) given a threshold a, those "less than or equal to a" and those "greater than a", for a quantitative feature. This approach naturally comes when the data are preprocessed by "enveloping" categories into the corresponding "quantitative" dummy features, that assign a unity to every object falling into the category, and a zero to all the rest. Indeed, at $a=0$, such a dummy feature would split the set in two parts – that for the corresponding category and the rest. Given a cluster, the choice of feature and threshold a for doing the split is driven by a correlation scoring function, be it Information gain, Pearson chi-squared, Gini index or anything else.

A cluster is not to be split anymore if it is smaller than a user defined threshold TS (TS = 10 is set further on) or is homogeneous enough. We use two different homogeneity tests: (a) large enough proportion of a target category in the cluster, say, above 80%, and (b) small enough value of the scoring function which is set to be 0.03 for Gini index, 0.08 for Pearson chi-squared, and 0.15 for Information gain. These levels of magnitude reflect the functions' ranges: Gini index is very close to 0 hardly reaching 0.5 at all, Pearson chi-squared, related to N, changes between 0 and 1 because it cannot be greater than the number of split parts minus 1, and Information gain can have larger values when the number of target categories is 3 or more. This sets the stopping conditions.

Worked example 4.3. Classification tree for Iris dataset

At Iris dataset with its three taxa, Iris setosa and Iris versicolor and Iris virginica, taken as target categories, all the three scoring functions – Impurity (Gini) function, Pearson chi-squared and Information gain – lead to the same classification tree, presented on Fig. 4.11.

The tree of Fig. 4.11 was found with program clatree.m, see p. 380 in Appendix. It comprises three leaf clusters: A, consisting of all all 50 Iris setosa specimens; B, containing 54 entities of which 49 are of Iris versicolor and 5 of Iris virginica;

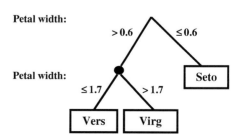

Fig. 4.11 Classification tree for the three-taxon partition at Iris dataset found by using each of Gini, Pearson chi-squared and Information gain scoring functions

4.5 Decision Trees

Table 4.15 Values of Gini index at the best split of each feature on Iris dataset clusters in Fig. 4.11

Feature	First split		Second split	
	Value	Gini	Value	Gini
w1	5.4	0.228	6.1	0.107
w2	3.3	0.127	2.4	0.036
w3	1.9	0.333	4.7	0.374
w4	0.6	0.333	1.7	0.390

C, containing 46 entities of which 45 are of Iris virginica and 1 of Iris versicolor. Altogether, this misplaces 6 entities leading to the accuracy of 96%. Of course, the accuracy would somewhat diminish if a cross-classification scheme is applied (see Loh and Shih, 1997, who draw a slightly different tree for Iris dataset).

Let us take a look at the action of each variable at each of the two splits in Table 4.15. Each time features w3 and w4 appear to be most contributing, so that at the first split, at which w3 and w4 give the same impurity value, w4 made it through just because it is the last maximum which is remembered by the program.

The tree involves just one feature, w4: Petal width, split twice, first at $w4 = 0.6$ and then at $w4 = 1.7$. The Pearson chi-squared value (related to N of course) is 1 at the first split and 0.78 at the second. The Impurity function grows by 0.33 at the first split and 0.39 at the second. The fact that the second value is greater than the first one may seem to be somewhat controversial. Indeed, the first split is supposed to be the best, so that it is the first value that ought to be maximum. Nevertheless, this opinion is wrong: if the first split was at $w4 = 1.7$ that would generate just 0.28 of impurity value, less than the optimal 0.33 at $w4 = 0.6$. Why? Because the first taxon has not been extracted yet and grossly contributes to a higher confusion (see the top part in Table 4.16).

Table 4.16 Confusion tables between a split and target partition on Iris dataset

Target partition classes	Iris setosa	Iris versicolor	Iris virginica	Total
Full set				
w4≤1.7	50	49	5	104
w4>1.7	0	1	45	46
Total	50	50	50	150
First cluster removed				
w4≤1.7	0	49	5	54
w4>1.7	0	1	45	46
Total	0	50	50	100

Project 4.1. Prediction of learning outcome at Student data

Consider the Student dataset and ask whether students' learning successes can be predicted from other features available (Occupation, Age, Number of children)? By looking at Table 1.5, it is hardly can be expected that marks can be predicted in this way. Therefore, let us divide students in three groups: I – not so good performers (average mark is less than 50), II – good performers (average mark between 50 and 70 inclusive), and III – excellent performers (average mark higher than 70). To do this, we compute the average mark over the three subjects (SE, OOP, and CI) and create a partition of students T as described; the distribution of T appears to be I–25, II–58, III–17.

We have a 100×5 matrix X to explore the correlation between X and T, the three columns, 1, 2, 3, being dummy variables for Occupation categories (IT, BA, AN), column 4 for Age, and column 5 for Number of children. The two conventional stopping criteria, the cluster's size and prevalence of a target class, are not sufficient at this data, because after one or two splits, the program just chips away small fragments of clusters without much improving them. This corresponds to the situations at which the scoring function does not show much improvements either. Therefore, we utilize one more criterion – the minimum value of the scoring function below which there is no splitting. Since the three scoring functions we use have different ranges, the thresholds must be different too. At this study, the threshold is set at 0.03 for Gini index, 0.08 for Pearson chi-squared, and 0.15 for Information gain. The minimum cluster size is taken at 10, and the prevalence of a target class at 80%.

The classification tree found with Gini index is presented on Fig. 4.12. The distributions of target categories in clusters on Fig. 4.12 are presented in Table 4.17. Bold font highlights four terminal clusters as well as high or low proportions of target classes in clusters. High proportions here are those greater than 70% and low proportions are those smaller than 5%.

Tree on Fig. 4.12 is driven by two features: AN Occupation, that structures the set rather well – one split part, those of AN occupation, get more than 70% of category I, and none of category III, and the other of category II. All further divisions are over feature Age; the 12 students in cluster 8 are rather specific – these are of AN occupation aged between 22 and 28, so that 75% of them are in category I, an improvement over parental cluster 6. Cluster 4 of younger not-AN students seems an attempt at drawing a cluster to predict category III – it has a highest jump in its proportion, to 36.4 from 17% in the entire set (cluster 1). The 25 older people

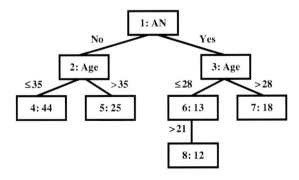

Fig. 4.12 Classification tree on students data targeting partition T of students in three categories found using Gini index. The legend *Number: A* presents, at a split cluster, *A* as the split variable or, at an unsplit cluster, *A* as the size (the number of students in it)

4.5 Decision Trees

Table 4.17 Distributions of target classes in clusters of tree on Fig. 4.12, per cent

Target categories	Clusters in tree on Fig. 4.12							
	1	2	3	4	5	6	7	8
I	25.0	2.9	**74.2**	**2.3**	**4.0**	69.2	**77.8**	**75.0**
II	58.0	**72.5**	25.8	61.4	**92.0**	30.8	22.2	25.0
III	17.0	24.6	0	36.4	**4.0**	0	0	0
Gini index at split	0.168	0.046	0.035			0.048		
Cluster size	100	69	31	44	25	13	18	12

among not-AN students are overwhelmingly, 92%, in category II. More splits would have followed if we had decreased the minimum acceptable value of Gini index, say from 0.03 to 0.01.

How well this tree would fare at prediction? To address this question properly, one should either conduct a cross-classification test as explained in the end of Section 4.5.2 or set aside a random testing set before using the rest for building a tree, after which see the levels of errors on the testing set.

Yet for the illustrative purposes, let us calculate the prediction error by using tree on Fig. 4.12. This is done by using the terminal clusters 4, 5, 7, 8 comprising 44, 25, 18, 12 elements, respectively. They total to 99, not 100, because of chipping off an element from cluster 6 to make it into cluster 7. That means: for students in AN category aged 21 or less, no prediction of their learning success level will be made; the classifier takes what is referred to as reject option (comprising approximately 1% of future cases if our sample is representative). According to the data in Table 4.17, the optimal prediction rule would predict then category II at cluster Cluster 4 (with error $100-61.4 = 38.6\%$), category II again, at cluster 5 (with error $100-92 = 8\%$), and category I at clusters 7 and 8 (with errors 22.2% and 25.0%, respectively). The average error is the sum of the individual cluster errors weighted by their relative sizes, $(38.6*44 + 8*25 + 22.2*18 + 25*12)/99 = 26.2\%$.

What happens, if we use the parental cluster 6 instead of the chipped cluster 8? First thing – no reject option is involved then. Second, the error somewhat increases as should be expected: $(38.6*44 + 8*25 + 22.2*18 + 30.8*13)/100 = 27.0\%$.

Figure 4.13 presents trees found by using Pearson chi-squared (a) and Information gain (b). In contrast to Gini index, decreasing the increment threshold does not much help at Information gain: chipping here and there rather than splits will be added. The change of splitting Age value to 30 at cluster 2 on tree (a) does lead to some improvements: the 45 older students are 82.2% in category II. Yet among the 24 younger students, 45.8% belong to category III (leaving 54.2% in category II and 0 in category I).

With this example, one can see that the 90–100% precision is not easy to achieve. That is, a terminal node may have rather modest proportions of target categories, like cluster 5 on Figure 4.13a: about 54% of II category and 46% of III category. Conventional thinking would label the node as an II category predictor because the share of II is greater than half.

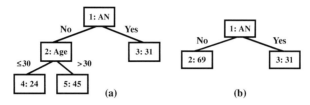

Fig. 4.13 Classification trees on Student dataset targeting partition T of students in three occupation categories found using Pearson chi-squared (**a**) and Information gain (**b**). The legends are of format "Number: A" where "A", at a split cluster, is the split variable or, at an unsplit cluster, the cluster's size

Yet, one should note that, in fact, the proportion 54% is smaller than that, 58%, in the entire set, which means that in fact these conditions, Not_AN and younger age, less than 31, wash out some of II category. It is a case when the style of Quetelet's thinking may produce a better description. This thinking goes beyond proportions in the terminal node and requires comparing the category shares at the node with that in the whole sample. In contrast to a reduction of II category, this cluster boasts a dramatic increase of III category – from 17% in the entire set to 46% in the cluster, 29%. This difference would be picked up by the absolute Quetelet coefficient which is equal to Gini index. Even more dramatic is the relative increase, $(45.8-17)/17 = 170\%$. It is this increase that has been picked up by Pearson chi-squared scoring function, because it is driven by the relative Quetelet coefficient.

Q.4.13. Drawing a lift chart in marketing research. Consider a marketing campaign advertising a product. There is a 1,000 strong sample from the set of targeted customers whose purchasing behavior is known because of prior campaigns. The sample is composed of clusters of a classification tree with different response (that is, purchasing) rates (see Table 4.18). To plan an effective campaign, marketing researchers use what is called a lift chart – a visual representation of the response rate. The x-axis of a lift chart shows the percentiles of the sample, say, from 10 to 100%. On y-axis, the so-called lifts are presented. Given a group of customers, the lift is defined as the ratio of the group's response rate to the baseline response rate, which is the response rate for the entire sample. On the lift chart, the percentiles of the sample are taken in the descending order of the lift. Both baseline and percentile lifts are presented on the chart. Build a lift chart for the sample. **A.** First, we calculate the baseline rate which is the average of the response rates in Table 4.18 weighted by the cluster proportions: $r = 0.1*30 + 0.4*10 + 0.25*4 + 0.25*0 = 8\%$.

Table 4.18 Proportions of four clusters in a sample of 1,000 customers and their purchasing behavior (response rate)

Cluster share, %	10	40	25	25
Response rate, %	30	10	4	0

4.5 Decision Trees

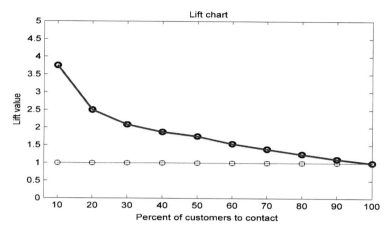

Fig. 4.14 Lift chart for data in Table 4.18

Now we take the most responsive 10% of the customers and calculate their lift value: $30/8 = 3.75$. Next, we take the most responsive 20% of the sample, that is the first cluster plus a hundred customers from the second cluster and see their response rate – there should be 30 customers from the first cluster plus 10 from the second who have purchased the product, which gives $40/200 = 20\%$ response rate leading to the lift value of $20/8 = 2.5$. Next percentile, 30% of the sample is composed of the first cluster plus 200 customers from the second cluster leading to $50/300 = 17.7\%$ response rate and lift 2.2. In this way, chart presented on Fig. 4.14 is computed.

C4.5.5 Building Classification Trees: Computation

Consider an entity set I along with a nominal target feature represented by partition T of I as well as a set of quantitative input features X (some or all of X-features may be binary dummy variables corresponding to categories). At each step of the process of building a classification tree a cluster $J \subseteq I$ is to be split according to a feature x_v from X in two clusters, S_1 and S_2 so that $S_1 = \{i | i \in J \text{ and } x_{iv} \leq y\}$ and $S_2 = \{i | i \in J \text{ and } x_{iv} > y\}$ where y is a value of x_v. The choice of x_v and y is guided by a scoring function $W(S,T)$ defined over the contingency table P cross-classifying T by S. That implies that a cluster, as an element of the hierarchical structure being built, should maintain at least the following data: (i) its entity set, (ii) its parental cluster, (iii) feature x_v over which it has been split, (iv) splitting value y, (v) the inequality, $\leq$ or $>$, in the cluster defining predicate. The process starts at the universal cluster consisting of the entire set I. The process stops if either of two conditions holds: (a) $|J| < n$, where n is a pre-specified threshold on the minimum number of entities in a cluster, and (b) if the frequency of a T-cluster is greater than a pre-specified threshold α. To make testing of (b) easier, each cluster should bear one more feature – (vi) the distribution of T in it. One more useful piece of data supplied with a cluster would be (vii) a signal of whether it may or may not be split again.

The recursive nature of the process, as well as the presence of a set of data to accompany each cluster, would make it a fitting subject of an object oriented code. Yet since the object oriented part of MatLab is not quite native in it, a procedural construction will be described in this section. This construction involves two parts, provided that computing scoring function $W(T,S)$ over contingency table P, has been implemented: (A) finding the best split over a feature, and (B) building a hierarchy of the best splits.

4.5.5.1 Finding the Best Split Over a Feature: Computation

A pseudocode, or MatLab function, msplit.m, takes in a column-feature x, a partition of the set of its indices, t, as a cell array of t-classes, and a string with the name of a scoring method. It produces partition s, the feature splitting value y, and the value of scoring function ma. The stages of computation are annotated within the code.

```
function [g,ma,y]=msplit(x,t,method)
n=length(x);
%-------preparing the set of split value candidates
xv=union(x,x);%set of x values sorted
ll=length(xv);
rl=length(t);
if ll==1 %feature x is constant
    g{1}=[1:n];
    ma=0;
    y=max(x);
else
    for k=1:(ll-1) %loop over splitting values
        f{1}=find(x<=xv(k)); %first split set
        f{2}=setdiff([1:n],f{1}); % the rest
        for ik=1:2; for il=1:rl
            p(ik,il)=length(intersect(f{ik},t{il}));
            end
        end %  contingency table p
        switch method
            case 'gini'
                res=gini(p);
            case 'chi'
                res=chi(p);
            case 'ing'
                res=ing(p);
            otherwise
                disp('The method is wrong ');
                pause(10);
        end
```

```
    %----------looking for the best split
        if res>ma
            ma=res;
            g=f;
            y=xv(k);
        end
    end
end
```

4.5.5.2 Organizing a Recursive Split Computation and Storage

The computation is organized in code clatree.m printed in the Appendix p. 380. Here are just a few comments on its structure. Consider a set of ss clusters stored in a cell structure indexed from 1 to ss; in the beginning, the structure stores just the universal cluster I and its features at ss $= 1$. Of these clusters, those in the end, starting from index tt $\geq$ ss are eligible for splitting. The newly split clusters are indexed by index bb starting from bb $=$ ss+1. (Note that with this system of indexing, there is no need to assign clusters with a label informing that they should not be split anymore: the clusters to split can only be fresh ones!) After split parts are put in the structure, the indices are updated.

There can be a number of stopping criteria that are to be set in the very beginning of the program: it stops when no clusters eligible for splitting remain. In the current version of program clatree.m, three types of stopping criteria are employed. First is the number of entities, TS: a cluster with a smaller number of entities cannot make it into the tree and of course cannot be split further. Second, the dominant proportion of the target classes, ee: a cluster is not split anymore if this has been reached. And the third stopping criterion is tin, a threshold on the scoring function value: if it is less then tin at a split, the cluster is not split.

4.6 Learning Correlation with Neural Networks

4.6.1 General

P4.6.1.1 Artificial Neuron and Neural Network: Presentation

Neural network is one of the most popular structures used for predictions of target features. It is a network of artificial neurons modeling the neuron cell in a living organism. A neuron cell fires an output when its summary input becomes higher than a threshold. Dendrites bring signal in, axons pass it out, and the firing occurs via synapse, a gap between neurons, that makes the threshold (see Fig. 4.15).

This is modeled in an artificial neuron as follows (see Fig. 4.16). A neuron model is drawn as a set of input elements connected to an output. The connections are assigned with wiring weights.

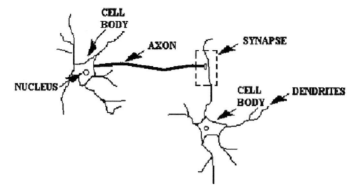

Fig. 4.15 Structure of neuron cells (from http://www.compeng.dit.ie/staff/tscarff/Music_technology/music/neuron_structure.htm visited 17 December 2010)

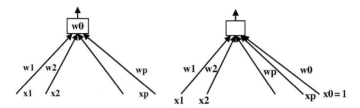

Fig. 4.16 A scheme of an artificial neuron, *on the left*. The same neuron with the firing threshold shown as a wiring weight on the fictitious input always equal to 1 is *on the right*

The input signals are data features or other neurons' outputs. The output element receives a combined signal, the sum of feature values weighted by the wiring weights. The output compares this with a firing threshold, otherwise referred to as a bias, and fires an output depending on the results. Ideally, the output is 1 if the combined signal is greater than the threshold, and −1 if it is smaller. This is, in fact, what is called the sign function of the difference, *sign(x)*, which is 1, 0 or −1 if x is positive, zero or negative, respectively. This activation function is overly straightforward sometimes. Instead, the so-called sigmoid and symmetric sigmoid functions are considered as smooth exponent-based counterparts to sign(x). Their graphs are shown alongside with that for *sign(x)* on Fig. 4.17. Sometimes the output element is assumed as doing no transformation at all, just passing the combined signal as the neuron's output, which is referred to as a linear activation function.

The firing threshold, or bias, hidden in the box in neuron on the left on Fig. 4.16, can be made explicit if one more, fictitious, input is added to the neuron.

Fig. 4.17 Graphs of sign (**a**), sigmoid (**b**) and symmetric sigmoid (**c**) functions

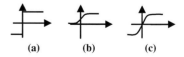

4.6 Learning Correlation with Neural Networks

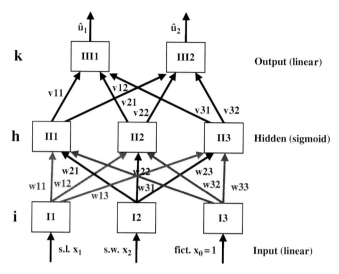

Fig. 4.18 A feed-forward network with two input and two output features (no feedback loops). Layers: Input (I, indexed by i), Output (III, indexed by k) and Hidden (II, indexed by h)

This input is always equal to 1 so that its wiring weight is always added to the combined input to the neuron. It is assumed to be equal to minus the bias so that the total sum is the difference between the combined signal and the bias. In the remainder, we assume that the bias, with the minus sign, is always explicitly present among the wiring weights in this way (see Fig. 4.16 on the right).

Artificial neurons can be variously combined in neural networks. There have been defined many specific types of neural network structures, referred to as architectures, of which the most generic is a three-layer structure with no feedback connections, such as presented on Fig. 4.18 in the next section. There are two outbound layers, the input and output ones, and one intermediate layer which is referred to as a hidden layer. This is why such a structure is referred to as a one hidden-layer neural network (NN).

Network on Fig. 4.18 is designed as a one-hidden-layer NN for predicting petal sizes of Iris features from their sepal sizes. Recall that in Iris data set, each of 150 specimens is presented with four features which are the length and width of petals (features w3 and w4) and sepals (features w1 and w2). It is likely that the sepal sizes and petal sizes are related.

In fact, the further material can be used for building an NN for modeling correlation between any inputs and outputs – the only possible difference, in numbers of input and/or output units, plays no role in the organization of computations.

This neural network consists of the following layers:

(a) Input layer that accepts three inputs: a bias input $x_0 = 1$ as explained above (see Fig. 4.16 on the right) as well as sepal length and width; these are combined to be inputs to each of the neurons at the hidden layer.

(b) Output layer producing an estimate for petal length and width with a linear activation function. Its input is the output signals from the hidden layer. No fictitious input $x_0 = 1$ is assumed here because the activation function here just passes the combined signal through without a threshold.
(c) Hidden layer consisting of three neurons. Each of them takes a combined input from the first layer and applies to it its sigmoid activation function. The output signals of these three neurons constitute inputs to the output layer. The architecture allows for any number of hidden neurons with no changes in the computations.

The one-hidden-layer structure is generic in NN theory. It has been proven, for instance, that such a structure can exactly learn any subset of the set of entities. Moreover, any pre-specified mapping of inputs to outputs can be approximated up to a pre-specified precision with such a one-hidden-layer network, if the number of hidden neurons is large enough (Cybenko 1989).

F4.6.1.2 Activation Functions and Network Function: Formulation

Two popular activation functions, besides the *sign* function $\overset{\circ}{u}_i = sign(\hat{u}_i)$, are the *linear* activation function, $\overset{\circ}{u}_i = \hat{u}_i$ and *sigmoid* activation function $\overset{\circ}{u}_i = s(\hat{u}_i)$ where

$$s(x) = (1 + e^{-x})^{-1} \tag{4.16}$$

is a smooth analogue to the sign function, except for the fact that its output is between 0 and 1 rather than −1 and 1 (see Fig. 4.17b). To imitate the *sign* function, we first double the output interval and then subtract 1 to obtain what is referred to as a symmetric sigmoid or hyperbolic tangent:

$$th(x) = 2s(x) - 1 = 2(1 + e^{-x})^{-1} - 1 \tag{4.17}$$

This function, illustrated on Fig. 4.17c, in contrast to sigmoid $s(x)$, is symmetric: $th(-x) = -th(x)$, like $sign(x)$, which can be useful in some contexts.

The sigmoid activation functions have nice mathematical properties; they are not only smooth, but their derivatives can be expressed through the functions themselves, see Q.4.14 and (4.24).

Let us express now the function of the one-hidden-layer neural network presented on Fig. 4.18. Its wiring weights between the input and hidden layer form a matrix $W = (w_{ih})$, where i denotes an input, and h a hidden neuron, $h = 1, 2, \ldots, H$ where H is the number of hidden neurons. The wiring weights between the hidden and output layers form matrix $V = (v_{hk})$, where h denotes a hidden neuron and k an output.

Layers I and III are assumed to be linear giving no transformation to their inputs; all of the hidden layer neurons will be assumed to have a symmetric sigmoid as their activation function.

4.6 Learning Correlation with Neural Networks

To find out an analytic expression for the network, let us work it out layer by layer. Neuron h in the hidden layer receives, as its input, a combined signal

$$z_h = w_{1h}x_1 + w_{2h}x_2 + w_{3h}x_0$$

which is h-th component of vector $z = \Sigma_i x_i * w_{ih} = x*W$ where x is a 1×3 input vector. Then its output will be $th(z_h)$. These constitute an output vector $th(z) = th(x*W)$ that is input to the output layer. Its k-th node receives a combined signal $\Sigma_h v_{hk} * th(z_h)$ which is k-th component of the matrix product $th(z)*V$, that is passed as the NN output $\hat{u}$. Therefore, the NN on Fig. 4.18 transforms input x into output $\hat{u}$ according to the following formula

$$\hat{u} = th(x*W)*V \qquad (4.18)$$

which combines linear operations of matrix multiplication with a nonlinear symmetric sigmoid transformation. If matrices W, V are known, (4.18) computes the function $u = F(x)$ in terms of th, W, and V. The problem is to fit this model with training data provided, at this instance, by the Iris data set.

4.6.2 Learning a Multi-layer Network

Given all the wiring weights W, between the input and hidden layers, and wiring weights V, between the hidden and output layers, as well as pre-specified hidden layer activation functions, the NN on Fig. 4.18 takes an input of the sepal length and width and transforms it into estimates of the corresponding petal length and width.

The quality of the estimates can be measured by the average squared error. The better adapted weights W and V are, the smaller the error. Where the weights come from? They are learnt from the training data.

Thus the problem is to estimate weight matrices W and V at the training data in such a way that the average squared error is minimized.

The machine learning paradigm is based on the assumption that a learning device adapts itself incrementally by facing entities one by one. This means that the full sample is assumed to be never known to the device so that global solutions, such as the orthogonal projection used in linear discrimination, are not applicable. In such a situation an optimization algorithm that processes entities one by one should be applied. Such is the gradient method, also referred to as the steepest descent.

This method relies on the so-called gradient of the function to be optimized. The gradient is a vector that can be derived or estimated at any admissible solution, that is, matrices W and V. This vector shows the direction of the steepest ascent over the optimized function considered as a surface. Its elements are the so-called partial derivatives of the optimized function that can be derived according to rules of calculus. The gradient is useful for maximizing a criterion, but how one can do minimization with the steepest ascent? Easily, by moving in the opposite direction, that is, taking minus gradient.

Fig. 4.19 The importance of properly choosing the step in the steepest descent process: if the leap is too big, the new state may be worse than the old one

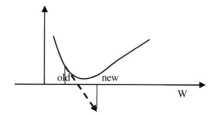

Assume, we have some estimates of matrices W and V as well as their gradients, that is, matrices gW and gV, whose components express the steepest ascent direction of changes in W and V. Then, according to the method of steepest descent, the matrices V and W should be moved in the direction of $-gW$ and $-gV$ with the control of the length of the step by a factor referred to as the learning rate. The equations expressing the move from the old state to the new one are as follows:

$$V(new) = V(old) - \mu * gV, \quad W(new) = W(old) - \mu * gW \quad (4.19)$$

where μ is the learning rate (step size). The importance of properly choosing the step size is illustrated on Fig. 4.19.

The gradient of the criterion of squared error is defined by: (a) matrices W and V, (b) error value itself, and (c) input feature values. This is why it is convenient to apply this approach when entities come in a sequence so that each individual entity gives an estimate of the gradient and, accordingly, the move to a new state of matrices W and V according to Equation (4.19). The sequence of entities is natural when the learning is done on the fly by processing entities in the order of their arrival. In the situations when all the entities have been already collected in a data set, as the Iris data set, the sequence is organized artificially in a random order. Moreover, as the number of entities is typically rather small (as it is in the case of just 150 Iris specimens) and the gradient process is rather slow, it is usually not enough to process all the entities just once. The processing of all the entities in a random order constitutes *an epoch*. A number of epochs need to be executed until the matrices V and W are stabilized.

Worked example 4.4. Learning Iris petal sizes

Consider, at any Iris specimen, its two sepal sizes as the input and its two petal sizes as the output. We are going to find a decision rule relating them in the format of a one-hidden-layer NN.

At the Iris data, the architecture presented on Fig. 4.18 and program nnn.m implementing the error back propagation algorithm leads to the average errors at each of the output variables presented in Table 4.19 at different numbers of hidden neurons h. Note that the errors are given relative to feature ranges.

The number of elements in matrices V and W here are five-fold of the number of hidden neurons, thus ranging from 15 at the current setting of three hidden neurons

4.6 Learning Correlation with Neural Networks

Table 4.19 Relative error values in the predicted petal dimensions with full Iris data after 5,000 epochs

Number of hidden neurons	Relative error, per cent	
	Petal length	Petal width
3	5.36	8.84
6	4.99	8.40
10	4.98	8.15

to 50 when this grows to 10. One can see that the increase in the numbers of hidden neurons does bring some improvement, but not that great – probably not worth doing.

Here are a few suggestions for further work on this example:

1. Find values of E for the errors reported in Table above.
2. Take a look at what happens if the data are not normalized.
3. Take a look at what happens if the learning rate is increased, or decreased, ten times.
4. Extend the table above for different numbers of hidden neurons.
5. Try petal sizes as input with sepal sizes as output.
6. Try predicting only one size over all input variables.

Worked example 4.5. Predicting marks at Student dataset

Let us embark on an ambitious task of predicting students mark at the Students data – we partially dealt with this in Section 4.4. The nnn.m program leads to the average errors in predicting student marks over three subjects, as presented Table 4.20 at different numbers of hidden neurons h. Surprisingly, the prediction works rather well: the errors are on the level of 3 points only, more or less independently on the number of hidden neurons utilized.

F4.6.2.1 Fitting Neural Networks and Gradient Optimization: Formulation

Steepest Descent for the Square Error Criterion with Linear Rules

In machine learning, the assumption is that the decision rule is learnt incrementally by using entities one by one. That is, the global solutions involving the entire sample

Table 4.20 Average absolute error values in the predicted student marks over all three subjects, with full Student data after 5,000 epochs

H	$|e1|$	$|e2|$	$|e3|$	# param.
3	2.65	3.16	3.17	27
6	2.29	3.03	2.75	54
10	2.17	3.00	2.64	90

are not applicable. In such a situation an optimization algorithm that processes entities one by one should be applied. The most popular is the gradient method, also referred to as the steepest descent.

This method relies on the gradient of the function to be optimized. If we are to minimize function $f(x)$ over x spanning a subspace D of the n-dimensional vector space R^n, we can utilize its gradient gf for this purpose. The gradient gf at $x \in D$ is a vector consisting of the f's partial derivatives over all components of x, under the assumption that a full derivative, geometrically corresponding to the tangential hyperplane, does exist. This vector shows the direction of the steepest ascent of $f(x)$, so that its opposite vector $-gf$ shows the opposite direction which is considered as that of the steepest descent of $f(x)$. The method of steepest descent produces a sequence of points $x(0), x(1), x(2), \ldots$ starting from an arbitrary $x(0)$ by using recursive equation

$$X(t+1) = x(t) - \mu_t * gf(x(t)) \qquad (4.19')$$

where parameter μ_t denotes the length of the step to go from $x(t)$ in the direction of the steepest descent, referred to as the learning rate in machine learning. The sequence $x(t)$ is guaranteed to converge to the minimum point at a constant $\mu_t = \mu$ if $f(x)$ is strictly convex, so that there is a sphere of a finite radius such that $f(x)$ is always greater than its lower part, as shown on the right of Fig. 4.20 (see B. Polyak 1987).

The process converges if $f(x)$ is a convex function and μ_t tends to 0 when t grows to infinity, but not too fast so that the sum of the series $\Sigma_t \mu_t$ is infinity. This guarantees that the moves from $x(t)$ to $x(t+1)$ are small enough to not over-jump the point of minimum but not that small to stop the sequence short of reaching the optimum by themselves.

If $f(x)$ is not convex however, the sequence reaches just one of the local optima depending on the starting point $x(0)$ (see Fig. 4.21). Luckily, the square error in

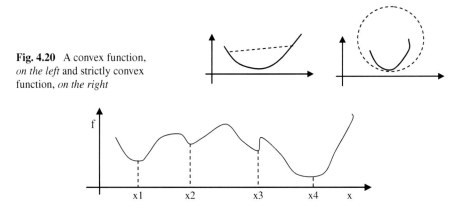

Fig. 4.20 A convex function, *on the left* and strictly convex function, *on the right*

Fig. 4.21 Points x1 to x4 are points of local minimum for the function whose graph is drawn with the *line*. The global minimum is only one of them, x4

4.6 Learning Correlation with Neural Networks

the problem of linear discriminant analysis is strictly convex so that the steepest descent sequence converges to the optimum from any initial point. This gives rise to the algorithm described in the following section.

Learning Wiring Weights with Error Back Propagation

The problem of learning a neuron network is to find weight matrices W and V minimizing the squared difference between u observed and $\hat{u}$ computed according to (4.20):

$$E = d(u, \hat{u}) = <u - th(x*W)*V, \ u - th(x*W)*V> /2 \qquad (4.20)$$

over the training entity set. The division by 2 is made to avoid factor 2 in the derivatives of E that has been already encountered in Section 4.4.2.2 (see Haykin 1999).

Specifically, with just two outputs on Fig. 4.18, the error function is

$$E = [(u_1 - \hat{u}_1)^2 + (u_2 - \hat{u}_2)^2]/2 \qquad (4.20')$$

where $u_1 - \hat{u}_1$ and $u_2 - \hat{u}_2$ are differences between the actual and predicted values of the two outputs.

Steepest descent Equation (4.19) for learning V and W can be written componentwise:

$$v_{hk}(t+1) = v_{hk}(t) - \mu * \partial E / \partial v_{kh}, \ w_{ih}(t+1) = w_{ih}(t) - \mu * \partial E / \partial w_{ih} \ (i \in I, \ h \in II, \ k \in III) \quad (4.21)$$

To make these computable, let us express the derivatives explicitly; first those at the output, over v_{hk}:

$$\partial E / \partial v_{hk} = -(u_k - \hat{u}_k) * \partial \hat{u}_k / \partial v_{hk}$$

To advance, notice that $\partial \hat{u}_k / \partial v_{hk} = th(zh)$, since $\hat{u}_k = \Sigma_j th(zh) * v_{hk}$. Putting this into equation above makes

$$\partial E / \partial v_{hk} = -(u_k - \hat{u}_k) * th(zh) . \qquad (4.22)$$

Regarding the second layer, of W, let us find the derivative $\partial E / \partial w_{ih}$ which requires more chain based derivations. Specifically,

$$\partial E / \partial w_{ij} = \Sigma_k [-(u_k - \hat{u}_k) * \partial \hat{u}_k / \partial w_{ij}].$$

Since $\hat{u}_k = \Sigma_j th(\Sigma_i x_i * w_{ij}) * v_{jk}$, this can be expressed as

$$\partial \hat{u}_k / \partial w_{ij} = v_{jk} * th'(\Sigma_i x_i * w_{ij}) * x_i.$$

The derivative $th'(z)$ can be expressed through $th(z)$ as explained in Q.4.14 later, which leads to the following final expression for the partial derivatives:

$$\partial E/\partial w_{ij} = -\Sigma_k[(u_k - \hat{u}_k)*v_{jk}]*(1 + th(z_j))(1 - th(z_j)) *x_i/2 \qquad (4.23)$$

Equations (4.19), (4.22) and (4.23) lead to the following rule for processing an entity, or instance, in the error back-propagation algorithm as applied to neural network on Fig. 4.18.

1. *Forward computation (of the output $\hat{u}$ and error)*. Given matrices V and W, upon receiving an instance (x, u), the estimate $\hat{u}$ of vector u is computed according to the neural network as formalized in Equation (3.18), and the error $e = u - \hat{u}$ is calculated.
2. *Error back-propagation (for estimation of the gradient elements)*. Each neuron receives the relevant error estimate, which is

   $-e_k = -(u_k - \hat{u}_k)$ for (4.22) for output neurons $k(k = III1, III2)$ or
   $-\Sigma_k[(u_k - \hat{u}_k)*v_{hk}]$, for (4.23) for hidden neurons $h(j = II1, II2, II3)$

   [the latter can be seen as the sum of errors arriving from the output neurons according to the corresponding wiring weights].

   These are used to adjust the derivatives (4.22) and (4.23) by multiplying them with local data depending on the input signal, which is $th(z_h)$, for neuron k's source h in (4.22), and $th'(z_h)x_i$ for neuron h's source i in (4.23).
3. *Weights update*. Matrices V and W are updated according to formula (4.19).

What is nice in this procedure is that the computation can be done locally, so that every neuron processes only the data that are available to this neuron, first from the input layer, then backwards, from the output layer. In particular, the algorithm does not change if the number of hidden neurons is changed from $h = 3$ on Fig. 4.18, to any other integer $h = 1, 2, \ldots$; nor it changes if the number of inputs and/or outputs changed.

C4.6.2.2 Error Back Propagation: Computation

For a data set available as a whole, "offline", due to the specifics of the binary target variables and activation functions, such as *th(x)* and *sign(x)*, which have −1 and 1 as their boundaries, the data in the NN context are frequently pre-processed to make every feature's range to lie between −1 and 1 and the midrange to be 0. This can be done by using the conventional shifting and rescaling formula for each feature v, $y_{iv} = (x_{iv} - a_v)/b_v$, at which b_v is equal to the half-range, $b_v = (M_v - m_v)/2$, and shift coefficient a_v, to the mid-range, $a_v = (M_v + m_v)/2$. Here M_v denotes the maximum and m_v the minimum of feature v.

The practice of digital computation, with a limited number of digits used for representation of reals, shows that it is a good idea to further expand the ranges into a [−10, 10] interval by multiplying afterwards all the entries by 10: in this range, digital numbers stored in computer arguably lead to smaller computation errors than in the range [−1, 1] if they are closer to 0.

The implementation of the method of gradient descent for learning neural networks cannot be straightforward because the minimized squared error depends both

4.6 Learning Correlation with Neural Networks

on the wiring weight matrices V and W and input/output pairs (x,u), yet there is no way to freely change the latter – the process is bound by the set of observations. This is why the observed pairs (x_i, u_i), the instances, are used as triggers to the steepest descent changes in matrices V and W. Specifically, given V and W, the instances are put one by one, in a random order, to see what are the discrepancies between the observed u and computed $\hat{u}$. When all of the instances have been entered, their order is randomly changed and they are ready to be put all over again – this is referred to as a new "epoch". The matrices V and W are changed either at each (x_i, u_i) instance, using the errors $\hat{u}-u$ locally, or after an epoch, using the accumulated errors.

The error back propagation algorithm, with the local changes of matrices V and W, can be formulated as follows.

A. Initialize weight matrices $W = (w_{ih})$ and $V = (v_{hk})$ by using random normal distribution N(0, 1) with the mean at 0 and the variance 1.
B. Standardize data to [–10, 10] ranges and 0 averages as described above.
C. Formulate halting criterion as explained below and run a loop over epochs.
D. Randomize the order of entities within an epoch and run a loop of the error back-propagation *instance processing procedure*, below, in that order.
E. If Halt-criterion is met, end the computation and output results: W, V, $\hat{u}$, e, and E. Otherwise, execute D again.

The best halting criterion, according to the nature of the steepest descent process should be at

(i) Matrices V and W stabilized. Unfortunately, in real world computations this criterion requires by far too many iterations, so that in practice the matrices fail to converge. Thus, other stopping criteria are used.
(ii) The difference between the average values (over iterations within an epoch) of the error function becomes smaller than a pre-specified threshold, such as 0.0001.
(iii) The number of epochs performed reaches a pre-specified threshold such as 5,000.

Instance Processing Procedure

Specifics of the NN structure and function provide for simple and effective rules for processing individual entities in the procedure of the steepest descent. Before updating the wiring weights according to Equation (4.19), two following steps are executed:

1. *Forward computation of the estimated output and its error.* Upon receiving a training instance input feature values, they are processed by the neuron network to produce an estimate of the output, after which the error is computed as the difference between real and estimated output values.

2. *Error back-propagation for estimation of the gradient.* The computed error of the output is back-propagated through the network. Each neuron of the output layer corresponds to a specific output feature and, thus, receives the error in this feature. Each neuron of the hidden layer receives a combined error signal from all output neurons weighted by the corresponding wiring weights. These are used to adjust the gradient elements by using the hidden neuron activation function as described in section "Learning Wiring Weights with Error Back Propagation".

In the Appendix A4, a Matlab code nnn.m is presented for learning NN weights with the error back propagation algorithm according to the NN of Fig. 4.18. Two parameters of the algorithm, the number of neurons in the hidden layer and the learning rate, are its input parameters. The output is the minimum level of error achieved and the corresponding weight matrices *V* and *W*.

The code includes the following steps:

1. *Loading data.* It is assumed that all data are in subfolder Data. According to the task, this can be either iris.dat or stud.dat or any other dataset.
2. *Normalizing data.* This is done by shifting each column to its midrange with the follow-up dividing it by the half-range, after which all data set is multiplied by 10, to have them in [−10, 10] scale as described above.
3. *Preparing input and output* training sub-matrices. This is done after the decision has been made of what features fall in the former and what features fall in the latter categories. In the case of Iris data, for example, the target is petal data (features w3 and w4) and input is sepal measurements (features w1 and w2) as described. In the case of Students data, the target can be students' marks on all three subjects (CI, SP and OOP), whereas the other variables (occupation categories, age and number of children), input.
4. *Initializing the network.* This is done by: (a) specifying the number of hidden neurons *H*, (b) filling in matrices *W* and *V* with random (0,1) normally distributed values, and (c) setting a loop over epochs with the counter initialized at zero.
5. *Organizing a loop over the entities.* For setting a random order of entities to be processed, the Matlab command randperm(n) for making a random permutation of integers 1, 2,..., *n* can be used.
6. *Forward pass.* Given an entity, the output is calculated, as well as the error, using the current *V, W* and activation functions. The program uses the symmetric sigmoid (4.17) as the activation function of hidden neurons.
7. *Error back-propagation.* Gradient matrices for *V* and *W* according to formulas (3.22) and (3.23) are computed.
8. *Weights V and W update.* Having the gradients computed and learning rate accepted as the input, updated *W* and *V* are computed according to (4.19).
9. *Halt-condition.* This includes both the level of precision, say 0.01, and a threshold to the number of epochs, say, 5,000. If either is reached, the program halts.

Q.4.14. Prove that the derivatives of sigmoid (4.16) or hyperbolic tangent (4.17) functions appear to be simple polynomials of selves. Specifically,

$$s'(x) = \left(\left(1 + e^{-x}\right)^{-1}\right)' = (-1)\left(1 + e^{-x}\right)^{-2}(-1)e^{-x} = s(x)(1 - s(x)), \quad (4.24)$$

$$th'(x) = [2*s(x) - 1]' = 2*s(x)' = 2*s(x)*(1 - s(x)) = (1 + th(x))*(1 - th(x))/2 \quad (4.24')$$

Q.4.15. Find a way to improve the convergence of the process, for instance, with adaptive changes in the step size values.

Q.4.16. Use k-fold cross validation to provide estimates of variation of the results regarding the data change.

Q.4.17. Develop a scoring function for learning a category by using the contribution of the partition to be built to the category.

4.7 Summary

The goal of this chapter is to present a variety of techniques for learning correlation from data. Most popular concepts – Bayes classifiers, decision trees, neuron networks and support vector machine – are presented along with more generic linear regression and discrimination. Some of these are accompanied with concepts that are interesting on their own such as the bag-of-words model or kernel. The description, though, is rather fragmentary, except perhaps the classification trees for which a number of theoretical results is invoked to show their firm relations to bivariate analysis, first, summary Quetelet indexes in contingency tables and, second, normalization options for dummy variables representing target categories.

Overall, the chapter contents reflect the current state of the art on the subject of learning correlations from data. Perhaps the subject is too big and major advances are a matter of future rather than the past.

References

Breiman, L., Friedman, J.H., Olshen, R.A., Stone, C.J.: Classification and Regression Trees. Wadswarth, Belmont, CA (1984)

Bring, J.: How to standardize regression coefficients. Am. Stat. **48**(3), 209–213 (1994)

Cybenko, G.: Approximation by superposition of sigmoidal functions. Math. Control Signals Systems **2**(4), 303–314 (1989)

Duda, R.O., Hart, P.E., Stork D.G.: Pattern Classification. Wiley-Interscience, New York, NY, ISBN 0-471-05669-3 (2001)

Esposito, F., Malerba, D., Semeraro, G.: A comparative analysis of methods for pruning decision trees. IEEE Trans. Pattern Anal.Mach. Intell. **19**(5), 476–491 (1997)

Fawcett, T.: An introduction to ROC analysis. Pattern Recognition Letters **27**, 861–874 (2006)

Fisher, R.A.: The use of multiple measurements in taxonomic problems. Annu. Eugen. **7**, Part II, 179–188 (1936); also in "Contributions to Mathematical Statistics" (Wiley, New York, NY, 1950)

Fisher, D.H.: Knowledge acquisition via incremental conceptual clustering. Machine Learning, **2**, 139–172 (1987)

Green, S.B., Salkind, N.J.: Using SPSS for the Windows and Mackintosh: Analyzing and Understanding Data. Prentice Hall, Upper Saddle River, NJ (2003)

Groenen, P.J.F., Nalbantov, G.I., Bioch, J.C.: SVM-Maj: A majorization approach to linear support vector machines with different hinge errors. Adv. Data Anal. Classification **2**(1), 17–43 (2008)

Grünwald, P.D.: The Minimum Description Length Principle, MIT Press (2007)

Haykin, S. S.: Neural Networks, 2nd edn. Prentice Hall, ISBN 0132733501 (1999)

Loh, W.-Y., Shih, Y.-S.: Split selection methods for classification trees. Stat. Sin. **7**, 815–840 (1997)

Manning, C.D., Raghavan, P., Schütze, H.: Introduction to Information Retrieval. Cambridge University Press, Cambridge, England (2008)

Mirkin, B.: Methods for Grouping in Socioeconomic Research. Finansy i Statistika Publishers, Moscow (in Russian) (1985)

Mirkin, B.: Mathematical Classification and Clustering. Kluwer Academic Press, Dordrecht/Boston (1996)

Mirkin, B.: Clustering for Data Mining: A Data Recovery Approach. Chapman & Hall/CRC, London, ISBN 1-58488-534-3 (2005)

Mitchell, T.M.: Machine Learning. McGraw Hill, New York, NY (2010)

Polyak, B.: Introduction to Optimization. Optimization Software, Los Angeles, CA, ISBN: 0911575146 (1987)

Quinlan, J.R.: C4.5: Programs for Machine Learning. Morgan Kaufmann, San Mateo, CA (1993)

Schölkopf, B., A.J. Smola, A.J.: (2005) Learning with Kernels. The MIT Press, Cambridge, MA (2005)

Vapnik, V.: Estimation of Dependences Based on Empirical Data, 2d edn. Springer Science + Business Media Inc. (2006)

Chapter 5
Principal Component Analysis and SVD

5.1 Decoder Based Data Summarization

5.1.1 Structure of a Summarization Problem with Decoder

Summarization as a concept covers many activities from data compression to labeling a dataset with a phrase like "Archeology finds indicate no King David Palace at the time of King David". Principal component analysis lies somewhere between these two to summarize the observed features in somewhat sharper structures. In contrast to a correlation problem, the features are not divided here into those belonging to input or output of the phenomenon under consideration. It is a different situation. One may think of this as that all features available are target features so that those to be constructed as a summary are in fact "hidden input features".

In this way, the structure of a summarization problem may be likened to that of a correlation problem if a rule is provided to predict all the original features from the summary. That is, the original data in the summarization problem act as target data in the correlation problem. That implies that there should be two rules involved in a summarization problem: one for building the summary, the other to provide a feedback from the summary to the observed data.

Unlike in the correlation problem, though, here the feedback rule must be pre-specified so that the focus is on building a summarization rule rather than on using the summary for prediction; this is why we refer to the feedback rule as a "decoder" rather than a "predictor". In the machine learning literature, the issue of data summarization has not been given yet that attention it deserves; this is why the problem is usually considered somewhat simplistically as just deriving a summary from data without any feedback (see Fig. 5.1c).

A proper consideration of the structure of a summarization problem should rely on the existence of a decoder to provide the feedback from a summary back to the data and make the summarization process more or less similar to that of the correlation process (see Fig. 5.1a versus 5.1b). More exactly, a decoder is a device that translates the summary representation encoded in the chosen summarization rule back into the original data format. This allows us to utilize the same criterion of minimization of the difference between the original data and those output by

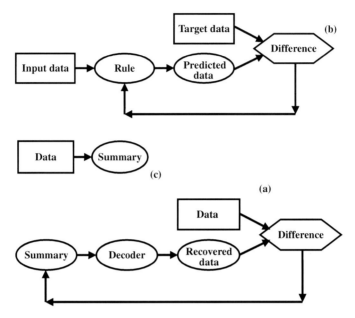

Fig. 5.1 A diagram for coder/decoder data summarization (**a**) versus learning input-target correlation (**b**) or summarization with no decoder (**c**). *Rectangles* are for observed data, *ovals* for computational constructions, *hexagons* for feedback comparisons

the decoder: the less the difference, the better. This text is largely concerned with methods that can be expressed in terms of decoder based criteria as presented on Fig. 5.1a.

P5.1.2 Data Recovery Criterion: Presentation

The data recovery approach in data summarization is based on the assumption that there is a regular structure in the phenomenon of which the observed dataset informs. This regular structure A is the summary to be found. When A is determined, this can feed back to the observed data Y in the format of the decoded data $F(A)$ that should coincide with Y up to residuals, that are due to possible flaws in any or all of the following three aspects:

(a) bias in entity sampling,
(b) selecting and measuring features,
(c) adequacy of the set of admissible A structures to the phenomenon in question.

Each of these three can drastically affect results. However, so far only the simplest of the aspects, (a) sampling bias, has been addressed scientifically, in statistics – as a random bias, due to the probabilistic nature of the data. The other two are subjects of much effort in specific domains but not in the general computational data analysis framework as yet. Rather than focusing on accounting for the causes of errors, let us consider the underlying equation in which the errors are looked at as a whole:

$$Observed\_Data\ Y = Model\_Data\ F(A) + Residuals\ E \tag{5.1}$$

This equation brings in an inherent data recovery criterion for the assessment of the quality of the model A in recovering data Y – according to the level of residuals E: the smaller the residuals, the better the model. Since a data model typically involves unknown parameters, this naturally leads to the idea of fitting these parameters to the data in such a way that the residuals become as small as possible.

In many cases this principle can be rather easily implemented as the least squares principle because of an extension of the Pythagoras theorem relating the square lengths of the hypotenuse and two other sides in a right-angle triangle connecting "points" Y, $F(A)$ and 0 (see Fig. 5.2). The least squares criterion requires fitting the model A by minimizing the sum of the squared residuals. Geometrically, it often means an orthogonal projection of the data set considered as a multidimensional point onto the space of all possible models represented by the x axis on Fig. 5.2. In such a case the dataset (pentagram), its projection (rectangle) and the origin (0) form a right-angle triangle for which a multidimensional extension of the Pythagoras' theorem holds. The theorem states that the squared length of the hypotenuse is equal to the sum of squares of two other sides. The squared hypotenuse translates into the data scatter, that is, the sum of all the data entries squared, being decomposed in two parts, the part explained by the summary model A, that is, the contribution of the line between 0 and rectangle, and the part left unexplained by A. The latter part is the contribution of the residuals E expressed as the sum of squared residuals, which is exactly the least squares criterion. This very decomposition was employed in the problems of linear and non-linear regression in Sections 3.2 and 4.3, classification trees in Section 4.5, and it will be used again in further described methods: Principal component analysis and K-Means clustering, as well as additive clustering.

When the data can be considered as a random sample from a multivariate Gaussian distribution, the least squares principle can be derived, under some simplifying assumptions, from a major statistical principle, that of maximum likelihood. In the data analysis framework, the data do not necessarily come from a probabilistic

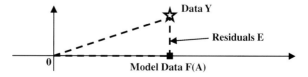

Fig. 5.2 Geometric relation between the observed data (*pentagram*), the fitted model data (*black rectangle*), and the residuals (*connecting line*)

population. Still, the least squares framework frequently provides for solutions that are both practically relevant and theoretically sound. The least squares will be the only criterion utilized in this text.

F5.1.3 Data Recovery Criterion: Formulation

A decoder based summarization problem can be stated as follows. Given N vectors forming a matrix $Y = \{(y_i)\}$ with rows $y_i = (y_{i1}, \ldots, y_{iV})$ of V features observed at entities $i = 1, 2, \ldots, N$ and a set of admissible summary structures A with decoder $D: A \Rightarrow R^p$, build a summary

$$A = F(Y), \ A \in \mathcal{A}$$

such that the error, which is the difference between the decoded data $D(A)$ and observed data Y, is minimal over the class of admissible rules F. More explicitly, one assumes that

$$Y = D(A) + E \qquad (5.2)$$

where E is matrix of residual values, or errors: the smaller the errors, the better the summarization A. According to the least-squares approach, the errors are minimized by minimizing the summary, or average, squared error:

$$E^2 = <Y - D(A), \ Y - D(A)> = <Y - D(F(Y)), \ Y - D(F(Y))> \qquad (5.3)$$

with respect to all admissible summarization rules F.

Expression (5.3) can be further decomposed into

$$E^2 = <Y, Y> - 2<Y, D(A)> + <D(A), D(A)>.$$

In many data summarization methods, such as the Principal component analysis and K-Means clustering described later in Sections 5.2 and 6.2, the set of all possible decodings $D(F(Y))$ forms a linear subspace. In this case, the data matrices Y and $D(A)$, considered as multidimensional points, form a "right-angle triangle" around the origin 0, as presented on Fig. 5.2 above. In such a case $<Y, D(A)> = <D(A), D(A)>$ and the square error (5.3) becomes part of a multivariate analogue to the Pythagoras equation relating the squares of the "hypotenuse", Y, and the "sides", $D(A)$ and E:

$$<Y, Y> = <D(A), D(A)> + E^2 \qquad (5.4)$$

or on the level of matrix entries,

5.1 Decoder Based Data Summarization

$$\sum_{i \in I} \sum_{v \in V} y_{iv}^2 = \sum_{i \in I} \sum_{v \in V} d_{iv}^2 + \sum_{i \in I} \sum_{v \in V} e_{iv}^2 \quad (5.4')$$

The data is an $N \times V$ matrix $Y = (y_{iv})$ that can be considered as either set of rows/entities y_i ($i = 1, \ldots, N$) or set of columns/features y_v ($v = 1, \ldots, V$) or both. The item on the left in (5.4') is usually referred to as the data scatter and denoted by $T(Y)$,

$$T(Y) = \sum_{i \in I} \sum_{v \in V} y_{iv}^2 \quad (5.5)$$

Why is this termed "scatter"? Indeed, $T(Y)$ is the sum of Euclidean squared distances from 0 to all entities, thus a measure of scattering them around 0. In fact, $T(Y)$ has a dual interpretation. On the one hand, $T(Y)$ is the sum of row-based entity contributions, the squared distances $d(y_i, 0)$ ($i = 1, \ldots, N$). On the other hand, $T(Y)$ is the sum of column-based feature contributions $t_v = \Sigma_{i \in I} y_{iv}^2$. In the case when the average c_v has been subtracted from all values of the column v, the summary contribution t_v is N times the variance, $t_v = N\sigma_v^2$.

Q.5.1. Prove that the summary contribution t_v is N times the variance, $t_v = N\sigma_v^2$ if feature v is centered. **A.** Indeed, $t_v = \Sigma_{i \in I} y_{iv}^2 = \Sigma_{i \in I}(y_{iv} - c_v)^2 = N[\Sigma_{i \in I}(y_{iv} - c_v)^2/N] = N\sigma_v^2$, where $c_v = 0$ is the mean of feature v.

5.1.4 Data Standardization

The least-squares solutions highly depend on the feature scales and may be highly affected by the scale changes, as decomposition (5.4') clearly demonstrates. This is not exactly the case in correlation problems, at least in those with only one target feature, because the least squares there are, in fact, just that feature's errors, thus all expressed in the same scale. The data standardization problem, which is rather marginal at learning correlations, is of a great importance in data summarization. The problem of data standardization can be reformulated as the issue of defining the relative relevance, or importance, among the features. The greater the range of v, the greater the contribution t_v, thus the greater the relevance of v. There can be no universal answer to the issue of feature importance, because the answer always depends on the goal of summarization.

The assumption of equal importance of features currently underlies all the efforts and makes the entire edifice of data analysis somewhat crippled – but there is nothing new in this. As the history of science clearly demonstrates, any breakthrough in the sciences starts with a rather shaky base.

To balance contributions of features to the data scatter, one conventionally applies the operation of data standardization comprising two transformations, shift of the origin and rescaling.

We have already encountered standardization while studying multivariate classifiers, decision trees and neural networks. In neural networks as well as in

Support Vector Machine, the standardization involves the scale shift to the midrange and rescaling by normalizing the feature values by the half-range. These parameters are distribution independent.

Another, much more popular, choice is the feature's mean for the scale shift and normalizing by the standard deviation for rescaling. This standardization is a cornerstone in mathematical statistics and it works very well if observations come from a Gaussian distribution, because the distribution becomes parameter-free if standardized by subtracting the mean followed by dividing over the standard deviation. In statistics, this transformation is frequently referred to as z-scoring. In the context of data analysis, though, distributions are rarely Gaussian and rarely of any popular family at all; moreover, observations are not necessarily random or independent. In these circumstances, the choice of shifting and rescaling needs a rethink.

First of all, there is no need in linking the two operations together: shifting the origin has nothing to do with balancing feature weights. The goal of the shifting is to position the data against a backdrop of a "norm" which is put to the origin by the shift. In this way, the analysis involves the differences of the data and the norm. The experimental evidence accumulated in the ever growing body of data analysis research suggests that it is much easier to find meaningful structures if the "normal background" has been removed from data. According to the least squares criterion, it is the mean that approximates the overall "norm" the best. Since this criterion underlies all the methods considered in this text, the mean – sometimes referred to as grand mean, to point out its position over the entire entity set – will be the choice for the origin.

The normalization seems to be better if done by half-range or, equivalently, the range, indeed. On the first glance, there is no advantage in normalization by the range. Z-scoring seems a better choice, especially since z-scoring satisfies the intuitively appealing equality principle – all features contribute to the data scatter equally after they have been divided by their standard deviations.

This view is, however, overly simplistic. In fact, the feature's contribution to the data scatter is affected by two unrelated factors: (a) the feature scale range and (b) the distribution shape. While reducing the effect of the former, normalization should not suppress the effect of the latter because the distribution shape is an important indicator of the data structure. But the standard deviation involves both and thus mixes them up. Take a look, for example, at distributions of two features presented on Fig. 5.3. One of them has one mode only (a), whereas the other has two modes (b). Since the features have the same range, the standard deviation is greater for the distribution (b), which means that its relative contribution to the data scatter decreases under z-scoring standardization. This means that its clear cut discrimination between two parts of the distribution will be stretched in while the unimodal structure, which is hiding the two-part structure, will be stretched out. This is not exactly what we want of data standardization. Data standardization should help in revealing the data structure rather than concealing it. Thus, normalization by the range helps in bringing forward multimodal features by assigning them relatively larger weights proportional to their variances.

5.1 Decoder Based Data Summarization

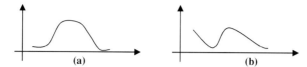

Fig. 5.3 One-modal distribution shape on (**a**) versus a two-modal distribution shape on (**b**): the standard deviation of the latter is greater, thus making it less significant under the z-scoring standardization

Therefore, in contrast to conventional wisdom, z-scoring standardization should be avoided unless there is a strong indication that the data come from a Gaussian distribution indeed. Any index related to the scale range can be used for normalization. In this text, the range is universally accepted. If, however, there is a strong indication that the range may be subject to outlier effects and, thus, unstable and random, more stable indexes could be used for normalization such as, for example, the distance between upper and lower 1% quantiles.

Worked example 5.1. Standardizing Iris dataset

Consider Iris dataset in Table 1.3. Its grand mean and midrange are presented in Table 5.1, along with its range and standard deviations.

These have been found by using the following MatLab commands:

```
≫ iris=load('Data\iris.dat');
≫ m=mean(iris); % grand mean
≫ ma=max(iris);% maximum
≫ ma=min(iris);% minimum
≫ mr=(ma+mi)/2; % midrange
≫ s−std(iris);% standard deviation
≫ ra=ma-mi;% range
```

In Table 5.1, midrange more or less follows the grand mean, but there are some discrepancies between the range and standard deviation. For example, ranges of w2 and w4 are the same, whereas standard deviations differ by almost 100%.

Table 5.1 Characteristics of Iris dataset

Characteristics	Features			
	w1	w2	w3	w4
Mean m	5.84	3.06	3.76	1.20
Midrange mr	6.10	3.20	3.95	1.30
Standard deviation s	0.83	0.44	1.77	0.76
Range ra	3.60	2.40	5.90	2.40

Let us take three different standardizations:

A – range related, $y \Leftarrow (x-mr)/ra$;
B – mean/range standardization, $y \Leftarrow (x-m)/ra$;
C – z-scoring, $y \Leftarrow (x-m)/s$;

and evaluate feature contributions to the data scatter after each of them (Table 5.2):

Feature contributions under A and B are similar, because both involve division by the range. According to these standardizations features w3 and w4 contribute most, because they are bimodal (see Q.2.24 and Fig. 2.19) and, thus play important role in further summarization methods, both Principal component analysis and cluster analysis. This concurs with the botanists' view that it is these sizes that determine the belongingness of an Iris specimen to a specific taxon (see references in Mirkin 2005). Moreover, at building a classification tree over Iris dataset, it was feature w4 that was involved in the splits according to three goodness criteria (see Fig. 4.11 in Section 4.5). In contrast, the first line assigns contributions according to feature values so that the lengths w1 and w3 get much larger contributions than the widths w2 and w4. And z-scoring (standardization C) makes all features contribute similarly, even in spite of the fact that two of them are bimodal.

The problem of standardization can be addressed by the user if they know the type of the distribution behind the observed data – the parameters of the distribution typically lead to a reasonable standardization. For example, the data should be standardized by z-scoring if the data is generated by independent one-dimensional Gaussian distributions. According to the formula for Gaussian density, a z-scored feature column would then fit the conventional $N(0,1)$ distribution making all features comparable to each other A similar strategy applies if the data is generated from a multivariate Gaussian density, just the data first needs to be transformed into mutually orthogonal singular vectors or, equivalently, principal components.

If no reasonable distribution can be assumed in the data, then there is no universal advice on standardization. However, with the summarization problems that we are going to address, the principal component analysis and clustering, some advice can be given in terms of the data scatter.

Table 5.2 Iris feature contributions to data scatter at different standardizations, per cent to the data scatter value

Standardization	Features			
	w1	w2	w3	w4
No standardization	54.76	15.00	27.07	3.17
A: midrange/range	20.61	12.70	31.48	35.66
B: mean/range	19.15	11.94	32.40	36.51
C: mean/std	25.00	25.00	25.00	25.00

5.1 Decoder Based Data Summarization

The data transformation effected by the standardization can be expressed as

$$y_{iv} = (x_{iv} - a_v)/b_v \tag{5.6}$$

where $X = (x_{iv})$ stands for the original and $Y = (y_{iv})$ for standardized data, whereas $i \in I$ denotes an entity and $v \in V$ a feature. Parameter a_v stands for the shift of the origin and b_v for normalizing factor at each feature $v \in V$. In other words, one may say that the transformation (5.6), first, shifts the data origin into the point $a = (a_v)$, after which each feature v is rescaled separately by dividing its values over b_v.

The position of the space's origin, zero point $0 = (0, 0, \ldots, .0)$, at the standardized data Y is unique because any linear transformation of the data, that is, matrix product AY, for any A, can be expressed as a set of rotations of the coordinate axes around the origin, so that the origin itself is invariant. The principal component analysis can be expressed mathematically as a set of linear transformations of the data features as becomes clear in Section 5.2, which means that all the action in this method occurs around the origin. Metaphorically, the origin can be likened to the eye through which data points are looked at by the methods below. Therefore, for the purposes of data analysis, the origin should be put somewhere in the center of the data set, for which the gravity center, the point of all within-feature averages, is a best candidate. What is nice about it is that the feature contributions to the scatter of the center-of-gravity standardized data (5.5) above are equal to $t_v = \Sigma_{i \in I} y_{iv}^2 (v \in V)$, which means that they are proportional to the feature variances. Indeed, after the average c_v has been subtracted from all values of the column v, the summary contribution satisfies equation $t_v = N\sigma_v^2$ so that t_v is N times the variance. Even nicer properties of the gravity center as the origin have been derived in the framework of the simultaneous analysis of the categorical and quantitative data, see in Sections 4.5 and 6.3.

As to the normalizing coefficients, b_v, their choice is underlied by the idea of balancing the features weights. A most straightforward expression of the principle of feature equal importance is the use of the standard deviations as the normalizing coefficients, $b_v = \sigma_v$. This standardization makes the variances of all the variables $v \in V$ equal to 1 so that all the feature contributions become equal to $t_v = N$, which is seen at Table 5.5.

A very popular way to take into account the relative importance of different features is by using weight coefficients of features in computing the distances. This, in fact, is equivalent to and can be achieved with a proper standardization. Take, for instance, the weighted squared Euclidean distance between arbitrary entities $x = (x_1, x_2, x_M)$ and $y = (y_1, y_2, y_M)$ which is defined as

$$D_w(x, y) = w_1(x_1 - y_1)^2 + w_2(x_2 - y_2)^2 + \ldots + w_M(x_M - y_M)^2$$

where w_v are a pre-specified weights of features $v \in V$. Let us define (additional) normalizing parameters $b_v = 1/\sqrt{w_v}$ $(v \in E)$ to transform x and y into $x'_v = x_v/b_v$ and $y'_v = y_v/b_v$. It is rather obvious that

$$D_w(x,y) = d(x', y')$$

where d is the unweighted Euclidean squared distance.

That is, the following fact holds: for the Euclidean squared distance, the feature weighting is equivalent to an appropriate normalization as described above.

Q.5.2. Is it true that the sum of feature values standardized by subtracting the mean is zero? **A.** Yes, because the sum is proportional to the mean which is zero after centering.

Q.5.3. Consider a reversal of the operations in standardizing data: the scaling to be followed by the scale shift. Is it that different from the conventional standardization? **A.** Denote the scale shift and rescaling factor by a and b. Then the conventional standardization produces $y = (x-a)/b = x/b - a/b$ from x, whereas that suggested gives $z = x/b - a$. These differ at $a \neq 0$. To make them equal, the scale shift in the latter case must be a/b.

C5.1.5 Data Standardization: Computation

For the N×V data set X, its V-dimensional arrays of averages, standard deviations and ranges can be found in MatLab with respective operations
    ≫ av=mean(X);
    ≫ st=std(X,1); % here 1 indicates that divisor at sigmas is N rather than N−1
    ≫ ra=max(X)−min(X);
To properly standardize X, these V-dimensional rows must be converted to the format of N×V matrices, which can be done with the operation repmat(x,m,n) that expands a $p \times q$ array x into an $mp \times nq$ array by replicating it n times horizontally and m times vertically as follows:
   ≫avm=repmat(av, N,1);
   ≫stm=repmat(st, N,1);
   ≫ram=repmat(ra, N,1);

These are N×V arrays, with the same lines in each of them – feature averages in avm, standard deviations in stm, and ranges in ram.

To range-standardize the data, one can use a non-conventional MatLab operation of the entry-wise division of arrays:
   ≫Y=(X−avm)./ram;

Project 5.1. Standardization of mixed scale data and its effect

Pr5.1.A Data table and its quantification

Consider the Company dataset described in Section 1.2 and copied here in Table 5.3. Let us convert it into a quantitative format. The table contains two categorical variables, EC, with categories Yes/No, and Sector, with categories Utility, Industrial

5.1 Decoder Based Data Summarization

Table 5.3 Data of eight companies producing goods A, B, or C related to the initial symbol of company's name

Company	Income	SharP	NSup	EC	Sector
Aversi	19.0	43.7	2	No	Utility
Antyos	29.4	36.0	3	No	Utility
Astonite	23.9	38.0	3	No	Industrial
Bayermart	18.4	27.9	2	Yes	Utility
Breaks	25.7	22.3	3	Yes	Industrial
Bumchist	12.1	16.9	2	Yes	Industrial
Civok	23.9	30.2	4	Yes	Retail
Cyberdam	27.2	58.0	5	Yes	Retail

and Retail. The former feature, EC, in fact represents just one category, "Using E-Commerce" and can be recoded as such by substituting 1 for Yes and 0 for No. The other feature, Sector, has three categories. To be able to treat them in a quantitative way, one should substitute each by a dummy variable.

Specifically, the three category features are:

(i) Utility: Is it Utility sector?
(ii) Industry: Is it Industrial sector?
(iii) Retail: Is it Retail sector?

Each of them admits Yes or No values, respectively substituted by 1 and 0. In this way, the original heterogeneous table will be transformed into a quantitative matrix in Table 5.4.

The first two features, Income and SharP, dominate the data in Table 5.4, especially with regard to the data scatter, that is, the sum of all the data entries squared, equal to 14833. As shown in Table 5.5, the two of them contribute more than 99% to the data scatter. To balance the contributions, features should be rescaled. Another

Table 5.4 Quantitatively recoded Company data table, along with summary characteristics

Company	Income	SharP	NSup	EC	Util	Indu	Reta
Aversi	19.0	43.7	2	0	1	0	0
Antyos	29.4	36.0	3	0	1	0	0
Astonite	23.9	38.0	3	0	0	1	0
Bayermart	18.4	27.9	2	1	1	0	0
Breaks	25.7	22.3	3	1	0	1	0
Bumchist	12.1	16.9	2	1	0	1	0
Civok	23.9	30.2	4	1	0	0	1
Cyberdam	27.2	58.0	5	1	0	0	1
Average	22.45	34.12	3.0	5/8	3/8	3/8	1/4
St deviation	5.26	12.10	1.0	0.48	0.48	0.48	0.43
Midrange	20.75	37.45	3.5	0.5	0.5	0.5	0.5
Range	17.3	41.1	3.0	1.0	1.0	1.0	1.0

Table 5.5 Within-column sums of the entries squared in Table 5.4

Contribution	Income	SharP	NSup	EC	Util	Ind	Retail	Data scatter
Absolute	4253	10487	80	5	3	3	2	14833
Per cent	28.67	70.70	0.54	0.03	0.02	0.02	0.01	100.00

important transformation of the data is the shift of the origin, because it affects the value of the data scatter and the decomposition of it in the explained and unexplained parts, as can be seen on Fig. 5.2.

Pr5.1.B Visualization of the data unnormalized

One can take a look at the effects of different standardization options. Table 5.6 contains data of Table 5.4 standardized by the scale shifting only: in each column, the within-column average has been subtracted from the column entries. Such standardization is referred to as centering.

The relative configuration of the 7-dimensional row-vectors in Table 5.6 can be captured by projecting them onto a plane, which is a two-dimensional, in an optimal way; this is provided by the two first singular values and corresponding singular vectors, as will be explained later in Section 5.2. This visualization is presented on Fig. 5.4 at which different product companies are shown with different shapes: squares (for A), triangles (for B) and circles (for C). As expected, this bears too much on features 2 and 1 that contribute 83.2% and 15.7%, respectively, here; a slight change from the original 70.7% and 23.7% according to Table 5.5. The features seem not related to products at all – the products are randomly intermingled with each other on the picture.

Table 5.6 The data in Table 5.4 standardized by the shift scale only, with the within-column averages subtracted. The values are rounded to the nearest two-digit decimal part, choosing the even number when two are the nearest. The bottom rows represent contributions of the columns to the data scatter as they are and per cent

Ave	−3.45	9.58	−1	−0.62	0.62	−0.38	−0.25
Ant	6.95	1.88	0	−0.62	0.62	−0.38	−0.25
Ast	1.45	3.88	0	−0.62	−0.38	0.62	−0.25
Bay	−4.05	−6.22	−1	0.38	0.62	−0.38	−0.25
Bre	3.25	−11.82	0	0.38	−0.38	0.62	−0.25
Bum	−10.4	−17.22	−1	0.38	−0.38	0.62	−0.25
Civ	1.45	−3.92	1	0.38	−0.38	−0.38	0.75
Cyb	4.75	23.88	2	0.38	−0.38	−0.38	0.75
Cnt	221.1	1170.9	8	1.9	1.9	1.9	1.5
Cnt %	15.7	83.2	0.6	0.1	0.1	0.1	0.1

5.1 Decoder Based Data Summarization

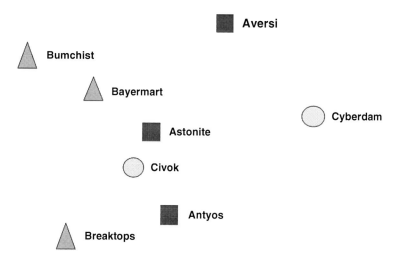

Fig. 5.4 Visualization of the entities in Companies data, centered only

Pr5.1.C Standardization by z-scoring

Consider now a more balanced standardization involving not only feature centering but also feature normalization over the standard deviations – z-scoring, as presented in Table 5.7.

An interesting property of this standardization is that contributions of all features to the data scatter are equal to each other, and moreover, to the number of entities, 8! This is not a coincidence but a property of z-scoring standardization.

The data in Table 5.7 projected on to the plane of two first singular vectors better reflect the products – on Fig. 5.5, C companies are clear-cut separated from the others; yet A and B are still intertwined.

Table 5.7 The data in Table 5.4 standardized by z-scoring. The bottom rows represent contributions of the columns to the data scatter as they are and per cent

Ave	−0.66	0.79	−1.00	−1.29	1.29	−0.77	−0.58	
Ant	1.32	0.15	0	−1.29	1.29	−0.77	−0.58	
Ast	0.28	0.32	0	−1.29	−0.77	1.29	−0.58	
Bay	−0.77	−0.51	−1.00	0.77	1.29	−0.77	−0.58	
Bre	0.62	−0.98	0	0.77	−0.77	1.29	−0.58	
Bum	−1.97	−1.42	−1.00	0.77	−0.77	1.29	−0.58	
Civ	0.28	−0.32	1.00	0.77	−0.77	−0.77	1.73	
Cyb	0.90	1.97	2.00	0.77	−0.77	−0.77	1.73	
Cnt	8	8	8	8	8	8	8	
Cnt, %	14.3	14.3	14.3	14.3	14.3	14.3	14.3	

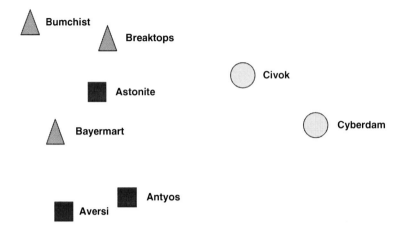

Fig. 5.5 Visualization of the entities in Companies data after z-scoring (Table 5.7)

Pr5.1.D Range normalization and rescaling of dummy features

We would like now to standardize the data in a mixed way by: (a) shifting the scales to the averages, as in z-scoring, but (b) dividing the results not by the feature's standard deviations but rather their ranges. However, when using both categorical and quantitative features, there is a catch here: each of the categories represented by dummy binary variables will have a greater variance than any of the quantitative counterparts after dividing by the ranges. Table 5.8 represents contributions of the range-standardized columns of Table 5.4.

Binary variables contribute much greater than the quantitative variables according to this standardization. The total contribution of the three categories of the original variable Sector looks especially odd – it is more than 55% of the data scatter, by far greater than should be assigned to one of the five original variables. This is partly because that variable Sector in the original table has been enveloped into three variables corresponding to its categories,, thus blowing out the contribution accordingly. To make up for this, the summary contribution of the three dummies should be decreased back three times. This can be done by making further normalization of them by dividing the normalized values by the square root of their number – 3 in our case. Why the square root is used, not just 3? Because contribution to the data scatter involves not the entries themselves but their squared values.

Table 5.8 Within-column sums of the entries squared in the data of Table 5.4 standardized by subtracting the averages and dividing the results by the ranges

Contribution	Income	SharP	NSup	EC	Util	Ind	Retail	Data scatter
Absolute	0.739	0.693	0.889	1.875	1.875	1.875	1.500	9.446
Per cent	7.82	7.34	9.41	19.85	19.85	19.85	15.88	100.00

5.1 Decoder Based Data Summarization

The data table after additionally dividing entries in the three right-most columns over the square root of 3 is presented in Table 5.9. One can see that the contributions of the last three features did decrease threefold from those in Table 5.8, though the relative contributions changed less. Now the most contributing feature is the binary EC that divides the sample along the product based lines. This probably has contributed to the structure visualized on Fig. 5.6. The product defined clusters, much blurred on the previous figures, are clearly seen here, which shows that the original features indeed are informative of the products – just a proper standardization has to be carried out.

Q.5.4. How to do a z-scoring in MatLab? **A.** Take [n,v]=size(X) where X is the data matrix. Then define Y=(X-repmat(mean(X,n,1))./ repmat(std(X,n,1).

Q.5.5. What are the feature contributions after z-scoring? **A.** They all are equal to the same value, the data scatter related to V, the number of features.

Table 5.9 The data in Table 5.4 standardized by: (i) shifting to the within-column averages, (ii) dividing by the within-column ranges, and (iii) further dividing the category based three columns by 3. The values are rounded to the nearest two-digit decimal part

Av	−0.20	0.23	−0.33	−0.63	0.36	−0.22	−0.14
An	0.40	0.05	0	−0.63	0.36	−0.22	−0.14
As	0.08	0.09	0	−0.63	−0.22	0.36	−0.14
Ba	−0.23	−0.15	−0.33	0.38	0.36	−0.22	−0.14
Br	0.19	−0.29	0	0.38	−0.22	0.36	−0.14
Bu	−0.60	−0.42	−0.33	0.38	−0.22	0.36	−0.14
Ci	0.08	−0.10	0.33	0.38	−0.22	−0.22	−0.14
Cy	0.27	0.58	0.67	0.38	−0.22	−0.22	0.43
Cnt	0.74	0.69	0.89	1.88	0.62	0.62	0.50
Cnt %	12.42	11.66	14.95	31.54	10.51	10.51	8.41

Note: only two different values stand in each of the four columns on the right – why? Moreover, the entries within every column sum to 0 (see Q.5.2).

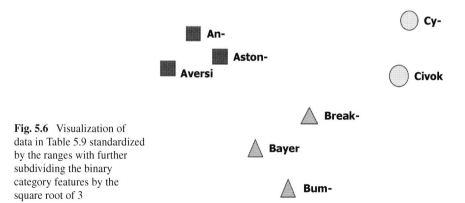

Fig. 5.6 Visualization of data in Table 5.9 standardized by the ranges with further subdividing the binary category features by the square root of 3

Q.5.6. How distances are affected if a different set of scale shifts is applied? **A.** Since the coordinates of two points are subtracted from each other in the distance, the scale shifts cancel each other and have no effect on the distances: the distances do not depend on the location of the origin of the space.

Q.5.7. How to do a distribution-free standardization by shifting to mid-range and normalizing by half-ranges? **A.** See Worked example 5.1, p. 179.

5.2 Principal Component Analysis: Model, Method, Usage

P5.2.1 SVD Based PCA and Its Usage: Presentation

The method of principal component analysis (PCA) has emerged in the research of "inherited talent" undertaken on the verge of nineteenth and twentieth centuries by F. Galton and K. Pearson, first of all to measure talent. For the time being, it has become one of the most popular methods for data summarization and visualization. The mathematical structure and properties of the method are based on the so-called singular value decomposition of data matrices (SVD); this is why in some publications terms PCA and SVD are used as synonymous. In the UK and USA, though, the term PCA frequently refers only to a technique for the analysis of inter-feature covariance or correlation matrix by extracting most contributing linear combinations of features, which utilizes no specific data models and is considered as purely heuristic. However, this method can be related to a genuine decoder based data summarization model that is underlied by the SVD equations – in the case when the data matrix has been centered beforehand. But the centering can hardly make a big difference to the method as such; this is why I refer to the method, even when the data matrix is not centered, as PCA.

There are many motivations for this method, of which we consider the following:

1. *Scoring a hidden factor*
2. *Data visualization*
3. *Feature space reduction*

P5.2.1.1 Scoring a Hidden Factor

Hidden Factor with a Multiplicative Decoder

Consider the following problem. Given student's marks at different subjects, can we derive from this their score at a hidden factor of talent that is supposedly reflected in the marks? Take a look, for example, at the first six students' marks over the three subjects in Table 5.10 extracted from Students data, Table 1.5.

To judge of the relative strength of a student, the average mark is used in practice. This ignores the relative work load that different subjects may impose on a student – can you see that CI mark is greater than SEn mark for each of the six students? – and

5.2 Principal Component Analysis: Model, Method, Usage

Table 5.10 Marks at three subjects for six students from Students data Table 1.5

#	SEn	OOP	CI	Average
1	41	66	90	65.7
2	57	56	60	57.7
3	61	72	79	70.7
4	69	73	72	71.3
5	63	52	88	67.7
6	62	83	80	75.0

in fact, is purely empirical and does not allow much theoretical speculation. Let us assume that there is a hidden factor, not measurable straightforwardly, the talent, that is manifested in the marks. Suppose that another factor manifested in the marks is subject load, and, most importantly, assume that these factors multiply to make a mark, so that a student's mark over a subject is the product of the subject's loading and the student's talent:

$$\text{Mark(Student, Subject)} = \text{Talent\_Score(Student)} * \text{Loading(Subject)}$$

One may point out two issues related to this model – one internal, the other external.

The external issue is that the mark, as observed, depends on many other factors differently affecting different students – the weather, a sleepless night or malady, level of interest in the subject, etc., which make the model as is overly simplistic and prone to errors. Well, a proponent would say, sure the model is simplistic – it takes on only most important factors. The others will cause errors indeed, but these can be tackled by minimizing them: the idea is that the hidden talent and loading factors can be found by minimizing the differences between the real marks and those derived from the model. The PCA method is based on the least-squares approach so that it is the sum of squared differences between the observed and computed marks that is minimized in PCA.

The internal issue is that the model as is admits no unique solution because it is the product of mark by loading that matters, not their individual values – if one multiples all the talent scores by a number, say Talent_Score(Student)*5, and simultaneously divides all the subject loadings by the same number, Loading(Subject)/5, the product will not change. How one is supposed to compute something which admits no definite representation? To make a solution unique, conventionally, a constant norm of one or both of the items is assumed so that one more item into the product is admitted – that expressing the product's magnitude. Then, as stated in the formulation part of this section, there is a unique solution indeed, with the magnitude expressed by the so-called maximum singular value of the data matrix with the score and load factors being its corresponding normed singular vectors.

Specifically, the maximum singular value of matrix in Table 5.10 is 291.4, and the corresponding normed singular vectors are $z = (0.40, 0.34, 0.42, 0.42, 0.41, 0.45)$, for the talent score, and $c = (0.50, 0.57, 0.66)$, for the loadings. That means that every mark in the matrix is product of three items. For example, to compute the model SEn value for student 6, one takes 291.4*0.45*0.50 = 65.6, which is not that far from the observed mark of 62. Yet our model involves the product of two items only. To get back to that, we need to distribute the singular value between the vectors. There is only one way to do it complying with the singular value equations (5.12) – by multiplying each of the vectors by the same value, the square root of the singular value which is 17.1. Thus, the denormalized talent score and subject loading vectors will be $z' = (6.85, 5.83, 7.21, 7.20, 6.95, 7.64)$ and $c' = (8.45, 9.67, 11.25)$. According to the model, the score of student 3 over subject SEn is the product of the talent score, 7.21, and the loading, 8.45, which is 60.9, quite close to the observed mark 61. Similarly, product 5.83*9.67 = 56.4 is close to 56, student 2's mark over OOP. The differences can be greater though: product 5.83*8.45 is 49.3, which is rather far away from the observed mark 57 for student 2 over SEn.

In matrix terms, the model can be represented by the following equation

$$\begin{bmatrix} 6.85 \\ 5.83 \\ 7.21 \\ 7.20 \\ 6.95 \\ 7.64 \end{bmatrix} * \begin{bmatrix} 8.45 & 9.67 & 11.25 \end{bmatrix} = \begin{bmatrix} 57.88 & 66.29 & 77.07 \\ 49.22 & 56.37 & 65.53 \\ 60.88 & 69.72 & 81.05 \\ 60.83 & 69.67 & 81.00 \\ 58.69 & 67.22 & 78.15 \\ 64.53 & 73.91 & 85.92 \end{bmatrix} \quad (5.7)$$

whereas its relation to the observed data matrix, by equation

$$\begin{bmatrix} 41 & 66 & 90 \\ 57 & 56 & 90 \\ 61 & 72 & 79 \\ 69 & 73 & 72 \\ 63 & 52 & 88 \\ 62 & 83 & 80 \end{bmatrix} = \begin{bmatrix} 57.88 & 66.29 & 77.07 \\ 49.22 & 56.37 & 65.53 \\ 60.88 & 69.72 & 81.05 \\ 60.83 & 69.67 & 81.00 \\ 58.69 & 67.22 & 78.15 \\ 64.53 & 73.91 & 85.92 \end{bmatrix} + \begin{bmatrix} -16.88 & -0.29 & 12.93 \\ 7.78 & -0.37 & -5.53 \\ 0.12 & 2.28 & -2.05 \\ 8.17 & 3.33 & -9.00 \\ 4.31 & -15.22 & 9.85 \\ -2.53 & 9.09 & -5.92 \end{bmatrix} \quad (5.8)$$

where the left-hand item is the observed mark matrix; that in the middle, the model-computed estimations of the marks; and the right-hand item comprises the differences between the real and decoded marks.

Error of the Model

Among questions that arise with respect to the matrix equation such as that in (5.8) are the following:

(i) Why are the differences appearing at all?
(ii) How can the overall level of differences be assessed?
(iii) Can any better fitting estimates for the talent be found?

We address them in turn.

5.2 Principal Component Analysis: Model, Method, Usage

(i) Differences between real and model-derived marks

The differences emerge because the model imposes significant constraints on the model-derived estimates of marks. They are generated as products of components of just two vectors, the talent score and the subject loadings. This means that every row in the model-based matrix (5.7) is proportional to the vector of subject loadings, and every column, to the vector of talent scores. Therefore, the rows are mutually proportional as well as the columns. Real marks, generally speaking, do not satisfy such a property: mark rows or columns are typically not proportional to each other. More formally, this can be expressed in the following way: 6 talent scores and 3 subject loadings together can generate not more than $6+3 = 9$ independent estimates. (One more degree of freedom may go because the norms of these two vectors are the same.) The number of marks however, is the product of these, $6*3 = 18$. The greater the size of the data matrix, $M \times V$, the smaller the proportion of the independent values, $M+V$, that can be generated from the model.

In other words, matrix (5.7) is one-dimensional. It is well recognized in mathematics in the concept of matrix rank which corresponds to the "inner" dimension bore by a matrix – matrices that are products of two vectors are referred to as matrices of rank 1.

(ii) Assessment of the level of differences

A conventional measure of the level of error of the model is the ratio of the scatters of the model derived matrix and the observed data matrix in (5.8). The scatter of matrix A, $T(A)$, is the sum of the squares of all of A-entries or, which is the same, the sum of the diagonal entries in matrix $A*A^T$, the *trace*$(A*A^T)$.

Worked example 5.2. Explained proportion of data scatter in Equation (5.8)

Consider scatters of three matrices in (5.8) in Table 5.11. The residual data scatter is rather small and accounts for only $\varepsilon^2 = 1183.2/86092 = 0.0137$, that is, 1.37%, of the original data scatter. Its complement to unity, 98.63%, is the proportion of the data scatter explained by the multiplicative model. This also can be straightforwardly derived from the singular value, 291.4: its square shows the part of the data scatter explained by the model, $291.4^2/86092 = 0.9863$.

Table 5.11 Scatters of matrices in Equation (5.8)

	Data matrix	Scatter of Model matrix	Residual matrix
Absolute	86092	84908.80	1183.20
Proportion	100	98.63	1.37

Q.5.8. In spite of the fact that some errors in (5.8) are rather high, the overall squared error is quite small, just about one per cent of the data scatter. Why is that?
A. Because the data values are far away from 0 – see Q.5.20 explaining the effect mathematically.

(iii) The singular vector estimates are the best

The squared error is the criterion optimized by the estimates of talent scores and subject loadings. No other estimates can give a smaller value to the error for data matrix in Table 5.10 than $\varepsilon^2 = 1.37\%$.

Formulaic Expression of the Hidden Factor Through the Data

The relations between singular vectors (see Eqs. (5.12) in Section 5.2.2.1 provide us with a conventional expression of the talent score as a weighted average of marks at different subjects. The weights are proportional to the subject loadings $c' = (8.45, 9.67, 11.25)$: weight vector w is the result of dividing of all entries in c' by the singular value, $w = c'/291.4 = (0.029, 0.033, 0.039)$. For example, the talent score for student 1 is the w weighted average of their marks, $0.029*41 + 0.033*66 + 0.039*90 = 6.88$.

The model-derived averaging allows one also to score the talent of other students, those not belonging to the sample being analyzed. If marks of a student over the three subjects are (50,50,70), their talent score will be the w-weighted average: $0.029*50 + 0.033*50 + 0.039*70 = 5.83$.

A final touch to the hidden factor scoring can be given by rescaling it in a way conforming to the application domain. Specifically, one may wish to express the talent scores in a 0-100 scale resembling that of the original mark scales. That means that the score vector z' has to be transformed into $z'' = \alpha^* z' + \beta$, where α and β are the scaling factor and shift coefficients, that can be found from two natural conditions: (a) z'' is 0 when all the marks are 0 and (b) z'' is 100 when all the marks are 100. Condition (a) means that $\beta = 0$, and condition (b) calls for calculation of the talent score of a student with all top marks. Summing three 100 marks with weights from w leads to the value $zM = 0.029*100 + 0.033*100 + 0.039*100 = 10.10$ which implies that the rescaling coefficient α must be $100/zM = 9.92$ or, equivalently, weights must be rescaled as $w' = 9.92*w(0.29, 0.33, 0.38)$. Talent scores found with these weights are presented in the right column of Table 5.12 – hardly a great difference from the average scores, except that the talent scores are slightly higher, due to a greater weight assigned to the mark-earning CI subject.

In spite of the fact that the original model does not assume any averaging of the marks, the optimal scoring is a form of averaging indeed. However, one should note that it is the model that provides us with both the weights, which are the optimal subject loadings, and the error – these are entirely out of the picture at the empirical averaging.

This line of thinking can be applied to any other hidden performance measures such as quality of life in different cities using scorings over its different aspects

5.2 Principal Component Analysis: Model, Method, Usage

Table 5.12 Marks and talent scores for six students

#	SEn	OOP	CI	Average	Talent
1	41	66	90	65.7	68.0
2	57	56	60	57.7	57.8
3	61	72	79	70.7	71.5
4	69	73	72	71.3	71.5
5	63	52	88	67.7	69.0
6	62	83	80	75.0	75.8

(housing, transportation, catering, pollution, etc.) or performance of different management sections in a big company or government.

Sensitivity of the Hidden Factor to Data Standardization

One big issue related to the multiplicative hidden factor model is its instability with respect to data standardization that has been clearly seen at different data normalization options in Project 5.1. Here is another example.

Worked example 5.3. Principal components after feature centering

Consider now the data set in Table 5.10 analyzed above. Take the means of marks over different disciplines in this table, 58.8 for SEn, 67.0 for OOP, and 78.2 for CI, and subtract them from the marks, to shift the data to the mean point (see Table 5.13). This would not much change the average scores presented in Tables 5.10, 5.12 – just shifting them back by the average of the means, $(58.8 + 67.0 + 78.2)/3 = 68$.

Everything changes, though, in the multiplicative model, starting from the data scatter, which is now 1729.7 – a 50 times reduction from the case of uncentered data. The maximum singular value of the feature centered matrix in Table 5.13,

Table 5.13 Centered marks for six students and corresponding talent scores, first, as found as explained in A.1, and, second, that rescaled to produce extreme values 0 and 100 if all subject marks are 0 or 100, respectively

#	SEn	OOP	CI	Average	Talent score	Talent rescaled
1	−17.8	−1.0	11.8	−2.3	−3.71	13.69
2	−1.8	−11.0	−18.2	−10.3	1.48	17.60
3	2.2	5.0	0.8	2.7	0.49	16.85
4	10.2	6.0	−6.2	3.3	2.42	18.31
5	4.2	−15.0	9.8	−0.3	−1.94	15.02
6	3.2	16.0	1.8	7.0	1.25	17.42

$$\begin{bmatrix} -17.8 & -1.0 & 11.8 \\ -1.8 & -11.0 & -18.2 \\ 2.2 & 5.0 & 0.8 \\ 10.2 & 6.0 & -6.2 \\ 4.2 & -15.0 & 9.8 \\ 3.2 & 16.0 & 1.8 \end{bmatrix} = \begin{bmatrix} -11.51 & -7.24 & 13.83 \\ 4.60 & 2.90 & -5.53 \\ 1.52 & 0.96 & -1.82 \\ 7.52 & 4.73 & -9.04 \\ -6.02 & -3.79 & 7.23 \\ 3.88 & 2.44 & -4.67 \end{bmatrix} + \begin{bmatrix} -6.33 & 6.24 & -2.00 \\ -6.44 & -13.90 & -12.63 \\ 0.65 & 4.05 & 2.66 \\ 2.65 & 1.27 & 2.87 \\ 10.18 & -11.21 & 2.60 \\ -0.72 & 13.56 & 6.50 \end{bmatrix} \quad (5.9)$$

is 27.37 so that the multiplicative model now accounts for only $27.37^2/1729.7 = 0.433 = 43.3\%$ of the data scatter. This goes in line with the idea that much of the data structure can be seen from the "grand" mean (see Fig. 5.10 on p. 204 illustrating the point), however, this also greatly increases the error. In fact, the relative order of errors does not change that much, as can be seen in formula (5.10) decomposing the centered data (in the box on the left) in the model-based item, the first on the right, and the residual errors in the right-hand item. What changes is the denominator. The model-based estimates have been calculated in the same way as those in formula (5.7) – by multiplying every entry of the new talent score vector $z^* = (-3.71, 1.48, 0.49, 2.42, -1.94, 1.25)$ over every entry of the new subject loading vector $c^* = (3.10, 1.95, -3.73)$.

Worked example 5.4. Rescaling the talent score from Worked example 5.3

Let us determine rescaling parameters α and β that should be applied to z^*, or to the weights c^*, in Worked example 5.3 so that at 0 marks over all three subjects the talent score would be 0 and at all 100 marks the talent score would be 100.

As in the previous section, we first determine what scores correspond to these situations in the current setting. All-zero marks, after centering, become minus the average marks, -58.8 for SEn, -67.0 for OOP, and -78.2 for CI. Averaged according to the loadings c from Worked example 5.3, they produce $3.10*(-58.8) + 1.95*(-67) - 3.73*(-78.2) = -21.24$. Analogously, all-hundred marks, after centering, become 41.2 for SEn, 33.0 for OOP, and 21.8 for CI to produce the score $3.10*(41.2) + 1.95*(33) - 3.73*(21.8) = 110.8$. The difference between these, $110.8 - (-21.2) = 132.0$ divides 100 to produce the rescaling coefficient a $= 100/132 = 0.75$, after which shift value is determined from the all-0 score as b $= -0.75*(-21.24) = 16.48$. Thus rescaled talent scores are in the last column of Table 5.13. These are much less related to the average scoring than it was the case at the original data. One can see some drastic changes such as, for instance, the formerly worst student 2 becoming second best, since their deficiency over CI has been converted to an advantage because of the negative loading at CI.

For a student with marks (50,50,70) that becomes $(-8.8, -17.0, -8.2)$ after centering, the rescaled talent score comes from the adjusted weighting vector $w = a*c = 0.75*(3.10, 1.95, -3.73) = (2.34, 1.47, -2.81)$ as the weighted

5.2 Principal Component Analysis: Model, Method, Usage

average $2.34^*(-8.8) + 1.47^*(-17) - 2.81^*(-8.2) = -22.73$ plus the shift value $b = 16.48$ so that the result is, paradoxically, -6.25 – less than at all zeros! This is again a result of the negative loading at CI.

This example illustrates not only the idea of a great sensitivity of the multiplicative model, but, also, that there should be no mark centering when evaluating performances.

P5.2.1.2 Data Visualization

For the purposes of *visualization* of the data entities on a 2D plane, the data set is usually first centered to put it against the backdrop of the center – we mentioned already that more structure in the dataset can be seen when looking at it from the center of gravity, that is, the grand mean location. What has been disastrous for the purposes of scoring in Worked example 5.4 is beneficial for the purposes of structuring. Solutions to the multiple factor model, that is, the hidden factor scoring vectors which are singular vectors of the data matrix, in this case, are referred to as principal components (PCs). Two principal components corresponding to the maximal singular values are needed for a 2D representation.

What is warranted in this arrangement is that the PC plane approximates the data, as well as the between-feature covariances and between-entity similarities, in the best possible way. The coordinates provided by the singular vectors/ principal components are not unique, though, and can be changed by rotating the axes, but they do hold a unique property that each next component maximally contributes to the residual data scatter.

Worked example 5.5. Visualization of a fragment of Students dataset

Consider four features in the Students dataset – the Age and marks for SEn, OOP and CI subjects. Let us center it by subtracting the mean vector $a = (33.68, 58.39, 61.65, 55.35)$ from all the rows, and normalize the features by their ranges $r = (31, 56, 67, 69)$. The latter operation seems a necessity because the Age, expressed in years, and subject marks, per cent, are not exactly comparable. Characteristics of all the four singular vectors, that are principal components (PCs), of these data for feature loadings are presented in Table 5.14.

The summary contribution of the first two PCs to the data scatter is $42.34 + 29.77 = 72.11\%$, which is not that bad for educational data and warrants a close representation of the entities on the principal components plane. The two principal components are found by multiplication of each of the corresponding left singular vectors z_1 and z_2 by the square root of the corresponding singular value. Each entity $i = 1, 2, \ldots, N$ is represented on the PC plane by the pair of the first and second PC values (z_{1i}^*, z_{2i}^*) (see Fig. 5.7). The left part is just the data with no labels. On the right part, two occupational categories are visualized using triangles (IT)

Table 5.14 Components of the normed loading parts of principal components for the standardized part of Student data set; corresponding singular values, along with their squares expressed both per cent to their total, the data scatter, and in real

Singular value	Singular value squared	Contribution, per cent	Singular vector			
			Age	SEn	OOP	CI
3.33	11.12	42.34	−0.59	0.03	0.59	0.55
2.80	7.82	29.77	0.53	0.73	0.10	0.42
2.03	4.11	15.67	−0.51	0.68	−0.08	−0.51
1.79	3.21	12.22	−0.32	0.05	−0.80	0.51

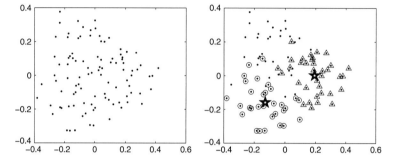

Fig. 5.7 Scatter plot of the student 4D data (Age and marks over SP, OO, CI) row points on the plane of two first principal components, after they have been centered and rescaled. *Pentagrams* represent the mean points of the occupation categories AN and IT

and circles (AN); remaining dots relate to category BA. In spite of the fact that the occupation has not been involved in building the PC space, its categories appear to occupy different parts of the plane, which will be explained later, in Worked example 5.6.

P5.2.1.3 Feature Space Reduction: Criteria of Contribution and Interpretability

The principal components provide for the best possible least-squares approximation of the data in a low dimension space. The quality of such a data compression is usually judged over (i) the proportion of the data scatter taken into account by the reduced dimension space and (ii) interpretability of the factors supplied by the PCs.

Contribution of the PCA model to the data scatter is reflected in the sum of squared singular values corresponding to the principal components in the reduced data. This sum should be related to the data scatter or, equivalently, to the total of all singular values squared, to see the impact. For example, the 2D representation of 4D student data on Fig. 5.7 contributes 72.11% to the data scatter, as found in

5.2 Principal Component Analysis: Model, Method, Usage

Worked example 5.5. In the example of marks for six students in Worked examples 5.2 and 5.4, the talent scoring factor contributed 98.6% to the data scatter at the original data and 43.3% after centering the data. Does that mean that marks should not be centered at all, to get a better approximation? Not necessarily. When all data entries are not negative, the large contribution of a principal component is an artifact of the very remoteness of the data set from the origin – the farther away you move the data from the origin, by adding a positive number to all the entries for example, the greater the contribution. This phenomenon follows a known property of positive fractions: if $0 < a/b < 1$, then adding a positive c to both the numerator and denominator may only increase the ratio; the greater the c, the greater the increase, so that $(a+c)/(b+c)$ converges to 1 when c tends to infinity (see Q.5.18). The analogy becomes clear if we consider b the data scatter and a, the principal component's contribution.

This example shows that the contribution, in spite of its firm mathematical footing, can be rather shaky an argument when data is not centered. One more criterion, of interpretability, gives a different perspective.

To interpret PCA results one should use the feature loadings according to the singular vectors related to features. These straightforwardly show how much a principal component is affected by a feature: the larger the value the greater the correlation. Features with relatively high positive or negative coefficients are used to interpret the component, as illustrated in the worked example below.

Worked example 5.6. Interpretation of principal components at the standardized Student data

Take a look at the first singular vector in Table 5.14 on p. 196 corresponding to the maximum singular value 3.33 at the standardized Students data. (Please note this 100×4 data differs from the 6×3 data of six students analyzed in the beginning.) One can see that the first component positively relates to marks over all subjects, perhaps except SEn at which the loading is almost zero, and negatively to the Age. That means that on average, the first factor is greater when a student gets better marks and is younger. Thus, the first component can be interpreted as the "Age-related Computer Science proficiency". The second component (the second line in Table 5.14) is positively related to all of the features, especially SEn marks, which can be interpreted as "Age defying inclination towards Software Engineering".Then the triangle and circle patterns on the right of Fig. 5.7 show that AN laborers are on the minimum side of the age-related CS proficiency, whereas IT occupations are high on that – all of which seem rather reasonable. Both are rather low on the second component, though, in contrast to students represented by dots, thus belonging to BA occupation category, that get the maximum values on it.

In the early days of the development of factor analysis, yet within the psychology community, researchers were trying to explore the possibility of achieving a more interpretable solution by rotating the axes of the PC space. The goal was to find a simple structure of the loadings, in which most of the loading elements are zero with a few non-zero values that should be as close to either 1 or -1 as possible. This goal, however, is subject to too much of arbitrariness and remains an open issue. Keeping singular vectors as they are, not rotated, has the advantage that each of them contributes to the residual data scatter as much as possible. This relates to frequently occurring real world situations in which factors underlying the phenomenon of interest contribute to it differently. The PCA factors express such a structure formally: that one contributing the most is followed by the second best contributing, then by the third best contributing, etc.

Q.5.9. Prove that the condition of statistical independence for a contingency data table can be equivalently reformulated as the contingency table being of rank 1. **A.** Indeed an $N \times V$ matrix of rank 1 is a matrix whose elements are products of components of two vectors, N- and V-dimensional ones. In the case of a relative contingency table, $P = (p_{kl})$, the statistical independence condition, $p_{kl} = p_{k+}p_{+l}$ for all rows k and columns l, shows exactly that: all elements of matrix P are products of components of two vectors, (p_{k+}) and (p_{+l}), which proves the statement.

Q.5.10. What could be a purpose to aggregate the features in the Market towns' data? **A.** Since all the features relate to the extent of development of a town, the aggregate feature perhaps would express the extent of the town's development.

F5.2.2 Mathematical Model of PCA-SVD and Its Properties: Formulation

F5.2.2.1 A Multiplicative Decoder

Let us consider a data matrix X with entries x_{iv} and standardize it into $Y = (y_{iv})$ ($i = 1, 2, \ldots, N$; $v = 1, 2, \ldots, V$). The PCA model assumes hidden factor scores z_i^* and feature loadings c_v^* such that their product $z_i^* c_v^*$ is the decoder for y_{iv}, which can be explicated, by using additive residuals e_{iv}, as

$$y_{iv} = c_v^* z_i^* + e_{iv} \qquad (5.10)$$

where the residuals are to be minimized using the least squares criterion

$$L^2 = \sum_{i \in I} \sum_{v \in V} e_{iv}^2 = \sum_{i \in I} \sum_{v \in V} (y_{iv} - c_v^* z_i^*)^2 \qquad (5.11)$$

The decoder in (5.10), as a mathematical model for deriving z_i^* and c_v^*, has a flaw from the technical point of view: its solution cannot be defined uniquely! Indeed, assume that we have got the talent score z_i^* for student i and the loading c_v^* at subject v, to produce $z_i^* c_v^*$ as the estimate for the student's mark at the subject. However, the same estimate will be produced if we halve the talent score vector and

5.2 Principal Component Analysis: Model, Method, Usage

simultaneously double the loading vector: $z_i^* c_v^* = (z_i^*/2)(2c_v^*)$. Any other real taken as the divisor/multiplier would, obviously, do the same.

A conventional remedy to this is following: specify the norms of vectors z^*, c^* to unities, and treat the multiplicative effect of the two as a real $\mu \geq 0$. Then put the product $\mu z_i c_v$ in (5.10) and (5.11) instead of $z_i^* c_v^*$ where z and c are normed versions of z^* and c^*, and μ is their multiplicative effect, the product of norms of z^* and c^*. The (Euclidean) norm $\|x\|$ of vector $x = (x_1,\ldots,x_N)$ is defined as its length, that is, the square root of $\|x\|^2 = x^T x = x_1^2 + x_2^2 + \ldots + x_N^2$. Thus a vector is referred to as normed if its length is 1, $\|x\| = 1$. After μ, z and c minimizing (5.11) are determined, return to the talent score vector z^* and loading vector c^* with formulas: $z^* = \mu^{1/2} z$, $c^* = \mu^{1/2} c$. It should be pointed out that a different norming condition such as say $|x_1| + |x_2| + \ldots + |x_N| = 1$ would lead to a different than the singular triplet solution – it seems no one ever explored such an opportunity.

The first-order optimality conditions to a triplet (μ, z, c) be the least-squares solution to (5.10) imply that $\mu = z^T Y c$ is maximum value satisfying equations

$$Y^T z = \mu c \quad \text{and} \quad Y c = \mu z \qquad (5.12)$$

These equations for the optimal scores give the transformation of the data leading to the summaries z^* and c^*. The transformation, denoted by $F(Y)$ in (5.1), appears to be linear, and combines optimal c and z so that each determines the other. It appears, this type of summarization is well known in linear algebra.

A triplet (μ, z, c) consisting of a non-negative μ and two vectors, c (size $M \times 1$) and z (size $N \times 1$) is referred to as to a singular triplet for Y if it satisfies (5.12); μ is referred to as a singular value and z, c the corresponding singular vectors. What can be proven immediately is the following:

Property 1 For any singular triplet (μ, z, c) satisfying (5.12) at $\mu \neq 0$ vectors z and c have the same norm.

Indeed, by multiplying the left-side equation in (5.12) by c^T, and the right-side equation by z^T, both from the left, one arrives at equations $c^T Y^T z = \mu c^T c$ and $z^T Y c = \mu z^T z$. Since $c^T Y^T z = (z^T Y c)^T$ and both are just real numbers, the equation $c^T c = z^T z$ holds because $\mu \neq 0$. Typically, the norms of c and z are taken to be unities. However, at the Principal components in (5.10), they are equal to the square root of the singular value μ, which proves the statement.

Any matrix Y can have only a finite number of singular values, and the number is equal to the rank of Y. Singular vectors z corresponding to different singular values are necessarily mutually orthogonal, as well as singular vectors c. When two or more singular values coincide, their singular vectors form a linear subspace and can be chosen to be orthogonal, which is the case in computational packages such as MatLab.

Therefore, $z^* = \mu^{1/2} z$ and $c^* = \mu^{1/2} c$ is a solution to the model (5.10) minimizing (5.11) defined by the maximum singular value of matrix Y and the corresponding normed singular vectors. Vectors z^* and c^* obviously also satisfy (5.12). This leads to other nice mathematical properties.

Property 2 The score vector z^* is a linear combination of columns of Y weighted by c^*'s components: c^*'s components are feature weights in the score z^*

Equations (5.12) allow mapping additional features or entities onto the other part of the hidden factor model. Consider, for example an additional N-dimemsional feature vector y standardized same way as Y. Its loading $c^*(y)$ is determined as $c^*(y) = <z^*, y>/\mu$ for the talent score z^*. Similarly, an additional standardized V-dimensional entity point h has its hidden factor score defined according to the other part of (5.12), $z^*(h) = <c^*, h>/\mu$.

Property 3 Pythagorean decomposition of the data scatter T(Y) relating the least squares criterion (5.11) and the singular value holds as follows:

$$T(Y) = \mu^2 + L^2 \tag{5.13}$$

This implies that the squared singular value μ^2 expresses the proportion of the data scatter explained by the principal component z^*.

F5.2.2.2 Extension of the PC Decoder to the Case of Many Factors

It is well known by now that there is not just one talent behind the human efforts but a range of them. Assume a relatively small number K of different hidden factors z^*_k and corresponding feature loading vectors c^*_k ($k = 1, 2, \ldots, K$; $K < V$), with students and subjects differently scored over them so that the observed marks, after standardization, are sums of those over the different talents:

$$y_{iv} = \sum_{k=1}^{K} c^*_{kv} z^*_{ik} + e_{iv}, \tag{5.14}$$

This is again a decoder that can be used for deriving a summary from the standardized marks matrix $Y = (y_{iv})$ so that the hidden score and loading vectors z^*_k and c^*_k are found by minimizing residuals, e_{iv}. To eliminate the mathematical ambiguity, we again assume that $z^*_k = \mu^{1/2} z_k$ and $c^*_k = \mu^{1/2} c_k$, where z_k and c_k are normed vectors.

Assume that the rank of Y is r and $K < r$. Assume that the singular values of Y are sorted so that $\mu_1 \geq \mu_2 \geq \ldots \geq \mu_r$. It can be proven that the least-squares solution to (5.14) is provided by the maximal singular values μ_k and corresponding normed singular vectors z_k and c_k ($k = 1, 2, \ldots, K$).

The underlying mathematical property is that any matrix Y can be decomposed over its singular values and vectors,

$$y_{iv} = \sum_{k=1}^{r} \mu_k c_{kv} z_{ik}, \tag{5.15}$$

5.2 Principal Component Analysis: Model, Method, Usage

which is referred to as the singular value decomposition (SVD). In matrix terms, SVD can be expressed as

$$Y = \sum_{k=1}^{r} \mu_k z_k c^T_k = ZMC^T \tag{5.15'}$$

where the right-hand item Z is $N \times r$ matrix with columns z_k and C is $M \times r$ matrix with columns c_k and M is an $r \times r$ diagonal matrix with entries μ_k on the diagonal and all other entries zero.

Equation (5.15') implies, because the singular vectors are mutually orthogonal, that the scatter of matrix Y is decomposed into the sum of the squared singular values:

$$T(Y) = \mu_1^2 + \mu_2^2 + \ldots + \mu_r^2 \tag{5.16}$$

This implies that the least-squares fitting of the PCA model in Equation (5.14) decomposes the data scatter into the sum of contributions of individual singular vectors and the least-squares criterion $L^2 = \Sigma_{i,v} e_{iv}^2$:

$$T(Y) = \mu_1^2 + \mu_2^2 + \ldots + \mu_K^2 + L^2 \tag{5.17}$$

This provides for the evaluation of the relative contribution of the model (5.14) to the data scatter as $(\mu_1^2 + \mu_2^2 + \ldots + \mu_K^2)/T(Y)$.

In particular, this part of decomposition (5.15) is used at 2D visualization:

$$y_{iv}^* \approx z_{i1}^* c_{1v}^* + z_{i2}^* c_{2v}^*$$

where the elements on the right come from the two first principal components. This equation holds not 100% but $100*(\mu_1^2 + \mu_2^2)/T(Y)$ percent. Every entity $i \in I$ is represented on a 2D Cartesian plane by pair (z_{i1}^*, z_{i2}^*). Moreover, because of the symmetry, every feature v can be represented, on the same plane by pair (c_{1v}^*, c_{2v}^*). Such a simultaneous representation of both entities and features is referred to as a joint display or a biplot. As a matter of fact, features are presented on a biplot by not just the corresponding points, but by lines joining them to 0. This reflects the fact, that projections of points, representing the entities (entity markers), to these lines are meaningful. For a variable v, the length and direction of the projection of an entity marker to the corresponding line reflects the value of v on the entity.

F5.2.2.3 Conventional Formulation of PCA Using Covariance Matrix

In the English-written literature, PCA is conventionally introduced in a different way: not via the decoder based model (5.10) or (5.14), but rather as a heuristic technique to build up most contributing linear combinations of features with the help of the data covariance matrix (see, for example, Kendall and Stewart 1973, Hair et al. 2010).

The covariance matrix is defined as $V \times V$ matrix $C = Y^T Y/N$, where Y is a centered version of the data matrix X, so that all its columns are centered. The (v', v'')-entry in the covariance matrix is the covariance coefficient between features v' and v''; and the diagonal elements are variances of the corresponding features. The covariance matrix is referred to as the correlation matrix if Y has been z-score standardized, that is, after shifting each column to its mean, it was further normalized by dividing by its standard deviation. In this case, elements of C are correlation coefficients between corresponding variables. (Note how a bivariate concept is carried through to multivariate data by using matrix multiplication.)

The conventional PCA problem formulation goes like this. Given a centered $N \times V$ data matrix Y, find a normed V-dimensional vector $c = (c_v)$ such that the sum of Y columns weighted by c, $f = Yc$, has the largest variance possible. This vector is the principal component, termed so because it shows the direction of the maximum variance in data. Vector f is centered for any c, since Y is centered. Therefore, its variance is $s^2 = <f,f>/N = f^T f/N$. The last equation comes under the convention that a V-dimensional vector is a $V \times 1$ matrix, that is, a column. By substituting Yc for f, this leads to equation $s^2 = c^T Y^T Yc/N$. Maximizing this with respect to all vectors c that are normed, that is, satisfy condition $c^T c = 1$, is equivalent to unconditionally maximizing the ratio

$$q(c) = \frac{c^T Y^T Yc}{c^T c} \tag{5.18}$$

over all V-dimensional vectors c. Expression (5.18) is well known in linear algebra as the Rayleigh quotient for matrix $NC = Y^T Y$ which is proportional to the covariance matrix of course. The maximum of Rayleigh quotient is reached at c being an eigen-vector, also termed latent vector, for matrix C, corresponding to its maximum eigenvalue (latent value) $q(c)$ (5.18).

Vector a is referred to as an eigenvector for a square matrix B if $Ba = \lambda a$ for some, possibly complex, number λ which is referred to as the eigenvalue corresponding to a. In the case of a covariance matrix all eigenvalues are not only real but non-negative as well. The number of eigenvalues of B is equal to the rank of B, and eigenvectors corresponding to different eigenvalues are orthogonal to each other.

Therefore, the first principal component, in the conventional definition, is vector $f = Yc$ defined by the eigenvector of the covariance matrix A corresponding to its maximum eigenvalue. The second principal component is conventionally defined as another linear combination of columns of Y, which maximizes its variance under the condition that it is orthogonal to the first principal component. It is defined, of course, by the second eigenvalue and corresponding eigenvector. Other principal components are defined similarly, in a recursive manner, under the condition that they are orthogonal to all preceding principal components; this implies that they correspond to other eigenvalues, in the descending order, and the corresponding eigenvectors.

This construction seems rather remote from how the principal components are introduced above. However, it is not difficult to prove that the two definitions are computationally equivalent. Indeed, take equation $Yc = \mu z$ from (5.12), express z from this as $z = Yc/\mu$, and substitute this z into the other Equation (5.12): $Y^T z = \mu c$ so that $Y^T Yc/\mu = \mu c$. This implies that μ^2 and c, defined by the multiplicative decoder model, satisfy equation

$$Y^T Yc = \mu^2 c, \qquad (5.19)$$

that is, c is an eigenvector of square matrix $Y^T Y$, corresponding to its maximum eigenvalue $\lambda = \mu^2$. Matrix $Y^T Y$, in the case when Y is centred, is the covariance matrix C up to the constant factor $1/N$. Therefore, c in (5.19) is an egenvector of C corresponding to its maximum eigenvalue. This proves that the two definitions are equivalent when the data matrix Y is centered. The given proof also establishes a simple relation between the eigen values λ of C and singular values μ of Y: $\lambda = \mu^2$.

In spite of the computational equivalence, there are some conceptual differences between the two definitions. In contrast to the definition in 5.2.2.1 based on the multiplicative decoder, the conventional definition is purely heuristic, assuming no underlying model whatsoever. It makes sense only for centered data because of its reliance on the concept of covariance. Moreover, the fact that the principal components are linear combinations of features is postulated in the conventional definition, whereas this is a derived property of the optimal solution to the multiplicative decoder model which involves no assumptions on a linear or nonlinear relation between features and hidden factors.

Q.5.11. Can you write equations defining μ^2 and z as an eigenvalue and corresponding eigenvector of matrix YY^T. Does this square matrix have any meaning of its own? **A.** By multiplying the left-side equation in (5.12) by Y on the left, we obtain $YY^T z = \mu Yc = \mu^2 z$, the latter equation following from the right-hand equation in (5.12). Elements of matrix YY^T are inner products of rows of matrix Y to express similarities between corresponding entities.

F5.2.2.4 Geometric Interpretation of Principal Components

Take all talent score points $z = (z_1, \ldots, z_N)$ that are normed, that is, satisfy equation $<z, z> = 1$ or $z^T z = 1$ or $z_1^2 + \ldots + z_N^2 = 1$: they form a sphere of radius 1 in the N-dimensional "entity" space (Fig. 5.8a). The image of these points in the feature space, $Y^T z$, forms a skewed sphere, an ellipsoid in the feature space, consisting of points μc where c is normed. The longest axis of this ellipsoid corresponds to the maximum μ, that is the first singular value of Y. [Indeed, the first singular value μ_1 and corresponding normed singular vectors c_1, z_1 satisfy equation $Yc_1 = \mu z_1$ and, thus, its transpose, $c_1^T Y^T = \mu_1 z^T$. Multiplying the latter by the former, one gets equation $c_1^T Y^T Yc_1 = \mu_1^2$, because $z^T z = 1$.]

Fig. 5.8 Sphere $z^T z = 1$ in the entity space (**a**) and its image, ellipsoid $c^T Y^T Y c = \mu_1^2$, in the feature space. The first component, c_1, corresponds to the maximal axis of the ellipsoid with its length equal to $2\mu_1$

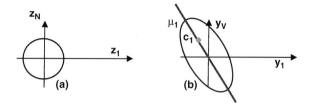

What the longest axis has to do with the data? This is exactly the direction which is looked for in the conventional definition of PCA. The direction of the longest axis of the data ellipsoid makes minimum of the summary distances (Euclidean squared) from data points to their projections on the line (see Fig. 5.9), because of the least-squares optimality of the decoder in (5.10) so that this axis is the best possible 1D representation of the data.

This property extends to all subspaces generated in the order of extraction of principal components: the first two PCs make a plane that is the best two-dimensional approximation of the dataset; the first three make a 3D space best approximating the dataset, etc.

One more illustration concerns the difference between SVD for the raw data and that after centering; see Fig. 5.10, which is basically extension of the previous Fig. 5.9 to the cases. One can see that at the raw data, Fig. 5.10a, the longest axis follows the direction between the origin and the points cloud, whereas after centering it follows the structure of the dataset, Fig. 5.10b.

Fig. 5.9 The direction of the longest axis of the data ellipsoid makes minimum the summary distances (Euclidean squared) from data points to their projections onto the line

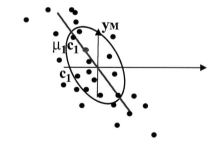

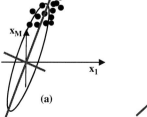 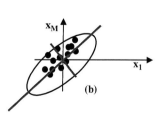

Fig. 5.10 (**a**) PCA at data not centered, (**b**) PCA at the data after centering

C5.2.3 Computing Principal Components

The SVD decomposition is found with MatLab's svd.m function

$$[Z, M, C] = svd(Y);$$

where Y is $N \times V$ data matrix after standardization whose rank is r. Typically, if all data entries come from observation, the rank $r = min(N, V)$.

The output consists of three matrices:

- Z – $N \times N$ matrix of which only r columns are meaningful, r factor score normed columns;
- C – $V \times V$ matrix of corresponding feature loading columns (normed) of which only r are meaningful;
- M – $N \times V$ matrix with $r \times r$ diagonal submatrix of corresponding singular values sorted in the descending order on the top left, the part below and to the right is all zeros.

Worked example 5.7. SVD for Six Students dataset

For the centered 6×3 matrix in Table 5.13 the SVD matrices are as follows:

```
      −0.7086   0.1783    0.4534    0.4659   0.1888    0.0888
       0.2836  −0.6934    0.4706    0.0552   0.2786    0.3697
z =    0.0935   0.1841   −0.0486   −0.1870   0.9048   −0.3184
       0.4629   0.0931   −0.1916    0.8513   0.0604   −0.1092
      −0.3705  −0.3374   −0.7293    0.1083   0.2279    0.3916
       0.2391   0.5753    0.0455   −0.0922   0.1116    0.7673

      −0.6566   0.0846    0.7495              27.37     0        0
c =   −0.2514   0.9123   −0.3232    M =        0      26.13      0
       0.7111   0.4006    0.5778                0       0      17.26
```

Worked example 5.8. Standardized Student data visualized

To produce the scatter-plot of Fig. 5.7 with 100×4 Student data matrix Y, the following MatLab commands can be used:

```
>> [z,m,c]=svd(y);
>> z1*=z(:,1)*sqrt(m(1,1));
>> z2*=z(:,2)*sqrt(m(2,2));
% z1*, z2* are first PCs defined on p. 201
>> subplot(1,2,1); plot(z1*,z2*,'k.'); %Fig. 5.7, picture on the left
>> subplot(1,2,2);
>> plot(z1*, z2*,'k.', z1*(1:35),z2*(1:35),'k^',...z1*(70:100),z2*(70:100),'ko',ad1,ad2,'kp');
```

In the last command, there are several items to be shown on the same plot:

(i) z1*, z2*, 'k.' – these are black dot markers for all 100 entities exactly as on the plot on the left;
(ii) z1*(1:35), z2*(1:35),'k^' – these are triangles to represent entities 1 to 35 – those in category IT;
(iii) z1*(70:100), z2*(70:100),'ko' – these are circles to represent entities 70 to 100 – those in category AN;
(iv) ad1, ad2, 'kp' – ad1 is a 2×1 vector of the averages of z1* over entities 1 to 35 (category IT) and 70 to 100 (category AN), and ad2 a similar vector of within-category averages of z2*. These are represented by pentagrams.

Q.5.12. Assume that a category covers subset S of entities and y(S) represents the feature mean vector over S. Prove that the supplementary introduction of y(S) onto the plain of singular vectors z via equation $z^* = \sqrt{\mu}z = Y^*y(S)/\sqrt{\mu}$ from (5.12) onto the 2D PCA display is equivalent to representing the category by the averages of the 2D points z_{1i}^* and z_{2i}^* over $i \in S$. **A.** Indeed, the operation of averaging involves but addition and dividing by a number, which are not affected by a linear operation of matrix multiplication.

Worked example 5.9. Evaluation of the quality of visualization of the standardized Student data

To evaluate how well the data are approximated by the PC plane such as that on Fig. 5.7, according to Equation (5.17), one needs to assess the summary contribution of the first two singular values squared in the total data scatter. To get the squares one can multiply the diagonal matrix of singular values by itself and then see the proportion of the first two values in the total:

```
>> mu=m(1:4,:); %no need in 4×100 matrix output, have a square size 4×4
>> la=diag(mu*mu);% make squares and put them as a vector
>> lar=la*100/sum(sum(la)) % vector of the relative contributions of each PC
>> lar(1)+lar(2) % contribution of the 2 first components
```

This prints to the screen:

lar =
   42.3426
   29.7719
   15.6664
   12.2191

ans =
   72.1145

The latter is the sum of two first elements of the former – the proportion of the data scatter taken into account by the 2D visualization on Fig. 5.7.

5.3 Application: Latent Semantic Analysis

The number of papers applying PCA to various problems – image analysis, information retrieval, gene expression interpretation, complex data storage, etc. – makes many hundreds published annually. Some of the applications are well established techniques of their own. We present two such techniques: Latent semantic indexing (analysis) in this section and Correspondence analysis, in the next section.

P5.3.1 Latent Semantic Analysis: Presentation

Latent semantic analysis is an application of PCA to document analysis – information retrieval, first of all, using document-to-keyword data (see Deerwester et al. 1990).

Information retrieval is an application that no computational data analysis may skip: given a set of records or documents stored, find out those related to a specific query expressed by a set of keywords. Initially, at the dawn of computer era, when all the documents were stored in the same database, the problem was treated in a hard manner – only documents containing the query words were to be given to the user. Currently, this is a much softer problem that is being constantly and efficiently solved by various search engines such as Google, for millions of World Wide Web users (see Manning et al. 2008).

In its generic format, the problem can be illustrated with data in Table 5.15, already utilized as Table 4.1 in Section 4.2. It refers to a number of newspaper

Table 5.15 Database of 12 newspaper articles along with 10 keywords and the conventional coding of term frequencies. The articles are labeled F for Feminism, E for Entertainment and H for Household. One line holds document frequencies of terms (df) and the other, inverse document frequency weights (idf)

Article	Keyword									
	Drink	Equal	Fuel	Play	Popular	Price	Relief	Talent	Tax	Woman
F1	1	2	0	1	2	0	0	0	0	2
F2	0	0	0	1	0	1	0	2	0	2
F3	0	2	0	0	0	0	0	1	0	2
F4	2	1	0	0	0	2	0	2	0	1
E1	2	0	1	2	2	0	0	1	0	0
E2	0	1	0	3	2	1	2	0	0	0
E3	1	0	2	0	1	1	0	3	1	1
E4	0	1	0	1	1	0	1	1	0	0
H1	0	0	2	0	1	2	0	0	2	0
H2	1	0	2	2	0	2	2	0	0	0
H3	0	0	1	1	2	1	1	0	2	0
H4	0	0	1	0	0	2	2	0	2	0
df	5	5	6	7	7	8	5	6	4	5
idf	0.88	0.88	0.69	0.54	0.54	0.41	0.88	0.69	1.10	0.88

articles related to subjects such as entertainment, feminism and households, conveniently coded with letters E, F and H, respectively. Columns correspond to keywords, or terms, listed in the first line of the table, and entries refer to term frequency in the articles, according to a conventional coding scheme:

0 – no occurrence,
1 – occurs once,
2 – occurs twice or more.

The user may wish to retrieve all the articles on the subject of households, but they are subjected to inquire by using the listed keywords only. For example, query "fuel" will retrieve all four of the household related articles, and, in fact more than that – E1 and E3 will show up too; query "tax" will get four items, three – H1, H3, and H4 – on the subjects of household and one – E3 – on the subject of entertainment. No combination of these two can improve the result.

This is very much a class description problem; just the decision rules, the queries, must be combinations of keywords. The error of such a query is characterized by two characteristics, precision and recall (see Section 4.2.3). For example, "fuel" query's precision is $4/6 = 2/3$ since only four of six are relevant and recall is 1 because all of the relevant documents have been returned. Similarly, for "tax" query both precision and recall are $3/4$.

The rigidity of the query format does not fit well into the polysemy of natural language – such words as "fuel" or "play" have more than one meaning – thus leading to impossibility of exact information retrieval in many cases.

The method of latent semantic indexing (LSI) utilizes the SVD decomposition of the document-to-term data to soften and thus improve the query system by embedding both documents and terms into a subspace of singular vectors of the data matrix.

Before proceeding to SVD, the data table sometimes is pre-processed, typically, with what is referred to Term-Frequency-Inverse-Document-Frequency (tf-idf) normalization. This procedure gives a different weight to any keyword according to the number of documents it occurs at (document frequency df). The intuition is that the greater the document frequency, the more common and thus less informative is the word. The idf weighting assigns each keyword with a weight inversely proportional to the logarithm of its document frequency. For Table 5.15, df and idf weights are in its last lines.

The term frequency, tf, value is that of the corresponding data matrix entry referring to the frequency of the occurrence of the column word in the row document related to the document size.

After the SVD of the data matrix is obtained, the documents are considered points of the subspace of a few first singular vectors. The dimension of the space is not very important here, though it still should be much smaller than the original dimension. Good practical results have been reported at the dimension of about 100-200 when the number of documents in tens and hundred thousands and the number of keywords in thousands. A query is also represented as a point in the same space. The

5.3 Application: Latent Semantic Analysis

principal components, in general, are considered as "orthogonal" concepts underlying the meaning of terms. This however, should not be taken too literally as the singular vectors can be quite difficult to interpret. Also, the representation of documents and queries as points in a Euclidean space is referred to sometimes as the vector space model in information retrieval.

The Euclidean space format allows to measure similarity between items using the inner product or even what is called cosine - the inner product between rows that have been pre-normalized. Then a query would return the set of documents whose similarity to the query point is greater than a threshold. This tool may provide for a better resolution in the problem of information retrieval, because it well separates different meanings of synonyms.

Worked example 5.10. Latent semantic space for article-to-term data

Let us illustrate the LSI at the data in Table 5.15. To apply tf-idf normalization, we assume that all the documents have the same length so that the absolute term frequencies in Table 5.15 can be used as tf estimates. Then the tf-idf coding of any entry is equal to the entry value multiplied by the corresponding idf value (the last line in Table 5.15).

Consider the combination fuel-price-relief-tax as a query Q that should relate to Household category. Table 5.16 contains data that are necessary for computing coordinates of a query Q in the concept space. The query is represented by 1/0 vector in the last line of Table 5.16. The first coordinate of Q image on the map is computed by summing all the corresponding components of the first left singular vector and dividing the result by the square root of the first singular value: $u1 = (-0.34 - 0.42 - 0.29 - 0.24)/8.6^{1/2} = -0.44$. The second coordinate is computed similarly from the second singular vector and value. $u2 = (-0.25 - 0.22 - 0.35 - 0.33)5.3^{1/2} = -0.48$. These correspond to the pentagram on the left part of Fig. 5.11. One can think that query Q corresponds to an additional row in the data table which is equal to the last line in Table 5.16, so that the visualization algorithm of PCA applies to that line.

Figure 5.11 represents the documents in the space of two first principal components; left part corresponds to the original term frequency codes and the right part to the data tf-idf normalized.

Table 5.16 Two first singular vectors of term frequency data in Table 5.15

	Ei.value	Contrib.,%	Left singular vectors normed									
1st comp.	8.6	46.9	−0.25	−0.19	−0.34	−0.40	−0.39	−0.42	−0.29	−0.32	−0.24	−0.22
2nd comp.	5.3	17.8	0.22	0.34	−0.25	−0.07	0.01	−0.22	0.35	0.48	−0.33	0.51
Query Q			0	0	1	0	0	1	1	0	1	0

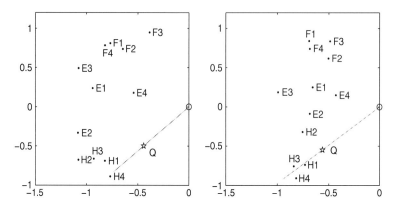

Fig. 5.11 Two first principal components plane for data in Table 5.15, both in the original format (*left*) and after tf-idf normalization (*right*). Query Q combining fuel-price- relief-tax keywords is mapped to the pentagram; the line connects it with the origin 0

As one can see, both representations separate the three subjects, F, E and H, more or less similarly, and provide the query Q with a rather good resolution. Taking into account the position of the origin of the concept space – the circle in the middle of the right boundary, the four H items are indeed have very good angular similarity to the pentagram representing the query Q.

The SVD representation of documents is utilized in other applications such as text mining, web search, text categorization, software engineering, etc.

F5.3.2 Latent Semantic Analysis: Formulation

The full SVD of data matrix F leads to equation $F = ZMC^T$ where Z and C are matrices whose columns are right and left normed singular vectors of F and M is a diagonal matrix with the corresponding singular values of F on the diagonal. By leaving only K columns in these matrices, we substitute matrix F by matrix $Z_K M_K C_K^T$ so that the entities are represented in the K-dimensional concept space by the rows of matrix $Z_K M_K^{1/2}$.

To translate a query presented as a vector q in the V-dimensional space into the corresponding point u in the K-dimensional concept space, one needs to take the product $g = C_K^T q$, which is equal to $g = z M_K^{1/2}$ according to the definition of singular values, after which z is found as $z = g M_K^{-1/2}$. Specifically, k-th coordinate of vector z is calculated as $z_k = <c_k, q>/\mu_k^{1/2}$ ($k = 1, 2, \ldots, K$).

The similarities between rows (documents), corresponding to row-to-row inner products in the concept space are computed as $Z_K M_K^2 Z_K^T$ and, similarly, the similarities between columns (keywords) are computed according to the dual formula $C_K M_K^2 C_K^T$. Applying this to the case of the K-dimensional point z representing the original V-dimensional vector q, its similarities to N original entities are computed as $z M_K^2 Z_K^T$.

C5.3.3 Latent Semantic Analysis: Computation

Let X be $N \times V$ array representing the original frequency data. To convert that to the conventional coding, in which all the entries larger than 1 are coded by 2, one can use this operation:

≫ Y=min(X,2*ones(N,V));

Computing vector *df* of document frequencies over matrix Y can be done with this line:

≫df=zeros(1,V); for k=1:V;df(k)=length(find(Y(:,k)>0));end;

and converting *df* to the inverse-document-frequency weights, with this:

≫ idf=log(N./df);

After that, it-idf normalization can be made by using command

≫YI=Y.*repmat(idf, N,1);

Given term frequency matrix Y, its K-dimensional concept space is created with commands:

≫ [z,m,c]=svd(Y);
≫zK=z(:, [1:K]); cK=c(:, [1:K]); mK=m([1:K], [1:K]);

Worked example 5.11. Drawing Figure 5.11

Consider that z is the matrix of normed document score singular vectors, c the matrix of normed keyword loading vectors, and m the matrix of singular values of the data in Table 5.15 as they are.

To draw the left part of Fig. 5.11, one can define the coordinates with vectors z1 and z2:

≫ z1=z(:,1)*sqrt(m(1,1)); %first coordinates of N entities in the concept space
≫ z2=z(:,2)*sqrt(m(2,2)); %second coordinates of N entities in the concept space

Then prepare the query vector and its translation to the concept space:
≫ q=[0 0 1 0 0 1 1 0 1 0]; % "fuel, price, relief, tax" query vector
≫ d1=q*c(:,1)/sqrt(m(1,1)); %first coordinate of query q in the concept space
≫ d2=q*c(:,2)/sqrt(m(2,2)); %second coordinate of query q in the concept space

After this, an auxiliary text data should be put according to MatLab requirements:
≫ tt={'E1','E2',..., 'H4'}; % cell of 12 names of the items in data matrix
≫ll=[0:.04:1.5]; zd1=d1*ll; zd2=d2*ll;
% pair zd1, zd2 will draw a line through origin and point (d1,d2)

Now we are ready for plotting the left drawing on Fig. 5.11:
≫ subplot(1,2,1);
≫ plot(u1,u2,'k.',d1,d2,'kp',0,0,'ko',ud1,ud2);text(u1,u2,tt);

```
≫ text(d1,d2,' Q');
≫ axis([−1.5 0 −1 1.2]);
```

The arguments of plot command here are:
u1,u2,'k.' – black dots corresponding to the original entities;
d1,d2,'kp' – black pentagram corresponding to query q;
0,0,'ko' – black circle corresponding to the space origin;
ud1,ud2 – line through the query and the origin.

Command *text* provides for string labels at corresponding points. Command *axis* specifies the boundaries of the Cartesian plane box on the figure, which can be used for making different plot boxes uniform.

The plot on the right of Fig. 5.11 is coded similarly by using SVD of tf-idf matrix *YI* rather than *Y*.

5.4 Application: Correspondence Analysis

P5.4.1 Correspondence Analysis: Presentation

Correspondence Analysis is an extension of PCA to contingency tables taking into account the specifics of co-occurrence data: they are not only comparable across the table but also can be meaningfully summed up across the table. This leads to a unique way of standardization of such data – by using the Quetelet coefficients rather than the original frequencies, which is an advantage over the common situations in which the data standardization is rather arbitrary.

Correspondence Analysis (CA) is a method for visually displaying both row and column categories of a contingency table $P = (p_{ij}), i \in I, j \in J$, in such a way that distances between the presenting points reflect the patterns of co-occurrences in P. This method is usually introduced as a set of dual heuristics applied simultaneously to rows and columns of the contingency table (see, for example, Lebart, Morineau and Piron 1995). Yet there is a way for introducing CA as a decoder based data recovery technique similar to that used for introducing PCA above. According to this perspective (Mirkin 1996), CA is a version of PCA differing from PCA, due to the specifics of contingency data, in the following aspects:

(i) CA decoder applies to the relative Quetelet coefficients rather than to the original frequency data;
(ii) Both rows and columns are assigned with weights equal to the marginal frequencies – these weights are used in the least-squares criterion as well as the orthogonality conditions;
(iii) Both rows and columns are visualized on the same display in such a way that the geometric distances between the representing them points reflect the so-called chi-square distances between row and column conditional frequency profiles (see (5.24) in 5.4.2);
(iv) Data scatter is measured by the Pearson chi-square association coefficient rather than just the sum of squares of Quetelet coefficients.

5.4 Application: Correspondence Analysis

Worked example 5.12. Correspondence analysis of Protocol/Attack contingency table

Consider Table 5.17, a copy of Table 3.6 representing the distribution of protocol types and attack types according to Intrusion data in Section 1.2: totals on its margins show separate summary distributions of protocols and attacks.

To apply the method to data in Table 5.17, we first transform it into Quetelet coefficients (see Table 5.18).

This standardization does make the data structure somewhat sharper, as has been explained in Section 3.4. One can see, for example, that $q = 9$ ($= 900\%$) for the equivalent pair *(Icmp, surf)*. But the transformation $p \Rightarrow q$ does not work alone in CA. It is coupled with the weighting of each row and column by its corresponding marginal probability so that the squared errors in the criterion are weighted by products of the marginal probabilities. Moreover, the vector norm is weighted by them too.

Figure 5.12 represents a CA visualization of Table 5.17 derived as described in Section 5.4.3, which shows indeed that the equivalent Smurf and Icmp fall in the same place; Norm is much associated to Udp, and Apache and Saint are between Tcp and Icmp protocols. The Norm category slightly falls out: all points representing columns should be within the convex closure of the three protocol points because of Equation (5.23) relating row points and column points.

F5.4.2 Correspondence Analysis: Formulation

Correspondence Analysis (CA) is a method for visually displaying both row and column categories of a contingency table $P = (p_{ij})$, $i \in I$, $j \in J$, in such a way that distances between the presenting points reflect the pattern of co-occurrences in P.

Table 5.17 Protocol/attack contingency table for intrusion data

Category	Apache	Saint	Smurf	Norm	Total
Tcp	23	11	0	30	64
Udp	0	0	0	26	26
Icmp	0	0	10	0	10
Total	23	11	10	56	100

Table 5.18 Quetelet indexes, per cent, for the Protocol/Attack contingency Table 5.17

Category	Apache	Saint	Smurf	Norm
Tcp	56.25	56.25	−100.00	−16.29
Udp	−100.00	−100.00	−100.00	**78.57**
Icmp	−100.00	−100.00	**900.00**	−100.00

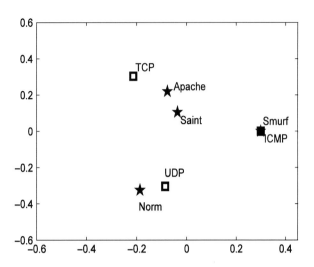

Fig. 5.12 Visualization of Protocol/Attack contingency data in Table 5.17 using Correspondence Analysis. *Squares* stand for protocol types and *stars* for attack categories

To be specific, let us take on the issue of visualization of P on a 2D plane so that we are looking for just two approximating factors, $u_1 = (v_1, w_1)$ where $v_1 = (v_1(i))$ and $w_1 = (w_1(j))$ and $u_2 = (v_2, w_2)$ where $v_2 = (v_2(i))$ and $w_2 = (w_2(j))$, with $I \cup J$ as their domain, such that each row $i \in I$ is displayed as point $u(i) = (v_1(i), v_2(i))$ and each column $j \in J$ as point $u(j) = (w_1(j), w_2(j))$ on the plane as shown in Fig. 5.12.

The $|I|$-dimensional vectors v_t and $|J|$-dimensional vectors w_t constituting the $u_t (t = 1, 2)$ are calculated to approximate the relative Quetelet coefficients $q_{ij} = p_{ij}/(p_{i+}p_{+j}) - 1$ rather than the co-occurences p_{ij} themselves, according to equations:

$$q_{ij} = \mu_1 v_1(i) w_1(j) + \mu_2 v_2(i) w_2(j) + e_{ij} \tag{5.20}$$

where μ_1 and μ_2 are positive reals, by minimizing the weighted least-squares criterion

$$E^2 = \sum_{i \in I} \sum_{j \in J} p_{i+} p_{+j} e_{ij}^2 \tag{5.21}$$

with regard to μ_t, v_t, w_t, subject to conditions of weighted orthonormality:

$$\sum_{i \in I} p_{i+} v_t(i) v_{t'}(i) = \sum_{j \in J} p_{+j} w_t(j) w_{t'}(j) = \begin{cases} 1, & \text{if } t = t' \\ 0, & \text{if not} \end{cases} \tag{5.22}$$

where $t, t' = 1, 2$.

The weighted criterion E^2 is equivalent to the unweighted least-squares criterion L^2 applied to the matrix R that has Pearson indexes $r_{ij} = q_{ij}(p_{i+}p_{+i})^{1/2} = (p_{ij} - p_{i+}p_{+i})/(p_{i+}p_{+i})^{1/2}$ as its entries. This implies that the factors v and w are

5.4 Application: Correspondence Analysis

determined by the singular-value decomposition of matrix $R = (r_{ij})$. More explicitly, the two maximal singular values μ_t and corresponding singular vectors $f_t = (f_{it})$ and $g_t = (g_{it})$ of matrix R, defined by equations $Rg_t = \mu_t f_t$, $R^T f_t = \mu_t g_t$ ($t = 1, 2$) determine the optimal values μ_t and optimal solutions to the problem of minimization of (5.21)–(5.22). Indeed, these singular triplets relate to the optimal solution according to equations $v_t(i) = f_{it}/(p_{i+}^{1/2})$ and $w_t(j) = g_{jt}/(p_{+j}^{1/2})$. The proof follows from the first-order optimality conditions for the Lagrange function of the problem (5.21)–(5.22).

The singular triplet equations can be rewritten in terms of v_t and w_t, as follows:

$$\sum_{j \in J} \frac{p_{ij}}{p_{i+}} w_t(j) = \mu_t v_t(i), \quad \sum_{i \in I} \frac{p_{ij}}{p_{+j}} v_t(i) = \mu_t w_t(j) \qquad (5.23)$$

To prove the left-hand equation, take equation $Rg_t = \mu_t f_t$ in its component-wise form, $\sum_{j \in J} r_{ij} g_j = \mu f_i$ (index t omitted for the sake of convenience) and substitute by vectors v and w defined above: $\sum_{j \in J} r_{ij} (\frac{p_{+j}}{p_{i+}})^{1/2} w(j) = \mu v(i)$. This is equivalent to $\sum_{j \in J} (\frac{p_{ij}}{p_{i+}} - p_{+j}) w(j) = \mu v(i)$. To complete the proof, equation $\sum_j p_{+j} w(j) = 0$ is to be proven. To do that, let us first prove that vector g_0 whose components are $p_{+j}^{1/2}$ is a singular vector of R corresponding to singular value 0 (the other singular vector is equal to $f_0 = (p_{i+}^{1/2})$). Indeed, $\sum_{j \in J} r_{ij} p_{+j}^{1/2} = (1/p_{i+}^{1/2}) \sum_{j \in J} (p_{ij} - p_{i+} p_{+j}) = (1/p_{i+}^{1/2})(p_{i+} - p_{i+}) = 0$. Then the equation $\sum_j p_{+j} w(j) = 0$ follows from the fact that all the singular vectors are mutually orthogonal so that singular vector g corresponding to w is orthogonal to g_0, which proves the statement. The right-hand equation can be proven in a similar way, from equation $R^T f_t = \mu_t g_t$.

Equations (5.23) are referred to as transition equations and considered to justify the joint display of rows and columns because the row-points $v_t(i)$ appear to be averaged column-points $w_t(j)$ and, vice versa, the column-points appear to be averaged versions of the row-points, up to the singular value of μ_t course.

The mutual location of the row-points is considered as justified by the fact that between-row-point squared Euclidean distances $d^2(u(i), u(i'))$ approximate the chi-square distances between corresponding rows of the contingency table. Specifically, chi-square distance is defined a weighted squared Euclidean distance:

$$\chi^2(i, i') = \sum_{j \in J} p_{+j} (q_{ij} - q_{i'j})^2 = \sum_{j \in J} (p_{ij}/p_{i+} - p_{i'j}/p_{i'+})^2 / p_{+j}. \qquad (5.24)$$

Here $u(i) = (v_1(i), v_2(i))$ for v_1 and v_2 rescaled in such a way that their norms are equal to μ_1 and μ_2, respectively. A similar property holds for columns j, j'. In fact, it is the right-hand item in (5.24) which is used to define the chi-squared distance (Lebart et al. 1995), but the definition in terms of Quetelet coefficients in the middle of (5.24) (Mirkin 1996) looks more natural. The distance is dubbed chi-square distance because of its links to the chi-square coefficient for table P. First of all if we take the weighted chi-square summary distance to 0, $\sum_{i \in I} p_{i+} \chi^2(i, 0)$ where 0 is put instead of $q_{i'j}$ in (5.24), it is easy to see that this is the Pearson chi-squared coefficient, without the factor N of course, which is simultaneously the expression for the data scatter according to criterion E^2 in (5.21):

$$\sum_{i \in I} p_{i+} \chi^2(i, 0) = \sum_{i \in I} \sum_{j \in J} p_{i+} p_{+j} e_{ij}^2 = X^2/N \quad (5.25)$$

The weighted data scatter is equal to the scatter of R, the sum of its squared entries $T(R)$, which can be easily calculated from the definition of R. Indeed, $\sum_{i \in I} \sum_{j \in J} (p_{ij} - p_{i+}p_{+j})^2/(p_{i+}p_{+j}) = X^2/N$. This implies that

$$X^2/N = \mu_1^2 + \mu_2^2 + E^2 \quad (5.26)$$

which can be seen as a decomposition of the contingency data scatter, expressed by X^2, into contributions of the individual factors, μ_1^2 and μ_2^2, and unexplained residuals, E^2. (Only two factors are considered here, but the number of factors to be found can be raised up to the rank of matrix R with no other changes).

In a common situation, the first two singular values account for a major part of X^2, thus justifying the use of the plane of the first two factors for visualization of the interrelations between I and J.

C5.4.3 Correspondence Analysis: Computation

Given a contingency table P, the computation of correspondence analysis factors can go in three steps: (a) computing Pearson index matrix R, (b) finding the singular decomposition of R and the two first correspondence analysis factors, and (c) visualization of the joint display of rows and columns of P. Here are MatLab commands for these.

(a) Computing Pearson index matrix R

```
>> Pc=sum(P); Pr=sum(P'); total=sum(Pc);
>> P=P/total; %relative frequencies
>> Pc=Pc/total; %column relative frequencies
>> Pr=Pr/total; %row relative frequencies
>> Prod=Pr'*Pc; % matrix of products
>> rProd=Prod.^(0.5); % square roots of products
>> r=(P−Prod)./rProd; % Pearson index matrix
```

(b) Finding the correspondence analysis factors:

```
>> [a,mu,b]=svd(r);
>> % finding first factor
>> x1=(a(:,1).*sqrt(Pr'))*sqrt(mu(1,1));
>> y1=(b(:,1).*sqrt(Pc'))*sqrt(mu(1,1));
>> % finding second factor
>> x2=(a(:,2).*sqrt(Pr'))*sqrt(mu(2,2));
>> y2=(b(:,2).*sqrt(Pc'))*sqrt(mu(2,2));
```

5.4 Application: Correspondence Analysis

As a bonus, one can estimate the proportion of data scatter, the chi-squared, taken into account by the factors, and display it on the screen:

```
>> yy=r.*r; chi=sum(sum(yy))% data scatter
>> ccn=(mu(1,1)^2+mu(2,2)^2)*100/chi;
     %contribution of the first two
>> disp('Contribution of the solution:'); ccn
```

(c) Visualization of the joint display of rows and columns of P. The plot is easy to do with command

```
>> plot(x1,x2,'ks', y1,y2,'kp');
```

Yet to make the points annotated with row and column names, which are to be available in a string cell termed say "names", the joint set of rows and columns should get their x-coordinate and y-coordinate vectors, z1 and z2 below:

```
>> z1=[x1' y1']; z2=[x2' y2']; text(z1,z2,names);
>> v=axis; axis(1.5*v);
```

The last line is to make the plot to look tighter by extending its boundaries.

Q.5.13. In many situations, (a) the first singular vector all positive and (b) the second singular vector half negative. Why can be that? **A.** A typical situation: (a) All features are positively correlated which implies that the first eigenvector is positive; (b) the second must be orthogonal to the first, to make 0 their inner product.

Q.5.14. For the data in Table 5.10, as well as many others, svd function in MatLab produces first singular vectors z and c negative, which contradicts the meaning of them as talent scores and subject loadings. Can anything be done about that? **A.** Yes, they can be changed to $-z$ and $-c$ without compromising their singular vector status.

Q.5.15. Is matrix

$$\begin{matrix} 1 & 2 \\ 2 & 1 \end{matrix}$$

of rank 1 or not? **A.** The rows are not proportional to each other, thus not.

Q.5.16. Prove that if matrix Y is symmetric then its eigenvalues and vectors (λ, z) are simultaneously its singular triplets (λ, z, z).

Q.5.17. Find a matrix of rank 1 that is the nearest to matrix in Q.5.15 according to the least-squares criterion. **A.** The solution is given by the first singular value and corresponding singular vectors which are the same as the first eigenvalue and and corresponding eigen-vector, $\lambda = 3$ and $z = (1/\sqrt{2}, 1/\sqrt{2})$, thus leading to matrix

$$\begin{matrix} 3/2 & 3/2 \\ 3/2 & 3/2 \end{matrix}$$

as the solution.

Q.5.18. For positive a and b, inequality $a < b$ can be equivalently expressed as $a/b < 1$. The difference between a and b does not change if $c > 0$ is added to both of them, but the ratio does. Prove that for any $c > 0$, $(a+c)/(b+c) > a/b$ – this would illustrate that the further away the positive data are from zero, the greater the contribution of the first principal component.

Q.5.19. There is another representation of a singular value problem as an eigenvalue problem. Given a rectangular $N \times V$ matrix Y, consider a $(N+V) \times (N+V)$ matrix Y^* that consists of four blocks, two of which, the diagonal $N \times N$ and $V \times V$ blocks, are all zeros:

$$Y^* = \begin{pmatrix} 0 & Y \\ Y^T & 0 \end{pmatrix}.$$

Prove that a triplet (μ, z, c) is singular for Y if and only if μ is eigenvalue for Y^* corresponding to eigenvector $y = (z, c)$ in which first N components are taken by z and the remaining V components, by c. **A.** Consider an arbitrary eigenvalue μ and corresponding eigenvector y of matrix Y^* and denote the vector of its first N component by z, and the rest by c so that $y = (z, c)$. The product $Y^* y$ will have its first N components equal to $0z + Yc = Yc$ and the next V components equal to $Y^T z + 0c = Y^T z$. Since $Y^* y = \mu y$, that means that $Yc = \mu z$ and $Y^T z = \mu c$, which proves the statement.

Q.5.20. Prove that if an eigenvector $y = (z, c)$ of Y^* is normed, then its components z and c both have norms of $0.5^{1/2}$. **A.** Indeed, $\|y\|^2 = \|z\|^2 + \|c\|^2$ because these are just sums of the squared components. On the other hand, as proven in Q.5.19, z and c are singular vectors of Y so that they must have equal norms because of Property 1 in Section 5.2.2. This leads to equation $\|z\|^2 = \|c\|^2 = 1/2$, which proves the statement.

Q.5.21. Prove that if μ is an eigenvalue of Y^* corresponding to its eigenvector $y = (z, c)$ then so is its negation $-\mu$ corresponding to eigenvector $y = (-z, c)$. **A.** Indeed, equations $Yc = \mu z$ and $Y^T z = \mu c$ hold if and only if $Yc = (-\mu)(-z)$ and $Y(-z) = (-\mu)c$.

5.5 Summary

This chapter introduces the concept of data summarization as a coder-decoder pair and describes the method of principal components (PCA) as a data-driven model in this framework. Luckily, this model is underlied with a well developed mathematical theory of singular value decomposition (SVD) for rectangular matrices. Unlike the conventional formulation of PCA, this model does not require to postulate that the principal components are to be linear combinations of features. This property is rather derived from the model. Yet the PCA model itself is rather simplistic and suggests that further thinking on better data summarization models should be undertaken.

Three applications of PCA – scoring hidden factors, data visualization, and feature space reduction are illustrated with further instructions and worked examples. Two more distant applications, Latent semantic analysis (for disambiguation in document retrieval) and Correspondence analysis (for visualization of contingency tables), are explained in full, too.

References

Deerwester, S., Dumais, S., Furnas, G.W., Landauer, T.K., Harshman, R.: Indexing by Latent Semantic Analysis, J. Am. Soc. Inf. Sci. **41**(6), 391–407 (1990)

Hair, J.F., Black, W.C., Babin, B.J., Anderson, R.E.: Multivariate Data Analysis, 7th edn, Prentice Hall, ISBN-10: 0-13-813263-1 (2010)

Kendall, M.G., Stewart, A.: Advanced Statistics: Inference and Relationship, 3rd edn. Griffin, London, ISBN: 0852642156 (1973)

Lebart, L., Morineau, A., Piron, M.: Statistique Exploratoire Multidimensionelle. Dunod, Paris, ISBN 2-10-002886-3 (1995)

Manning, C.D., Raghavan, P., Schütze, H.: Introduction to Information Retrieval. Cambridge, Cambridge University Press (2008)

Mirkin, B.: Mathematical Classification and Clustering. Dordrecht, Kluwer Academic Press (1996)

Mirkin, B.: Clustering for Data Mining: A Data Recovery Approach. London, Chapman & Hall/CRC, ISBN 1-58488-534-3 (2005)

Chapter 6
K-Means and Related Clustering Methods

6.1 General

Clustering is a set of methods for finding and describing cohesive groups in data, typically, as "compact" clusters of entities in the feature space.

Consider data patterns on Fig. 6.1: a clear-cut cluster structure on part (a), a blob on (b), and an ambiguous "cloud" on (c).

Some argue that term "clustering" applies only to structures of the type presented on Fig. 6.1a, c, moreover, depending on the resolution, one may distinguish 3 or 7 clusters on (c). Yet there are no "natural" clusters in the other two cases, Fig. 6.1a and 6.1b. Indeed, initially the term was used to express a clear-cut clustering. But currently clustering has become synonymous to building a classification over empirical data, and as such it embraces all the situations in which data is structured into cohesive chunks.

To serve as models of natural classes and categories, clusters need to be not only found but conceptually described as well. A class always expresses a concept embedded into a fragment of knowledge – this is what is referred to, in logics, as the class' "intension", in contrast to empirical instances of the class constituting what is referred to as the class' "extension", e.g., the concept of "tree" versus real plants growing here and there. Therefore, two dual intelligent activities – cluster finding and cluster describing – should be exercised both when clustering.

As Fig. 6.2 illustrates, a cluster is rather easy to describe by combining corresponding feature intervals when it is clear-cut. This knowledge-driven data analysis perspective can be reflected in dividing all cluster finding techniques in the following categories:

(a) clusters are to be found directly in terms of features – this is frequently referred to as conceptual clustering;
(b) clusters are to be found simultaneously with a transformation of the feature space making them clear-cut – this direction only started very recently and is not well shaped yet;
(c) clusters are to be found as subsets of entities first, so that the description comes as a follow-up stage – this is the genuine clustering which covers most of the clustering activities so far.

Fig. 6.1 Clear-cut cluster structures at (**a**) and (**c**); data clouds with no clear structure at (**b**) and (**d**)

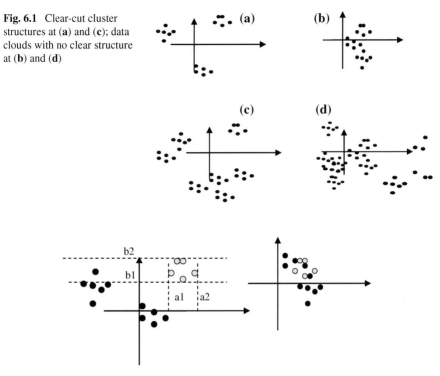

Fig. 6.2 Cluster of *blank circles* on the right is well described by the predicate a1<x<a2 & b1<y<b2. A similar cluster on the right cannot be accurately described by interval predicates without false positive and false negative errors

6.2 K-Means Clustering

P6.2.1 Batch K-Means Partitioning

K-Means is a major clustering technique, of type (c), that is present, in various forms, in all major statistical packages such as SPSS and SAS, as well as data mining packages such as Clementine, iDA tool and DBMiner. It is very popular in many application areas such as image analysis, marketing research, bioinformatics, and medical informatics.

In general, the cluster finding process according to K-Means starts from K tentative centroids and repeatedly applies two steps:

(a) collecting clusters around centroids,
(b) updating centroids as within cluster means,

 – until convergence.

6.2 K-Means Clustering

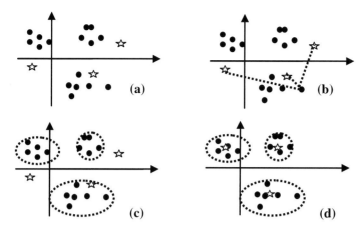

Fig. 6.3 Main steps of Batch K-Means: (**a**) initialization of centroids, (**b**) cluster update using Minimum distance rule (the pointed lines show distances from an entity to all centroids), (**c**) cluster update completed, (**d**) centroid update completed

This makes much sense – whichever centroids are suggested first, as hypothetical cluster tendencies, they are checked then against the real data and moved to the areas of higher density.

In its generic, so-called Batch mode, K-Means can be formulated as comprising the following steps 0–3 illustrated on Fig. 6.3 for $K = 3$ and entity set I:

0. *Initialization*: the user chooses the number K of clusters and puts K hypothetic cluster centroids among the entity points, see Fig. 6.3a;
1. *Cluster update*: Given K centroids c_k ($k = 1, 2, \ldots, K$), each of the entities $i \in I$ is assigned to one of the centroids according to Minimum distance rule: distances between i and each c_k are calculated, and i is assigned to the nearest c_k, see Fig. 6.3b. For each centroid c_k, the entities assigned to it form cluster S_k ($k = 1, 2, \ldots, K$), see Fig. 6.3c.
2. *Centroid update*: At each of the given K clusters S_k, its gravity center is computed and set as the new centroid c_k' ($k = 1, 2, \ldots, K$), see Fig. 6.3d.
3. *Halting test*: New centroids c_k' are compared with those from the previous iteration. If $c_k' = c_k$ for all $k = 1, 2, \ldots, K$, stop and output both c_k' and S_k for all $k = 1, 2, \ldots, K$. Otherwise, set c_k' as c_k and go to "1. Cluster update step".

The algorithm is appealing in several aspects. Conceptually it may be considered a model for the human process of typology making, with types represented by clusters S_k and centroids c_k. Also, it has nice mathematical properties. This method is computationally easy, fast and memory-efficient. However, researchers and practitioners point to some less desirable properties of K-Means. Specifically, they refer to lack of advice with respect to

(a) the initial setting, i.e. the number of clusters K and initial positioning of centroids,

(b) instability of clustering results with respect to the initial setting and data standardization, and
(c) insufficient interpretation aids.

These issues can be alleviated, to an extent, as will be explained later in this section.

A decoder based summarization model underlying the method is that the entities are assigned to clusters in such a way that each cluster is represented by its centroid, sometimes referred to as the cluster's standard point or prototype. This point expresses, intensionally, the typical tendencies of the cluster.

Worked example 6.1. K-Means clustering of Company data

Consider the standardized Company data in Table 5.9 copied here as Table 6.1.

This data set can be visualized with two principal components as presented on Fig. 6.4 (copied from Section 5.2.2).

For example, let entities An, Br and Ci be suggested centroids of three clusters. Now we can computationally compare each of the entities with each of the centroids to decide which centroid better represents an entity. To compare two points, Euclidean squared distance is a natural choice (see Table 6.2).

According to the Minimum distance rule, an entity is assigned to its nearest centroid (see Table 6.3 in which all distances between the entities and centroids are presented; those chosen according to the Minimum distance rule are highlighted in bold.) Entities assigned to the same centroid form a tentative cluster around it. Clusters found at Table 6.3 are

$S1 = \{Av, An, As\}$, $S2 = \{Ba, Br, Bu\}$, and $S3 = \{Ci, Cy\}$. These are the product classes already, but this is irrelevant to the computation. K-Means procedure has its own logic that needs to ensure that the new tentative clusters lead to the same centroids.

Table 6.1 The company data standardized by: (i) shifting to the feature averages, (ii) dividing by the feature ranges, and (iii) further dividing the category based three columns by 3. Contributions of the features to the data scatter are presented in the bottom

Av	−0.20	0.23	−0.33	−0.63	0.36	−0.22	−0.14
An	0.40	0.05	0	−0.63	0.36	−0.22	−0.14
As	0.08	0.09	0	−0.63	−0.22	0.36	−0.14
Ba	−0.23	−0.15	−0.33	0.38	0.36	−0.22	−0.14
Br	0.19	−0.29	0	0.38	−0.22	0.36	−0.14
Bu	−0.60	−0.42	−0.33	0.38	−0.22	−0.22	−0.14
Ci	0.08	−0.10	0.33	0.38	−0.22	−0.22	0.43
Cy	0.27	0.58	0.67	0.38	−0.22	−0.22	0.43
Cnt	0.74	0.69	0.89	1.88	0.62	0.62	0.50
Cnt %	12.42	11.66	14.95	31.54	10.51	10.51	8.41

6.2 K-Means Clustering

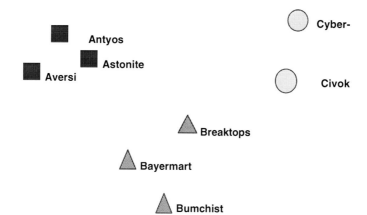

Fig. 6.4 Table 6.1 rows on the plane of two first principal components: it should not be difficult to discern clusters formed by products: distances within A, B and C groups are smaller than between them

Table 6.2 Computation of squared Euclidean distance between rows Av and An in Table 6.1 as the sum of squared differences between corresponding components

Points	Coordinates							d(An,Av)
An	0.40	0.05	0.00	−0.63	0.36	−0.22	−0.14	
Av	−0.20	0.23	−0.33	−0.63	0.36	−0.22	−0.14	
An-Av	0.60	−0.18	0.33	0	0	0	0	
Squares	0.36	0.03	0.11	0	0	0	0	0.50

Table 6.3 Distances between three company entities chosen as centroids and all the companies; each company column shows three distances between the company and centroids – the highlighted minima present best matches between centroids and companies

Point	Av	An	As	Ba	Br	Bu	Ci	Cy
An	**0.50**	**0.00**	**0.77**	1.55	1.82	2.99	1.90	2.41
Br	2.20	1.82	1.16	**0.97**	**0.00**	**0.75**	0.83	1.87
Ci	2.30	1.90	1.81	1.22	0.83	1.68	**0.00**	**0.61**

One needs to proceed further on and update centroids by using the information of the assigned clusters. New centroids are defined as centers of the tentative clusters, whose components are the averages of the corresponding components within the clusters; these are presented in Table 6.4.

The updated centroids differ from the previous ones. Thus we must update their cluster lists by using the distances between updated centroids and entities; the distances are presented in Table 6.5. As it is easy to see, the Minimum distance rule assigns centroids again with the same entity lists. Therefore, the process has stabilized – if we repeat it all over again, nothing new would ever come – the same

Table 6.4 Tentative clusters from Table 6.3 and their centroids

Av	−0.20	0.23	−0.33	−0.63	0.36	−0.22	−0.14
An	0.40	0.05	0	−0.63	0.36	−0.22	−0.14
As	0.08	0.09	0	−0.63	−0.22	0.36	−0.14
Centroid1	0.10	0.12	−0.11	−0.63	0.17	−0.02	−0.14
Ba	−0.23	−0.15	−0.33	0.38	0.36	−0.22	−0.14
Br	0.19	−0.29	0	0.38	−0.22	0.36	−0.14
Bu	−0.60	−0.42	−0.33	0.38	−0.22	0.36	−0.14
Centroid2	−0.21	−0.29	−0.22	0.38	−0.02	0.17	−0.14
Ci	0.08	−0.10	0.33	0.38	−0.22	−0.22	0.43
Cy	0.27	0.58	0.67	0.38	−0.22	−0.22	0.43
Centroid3	0.18	0.24	0.50	0.38	−0.22	−0.22	0.43

Table 6.5 Distances between the three updated centroids and all the companies; the highlighted column minima present best matches between centroids and companies

Point	Av	An	As	Ba	Br	Bu	Ci	Cy
Centroid1	**0.22**	**0.19**	**0.31**	1.31	1.49	2.12	1.76	2.36
Centroid2	1.58	1.84	1.36	**0.33**	**0.29**	**0.25**	0.95	2.30
Centroid3	2.50	2.00	1.95	1.69	1.20	2.40	**0.15**	**0.15**

centroids and the same assignments. The process stops at this point, and the found clusters along with their centroids are returned (they are, in the standardized format, in Table 6.4).

The result obviously depends on the standardization of the data performed beforehand, as the method heavily relies on the squared Euclidean distance and, thus, on the relative weighting of the features, just like PCA.

Case Study 6.1. Dependence of K-Means on Initialization: A Drawback and Advantage

The bad news is that the result of K-Means depends on initialization – the choice of the initial tentative centroids, even if we know, or have guessed correctly, the number of clusters K. Indeed, if we start from wrong entities as the tentative centroids, the result can be rather disappointing.

In some packages, such as SPSS (Green and Salkind 2003), K first entities are taken as the initial centroids. Why not start from rows for Av, An and As then? Taking these three as the initial tentative centroids will stabilize the process at wrong clusters S1 = {Av, Ba, Br}, S2 = {An}, and S3 = {As, Bu, Ci, Cy} (See Q.6.5). But what else can be expected if all centroids are taken from the same cluster?

However, even if centroids are taken from right clusters, this would not necessarily guarantee good results either. Start, for example, from Av, Ba and Ci (note, these produce different products!); the final result still will be rather disappointing – S1={Av,An,As}, S2={Ba, Bu}, and S3={Br, Ci, Cy} (see Q.6.6).

6.2 K-Means Clustering

Fig. 6.5 Case of two clear-cut clusters and two different initializations: (**a**) and (**b**). Case (**a**) results in a correct separation of the clusters, case (**b**) does not

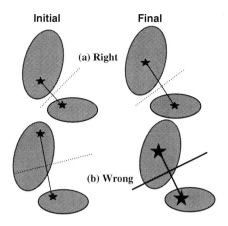

Figure 6.5 illustrates the fact that such instability is not because of a specially designed example but rather an ordinary phenomenon. There are two clear-cut clusters on Fig. 6.5, that can be thought of as uniformly distributed sets of points, and two different initializations, symmetric one on the (a) part and not symmetric one on (b) part. The Minimum distance rule at $K=2$ amounts to drawing a hyperplane that orthogonally cuts through the middle of line between the two centroids; the hyperplane is shown on Fig. 6.5 as the line separating the centroids. In Fig. 6.5, the case (a) presents initial centroids that are more or less symmetric so that the line through the middle separates the clusters indeed. In the case (b), initial centroids are highly asymmetric so that the separating line cuts through one of the clusters, thus distorting the position of the further centroids; the final separation still cuts through one of the clusters and, therefore, is utterly wrong.

There is one more property of K-Means clusters illustrated by Fig. 6.5: they are convex. Indeed, the Minimum Distance rule assigns each centroid with the intersection of half-spaces formed by the orthogonal cutting hyperplanes.

Another example of non-optimality of K-Means is presented on Fig. 6.6 which involves four points only.

Yet non-optimality of K-Means can be of an advantage, too – in those cases when K-Means criterion leads to solutions that are counterintuitive such as those in which the fact that K-Means favors equal cluster sizes brings good results.

Fig. 6.6 An example of K-Means failure: two clusterings at a four-point-set with K-Means – that intuitive on the *right* and that counter-intuitive on the *left*, with *stars* as centroids

Case Study 6.2. Uniform Clusters Can Be Too Costly

Here is an example when the square-error clustering criterion leads to a solution which is at odds with intuition, however that cannot be reached with batch K-Means algorithm because of its local nature.

Consider the case of Fig. 6.7 that presents three sets of points, two consisting of big clumps of say 100 entities each, around points A and B, and a small one around point C, consisting say of just one entity located at that point. Assume that the distance between A and B is 2, and between B and C, 10. There can be only two 2-cluster partitions possible: (I) 200 of A and B entities together in one cluster while the second cluster consists of just one entity in C; (II) 100 A entities for one cluster while 101 entities in B and C for the other. The third partition, consisting of cluster B and cluster A+C, cannot be optimal because cluster A+C is more outstretched than a similar cluster B+C in (II).

Let us compare the values of K-Means criterion using the squared Euclidean distance between entities and their centroids.

In case (I), centroid of cluster A+B will be located in the middle of the interval between A and B, thus on the distance 1 from each, leading to the total squared Euclidean distance $200*1 = 200$. Since cluster C contains just one entity, it adds 0 to the value of K-Means criterion, which is 200 in this case.

In case (II), cluster B+C has centroid, which is the gravity center, between B and C distanced from B by $d = 10/101$. Thus, the total value of K-Means criterion here is $100*d^2 + (10-d)^2$ which is less than $100*(1/10)^2 + 10^2 = 101$ because $d<1/10$ and $10-d<10$. Cluster A contributes 0 because all 100 entities are located in A which is, therefore, the centroid of this cluster.

Case (II) wins by a great margin: K-Means criterion, in this case, favors more equal distribution of entities between clusters in spite of the fact that case (I) is intuitive and case (II) is not: A and B are much closer together than B and C.

Yet Batch K-means algorithm leads to non-optimal, but intuitive, case (I) rather than optimal, but odd, case (II), if started from the most distant points A and C as initial centroids – an intuitively appealing option. Indeed, according to Minimal distance rule, all entities in B will join A, thus resulting in (I) clustering.

Fig. 6.7 Three sets of points subject to 2-Means clustering: which two will join together, A and B or B and C?

Case Study 6.3. Robustness of K-Means Criterion with Data Normalization

Let us generate two 2D clusters, of 100 and 200 elements. First cluster – a Gaussian spherical distribution with the mean in point (1,1) and the standard deviation 0.5. The second cluster, of 200 elements, is uniformly randomly distributed in the

6.2 K-Means Clustering

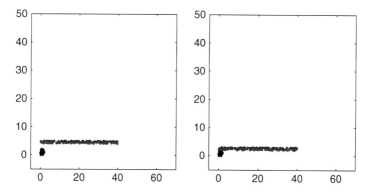

Fig. 6.8 A set of two differently shaped clusters, a *circle* and *rectangle*; the y-coordinates of their centers are 1 (*circle*) and 5.5 (*rectangle*) on the *left*, and 1 and 3.5, on the *right*

rectangle of the length 40 and width 1, put over axis x either at $y = 5$ (Fig. 6.8, on the left) or at $y = 3$ (Fig. 6.8, on the right). K-Means criterion of course cannot separate these two if applied in the original space; its criterion value will be minimized by dividing the set somewhere closer to one fourth of the strip of rectangular cluster so that the split parts have approximately 150 points each.

Yet after the data standardization, with the grand means subtracted and range-normalized, the clusters on the left part of Fig. 6.8 are perfectly separable with K-Means criterion, that does attain its minimum value, at $K = 2$, on the cluster-based partition of the set. This holds with z-scoring, too.

This tendency changes, though, at a less structured case on the right of Fig. 6.8. The best split indeed holds at about $x = 10$ in this case. At a random sample, 32 points of the rectangular cluster join circular cluster in the optimal split. Curiously, the z-scoring standardization, in this case, works towards a better recovery of the structure so that the optimal 2 cluster partition, at the same data sample, merges only 5 rectangular cluster elements into the circular cluster, thus splitting the rectangular cluster over a mark $x = 5$.

These results do not much change when we go to Gaussian similarities (affinity data) defined by formula $G(x, y) = e^{-d(x,y)/2\sigma^2}$ where $d(x,y)$ is the squared Euclidean distance between x and y if $x \neq y$, and σ^2 is equal to 10 at the original data and 0.5 at the standardized data, in the manner of spectral clustering (see Section 8.2) and apply algorithm AddRem from Section 8.3.

F6.2.2 Batch K-Means and Its Criterion: Formulation

F6.2.2.1 Batch K-Means as Alternating Minimization

The cluster structure in K-Means is specified by a partition S of the entity set in K non-overlapping clusters, $S = \{S_1, S_2, \ldots, S_K\}$ represented by lists of entities S_k, and cluster centroids $c_k = (c_{k1}, c_{k2}, \ldots, c_{kV})$, $k = 1, 2, \ldots, K$.

There is a model that can be thought of as that driving K-Means algorithm. According to this model, each entity, represented by the corresponding row of Y matrix as $y_i = (y_{i1}, y_{i2}, \ldots, y_{iV})$, belongs to a cluster, say S_k, and is equal, up to small residuals, or errors, to the cluster's centroid:

$$y_{iv} = c_{kv} + e_{iv} \text{ for all } i \in S_k \text{ and all } v = 1, 2, \ldots, V \qquad (6.1)$$

Equations (6.1) define as simple a decoder as possible: whatever entity belongs to cluster S_k, its data point in the feature space is decoded as centroid c_k.

The problem is to find such a partition $S = \{S_1, S_2, \ldots, S_K\}$ and cluster centroids $c_k = (c_{k1}, c_{k2}, \ldots, c_{kV})$, $k = 1, 2, \ldots, K$, that minimize the square error of decoding

$$L^2 = \sum_{i \in I} \sum_{v \in V} e^2_{iv} = \sum_{k=1}^{K} \sum_{i \in S_k} \sum_{v \in V} (y_{iv} - c_{kv})^2 \qquad (6.2)$$

Criterion (6.2) can be equivalently reformulated in terms of the squared Euclidean distances as the summary distance between entities and their cluster centroids (see (6.3)) (see Fig. 6.9). Please note that the number of distances in the sum is N and does not depend on the number of clusters.

$$L^2 = W(S, c) = \sum_{k=1}^{K} \sum_{i \in S_k} d(y_i, c_k) \qquad (6.3)$$

This is because the distance referred to as squared Euclidean distance is defined, for any V-dimensional $x = (x_v)$ and $y = (y_v)$ as $d(x, y) = (x_1 - y_1)^2 + (x_2 - y_2)^2 + (x_2 - y_2)^2 + \ldots + (x_V - y_V)^2$ so that the rightmost summation symbol in (6.2) leads to $d(y_i, c_k)$ indeed.

This criterion depends on two groups of variables, S and c, and thus can be minimized by the alternating minimization method which proceeds by repetitively applying the same minimization step: Given one group of variables, optimize criterion over the other group, and so forth, until convergence.

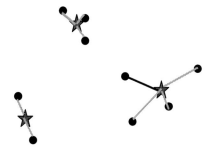

Fig. 6.9 The distances – intervals connecting centroids with entity points – in criterion $W(S,c)$

6.2 K-Means Clustering

Specifically, given centroids $c = (c_1, c_2, \ldots, c_K)$, find a partition S minimizing the summary distance (6.3). Obviously, to choose a partition S, one should choose, for each entity $i \in I$, one of K distances $d(y_i, c_1), d(y_i, c_2), \ldots, d(y_i, c_K)$. The choice to minimize (6.3) is according to the Minimum distance rule: for each $i \in I$, choose the minimum of $d(y_i, c_k), k = 1, \ldots, V$, that is, assign any entity to its nearest centroid. When there are several nearest centroids, the assignment is taken among them arbitrarily. In general, some centroids may be assigned no entity at all with this rule.

The other step in the alternating minimization would be minimizing (6.3) over c at a given S. The solution to this problem comes from the additive format of criterion (6.2) that provides for the independence of cluster centroid components from each other. As was indicated in Section 2.2.2, it is the mean that minimizes the square error, and thus the within-cluster mean vectors minimize (6.3) over c at given S.

Thus, starting from an initial set of centroids, $c = (c_1, c_2, \ldots, c_K)$, the alternating minimization method for criterion (6.3) will consist of a series of repeated applications of two steps: (a) clusters update – find clusters S according to the Minimum distance rule, (b) centroids update – make centroids equal to within cluster mean vectors. The computation stops when new clusters coincide with those on the previous step. This is exactly the K-Means in its Batch mode.

The convergence of the method follows from two facts: (i) at each step, criterion (6.3) can only decrease, and (ii) the number of different partitions S is finite.

F6.2.2.2 Various Formulations of K-Means Criterion

Let us consider any S_k and define c_k being within-cluster means ($k = 1, 2, \ldots, K$) so that $c_{kv} = \sum_{i \in S_k} y_{kv}/N_k$ where N_k is the number of entities in S_k. Multiply then each equation in (6.1) by itself and sum the resulting equations over both entities and features. The result will be the following equation:

$$\sum_{i \in I} \sum_{v \in V} y_{iv}^2 = \sum_{k=1}^{K} \sum_{v \in V} c_{kv}^2 N_k + \sum_{k=1}^{K} \sum_{i \in S_k} \sum_{v \in V} e_{iv}^2 \quad (6.4)$$

The three items in (6.4) come from summation of the products of the equations in (6.1) by themselves. The remaining sum $2\sum_v \sum_k c_{kv} \sum_{i \in S_k} e_{kv}$ is zero because $\sum_{i \in S_k} e_{kv} = \sum_{i \in S_k}(y_{iv} - c_{kv}) = c_{kv}N_k - c_{kv}N_k = 0$. This proves (6.4). Note that the item on the left in (6.4) is just the data scatter $T(Y)$, whereas the right-hand item is the least-squares criterion of K-Means (6.2). Therefore, Equation (6.4) can be reformulated as

$$T(Y) = B(S, c) + W(S, c) \quad (6.5)$$

where $T(Y)$ is data scatter, $W(S,c)$ the least-squares clustering criterion expressed as the summary within cluster distance (6.3) and $B(S,c)$ is clustering's contribution to the data scatter:

$$B(S,c) = \sum_{k=1}^{K} \sum_{v \in V} c_{kv}^{2} N_k \qquad (6.6)$$

Pythagorean Equation (6.5) decomposes data scatter $T(Y)$ in two parts: that one explained by the cluster structure (S,c), which is $B(S,c)$, and the unexplained part which is $W(S,c)$. The larger the explained part, the smaller the unexplained part, and the better the match between clustering (S,c) and data. Equation (6.5) is well known in the analysis of variance in statistics; items $B(S,c)$ and $W(S,c)$ are referred to in that other context as between-group and within-group variance.

Criteria $W(S,c)$ and $B(S,c)$ admit different equivalent reformulations that could lead to different systems of neighborhoods and local algorithms for minimization of $W(S,c)$ which may have been never attempted yet. Take, for example, the criterion of maximization of clustering's contribution to the data scatter (6.6). Since the sum of c_{kv}^2 over v is but the squared Euclidean distance between 0 and c_k, one has

$$B(S,c) = \sum_{k=1}^{K} \sum_{v \in V} c_{kv}^{2} N_k = \sum_{k=1}^{K} N_k d(0, c_k) \qquad (6.7)$$

The criterion on the right in formula (6.7) was first mentioned, under the name of "criterion of distant centers", by Mirkin (1996, p. 292). To maximize the criterion on the right in formula (6.7), the clusters should be as far away from 0 as possible. This idea may lead to a "parallel" version of the Anomalous Pattern method described later in Section 6.2.7.

Another expression of the cluster-explained part of the data scatter is

$$B(S,c) = \sum_{k=1}^{K} \sum_{i \in S_k} <y_i, c_k> \qquad (6.8)$$

which can be derived from (6.6) by taking into account that the internal sum $\Sigma_v c_{kv}^2$ in (6.6) is in fact the inner square $<c_k c_k>$ and substituting instead of one its expression as within cluster average $<c_k, c_k> = <c_k, \Sigma_{i \in S_k} y_{kv}/N_k> = \Sigma_{i \in S_k} <c_k, c_k>/N_k$. This expression shows that the K-Means criterion of minimizing within-cluster distances to centroids is equivalent to criterion of maximizing within-cluster inner products with centroids – they sum to the data scatter which does not depend on the clustering. Note that the distance based criterion makes sense at any set of centroids whereas the inner product based criterion makes sense only when centroids are within-cluster averages. As is well known, the distance does not depend on the location of the space origin whereas the inner product heavily depends on that – only special arrangements are suitable for the latter.

In this regard, it deserves to be mentioned that $W(S,c)$ can be reformulated in terms of entity-to-entity distances or similarities only – without any reference to centroids at all. One can prove that minimization of K-Means criterion is equivalent to minimization of $D(S)$ or maximization of $C(S)$ defined by

6.2 K-Means Clustering

$$D(S) = \sum_{k=1}^{K} \sum_{i,j \in S_k} d(y_i, y_j)/N_k \qquad (6.9)$$

$$C(S) = \sum_{k=1}^{K} \sum_{i,j \in S_k} a_{ij}/N_k \qquad (6.10)$$

where $d(y_i, y_j)$ is the squared Euclidean distance between i and j's rows and $a_{ij} = <y_i, y_j>$ is the inner product of them. Both follow from the expression $d(y_i, y_j) = <y_i - y_j, y_i - y_j> = <y_i, y_i> + <y_j, y_j> - 2<y_i, y_j>$ and the definition of the centroid of S_k as $\sum_{i \in S_k} y_i / N_k$. These formulations suggest algorithms for optimization based on exchanges and mergers between clusters.

A most unusual reformulation can be stated as a criterion of consensus among the features. Consider a measure of association $\zeta(S,v)$ between a partition S of the entity set I and a feature v, $v = 1, 2, \ldots, V$. Consider, for the sake of simplicity, that all features are quantitative and have been standardized by z-scoring, then $\zeta(S,v)$ is the correlation ratio η^2 defined by formula (3.15). Then maximizing the summary association

$$\zeta(S) = \Sigma_{v \in V} \eta^2(S, v) \qquad (6.11)$$

is equivalent to minimization of the K-Means least squares criterion $W(S,c)$ involving the squared Euclidean distances. A very similar equivalent consensus criterion can be formulated for the case when feature set consists of categorical or even mixed scale features. For the case of all features being categorical so that the categories are represented by dummy variables the total contribution to the data scatter in (6.4) can be formulated as

$$\varphi(S) = \Sigma_{v \in V} \varphi(S, v) \qquad (6.11')$$

where $\varphi(S,v)$ can be an association measure such as Pearson chi-squared or Gini index – see more detail in Section 4.5.3.

Since clusters are not overlapping, model in (6.1) can be rewritten differently in such a way that no explicit references are made over individual clusters. To do that, let us introduce N-dimensional membership vectors $z_k = (z_{ik})$ such that $z_{ik} = 1$ if $i \in S_k$ and $z_{ik} = 0$, otherwise. Using this notation allows us to use the following reformulation of the model. For any data entry, the following equation holds:

$$y_{iv} = \sum_{k=1}^{K} c_{kv} z_{ik} + e_{iv}, \qquad (6.12)$$

Indeed, since any entity $i \in I$ belongs to one and only one cluster S_k, only one of $z_{i1}, z_{i2}, \ldots, z_{iK}$ can be non-zero, that is, equal to 1, at any given i, which makes (6.12) equivalent to (6.1).

Yet (6.12) makes the clustering model similar to that of PCA in (5.14) except that z_{ik} in PCA are arbitrary values to score hidden factors, whereas in (6.12) z_{ik} are to be 1/0 binary values: it is clusters, not factors, that are of concern here. That is, clusters in model (6.12) correspond to factors in model (5.14).

The decomposition (6.4) of data scatter into explained and unexplained parts is similar to that in (5.17) making the contributions of individual clusters $\sum_{v \in V} c_{kv}^2 N_k$ akin to contributions μ_k^2 of individual principal components. More precisely, μ_k^2 in (5.17) are eigen-values of YY^T, that can be expressed thus with the analogous formula,

$$\mu_k^2 = z_k^T YY^T z_k / z_k^T z_k = \Sigma_v c_{kv}^2 |S_k| \quad (6.13)$$

in which the latter equation is due to the fact that vector z_k here consists of binary 1/0 entries.

Q.6.1. How many distances are summed up in $W(S, c)$? (**A:** This is equal to the number of entities N.) Does this number depend on the number of clusters K? (**A:** No.) Does this imply: the greater the K, the less the $W(S, c)$? (**A:** Yes.) Why?

Q.6.2. What is the difference between PCA model (5.14) and clustering model (6.12)? **A.** The scores z are constrained in (6.12) to take only binary values – either 1 or 0.

Q.6.3. Why is convergence guaranteed for K-Means? **A.** Because K-Means is alternating minimization process at which criterion $W(S,c)$ may only decrease at each step. Convergence follows from the fact that there are only a finite number of different partitions on I.

Q.6.4. Assume that $d(y_i, c_k)$ in $W(S, c)$ is city-block distance rather than Euclidean squared. Could K-Means be adjusted to make it alternating minimization algorithm for the modified $W(S,c)$? **A.** Yes, just use the city-block distance through, as well as within cluster median points rather than gravity centers.) Would this make any difference? (Yes, it will; especially at skewed distributions of the variables.)

Q.6.5. Demonstrate that, at Companies data, value $W(S,c)$ at product-based partition {1-2-3, 4-5-6, 7-8} is lower than at partition {1-4-6, 2, 3-5-7-8} found at seeds 1, 2 and 3. **A.** Indeed the sums of within-cluster distances to cluster centroids in the product based clusters are 0.7193, 0.8701, 0.3070, respectively, totaling to 1.8964, whereas the sums ot the second partition are 1.4411, 0, 2.1789 and sum to 3.62.

Q.6.6. Demonstrate that, at Companies data, value $W(S,c)$ at product-based partition {1-2-3, 4-5-6, 7-8} is lower than at partition {1-2-3, 4-6, 5-7-8} found at seeds 1, 4 and 7. **A.** Indeed the sums of within-cluster distances to cluster centroids in the product based clusters are 0.7193, 0.8701, 0.3070, respectively, totaling to 1.8964, whereas the sums ot the second partition are 0.7193, 0.4413, 1.1020 that total to 2.2626.

6.2 K-Means Clustering

Q.6.7. Can example of Fig. 6.6 or its modification lead to a similar effect for the case of least-modules criterion related to the city-block distance and median rather than average centroids? Can it be further extended to PAM method which uses city-block distance and median entities rather than coordinates?

Q.6.8. Formulate a version of K-Means to alternatingly maximize criterion (6.8) rather than to minimize (6.3) as the generic version.

Q.6.9. Formulate a version of K-Means to alternatingly maximize criterion (6.7) rather than to minimize (6.3) as the generic version (a "parallel" version of the Anomalous Pattern method). Take care of starting from a most distant set of centroids.

C6.2.3 A Pseudo-Code for Batch K-Means: Computation

To summarize, an application of K-Means clustering involves the following steps:

0. Select a data set.
1. Standardize the data.
2. Choose number of clusters K.
3. Define K hypothetical centroids (seeds).
4. Clusters update: Assign entities to the centroids according to Minimum distance rule.
5. Centroids update: define centroids as the gravity centers of thus obtained clusters.
6. Iterate 4. and 5. until convergence.

MatLab codes for the items 4 and 5 can be put as follows.

4. *Clusters update*: Assign points to the centroids according to Minimum distance rule.

   Given data matrix X and a KxV array of centroids cent, produce an N-dimensional array of cluster labels for the entities and the summary within cluster distance to centroids, wc:

```
function [labelc,wc]=clusterupdate(X,cent)
    [K,m]=size(cent);
    [N,m]=size(X);
    for k=1:K
            cc=cent(k,:); %centroid of cluster k
            Ck=repmat(cc,N,1);
            dif=X-Ck;
            ddif=dif.*dif; %Nxm matrix of squares
            dist(k,:)=sum(ddif');
    %distances from entities to cluster centroid
    end
```

```
[aa,bb]=min(dist); %Minimum distance rule
wc=sum(aa);
labelc=bb;
return
```

5. *Centroids update*: Put centroids in gravity centres of clusters defined by the array of cluster labels *labelc* according to data in matrix X, to produce KxV array centres of the centroids:

```
function  centres=ceupdate(X,labelc)
       K=max(labelc);
       for k=1:K
    clk=find(labelc==k);
    elemk=X(clk,:);
    centres(k,:)=mean(elemk);
  end
  return
```

Batch K-Means with MatLab, therefore, is to embrace steps 3–6 above and output a clustering in cell array termed, say, Clusters, along with the proportion of unexplained data scatter found by using preliminarily standardized matrix X and set of initial centroids, cent, as input. This can be put like this:

```
function [Clusters,uds]=k_means(X,cent)
 [N,m]=size(X);
 [K,m1]=size(cent);
 flag=0; %-- stop-condition
 membership=zeros(N,1);
 dd=sum(sum(X.*X)); %-- data scatter
 %--- clusters and centroids updates
 while flag==0
      [labelc,wc]=clusterupdate(Y,cent);
      if isequal(labelc,membership)
     %--stop-condition's working
             flag=1;
             centre=cent;
             w=wc;
         else
             cent=ceupdate(Y,labelc);
             membership=labelc;
         end
 end
 %-----preparing the output --------------
  uds=w*100/dd;
  Clusters{1}=membership;
  Clusters{2}=centre;
  return
```

6.2 K-Means Clustering

Q.6.10. Check the values of criterion (6.3) at each initial setting considered for Company data above. Find out which is the best among them.

Q.6.11. Prove that the square-error criterion (6.2) can be reformulated as the sum of within cluster variances $\sigma_{kv}^2 = \sum_{i \in S_k}(y_{iv} - c_{kv})^2/N_k$ weighted by the cluster cardinalities N_k.

Q.6.12. Prove that reformulation (6.9) of criterion (6.3) in terms of the squared Euclidean distances is correct.

Q.6.13. Prove that if Batch K-Means is applied to Iris data mean-range normalized with $K = 3$ and specimens 1, 51, and 101 taken as the initial centroids, the resulting clustering cross-classified with the prior three classes forms contingency table presented in Table 6.6.

In the following two sections we describe two approaches at reaching deeper minima of K-Means criterion (6.3): (a) an incremental version and (b) nature inspired versions.

6.2.4 Incremental K-Means

P6.2.4.1 Incremental K-Means: Presentation

An incremental version of K-Means uses the Minimum distance rule not for all of the entities but for one of them only. There can be two different reasons for doing so:

(**Ri**) The user is not able to operate over the entire data set and takes in the entities one by one, because of the data protocol, so that entities are to be clustered on the fly as, for instance, in an on-line application process.

(**Rii**) The user operates with the entire data set, but wants to smooth the action of the algorithm so that no drastic changes in the cluster contents may occur. To do this, the user may specify an order of the entities and run entities one-by-one in this order for a number of epochs like it is done in a neural network learning process.

The result of such a one-by-one entity processing may differ from that of Batch K-Means because each version finds a locally optimal solution on a different structure of locality neighborhoods.

Table 6.6 Cross classification of the original Iris taxa and 3-cluster clustering found starting from entities 1, 51 and 101 as initial seeds. The clustering does separate Iris Setosa but misplaces 14+3=17 specimens between two other taxa

Cluster	Setosa	Versicolor	Virginica	Total
S1	50	0	0	50
S2	0	47	14	61
S3	0	3	36	39
Total	50	50	50	150

Q.6.14. What is the difference in neighborhoods between Batch and incremental versions of K-Means?

Q.6.15. Consider a run of incremental K-Means at situation **Rii** on the Companies data, at which the order of entities follows the order of their distances to nearest centroids. Let $K = 3$ and entities Av, Ba and Ci initial centroids. **A.** Sequential steps of the incremental computation are presented in Table 6.7. In this table, cluster updates are provided as well as their centroids after each single update. The column on the right presents squared Euclidean distances between centroids and entities yet unclustered, with the minima highlighted in bold. The minimum distance determines, in this version, which of the entities joins the clustering next. One can see that on iteration 2 company Br switches to centroid Ba after centroid Ci of the third cluster had been updated to the mean of Ci and Cy – because its distance to the new centroid increased from the minimum 0.83–1.20. This leads to correct, product-based, clusters.

Q.6.16. Prove that the same initialization leads to wrong, that is, non-product based, clusters with Batch K-Means.

F6.2.4.2 Incremental K-Means: Formulation

When an entity y_i joins cluster S_k whose cardinality is N_k, centroid c_k changes to c'_k to follow the within cluster means, according to the following formula:

$$c'_k = N_k c_k/(N_k + 1) + y_i/(N_k + 1)$$

When y_i moves out of cluster S_k, the formula remains valid if all pluses are changed for minuses. To extend the formula so that it holds for both cases, let us introduce variable z_i which is equal to +1 when y_i joins the cluster and −1 when it moves out of it. Then the extended formula is:

$$c'_k = N_k c_k/(N_k + z_i) + y_i z_i/(N_k + z_i)$$

Accordingly, the distances from other entities change to $d(y_j, c'_k)$.

Because of the incremental setting, the stopping rule of the straight version (reaching a stationary state) may be not necessarily applicable here. In **Ri** case, the natural stopping rule is to end when there are no new entities observed. In **Rii** case, the process of running through the entities one-by-one stops when all entities remain in their clusters. The process may be stopped as well when a pre-specified number of runs through the entity set, that is, epochs, is reached.

6.2.5 Nature Inspired Algorithms for K-Means

P6.2.5.1 Nature Inspired Algorithms: Presentation

In real-world applications, K-Means typically does not move far away from the initial setting of centroids. Considered in the perspective of minimization of criterion

6.2 K-Means Clustering

Table 6.7 Iterations of incremental K-means on standardized company data starting with centroids Av, Ba and Ci

Incremental one-by-one entity clustering

Iteration	Cumulative clusters	Centroids							Distances					
									An	As	Br	Bu	Cy
0	Av	−0.20	0.23	−0.33	−0.62	0.36	−0.22	−0.14	0.51	0.88	2.20	2.25	3.01
	Ba	−0.23	−0.15	−0.33	0.38	0.36	−0.22	−0.14	1.55	1.94	0.97	0.87	2.46
	Ci	0.08	−0.10	0.33	0.38	−0.22	−0.22	0.43	1.90	1.81	0.83	1.68	0.61
1	Av, An	0.10	0.14	−0.17	−0.62	0.36	−0.22	−0.14		0.70	1.88	2.50	2.59
	Ba	−0.23	−0.15	−0.33	0.38	0.36	−0.22	−0.14		1.94	0.97	0.87	2.46
	Ci	0.08	−0.10	0.33	0.38	−0.22	−0.22	0.43		1.81	0.83	1.68	**0.61**
2	Av, An	0.10	0.14	−0.17	−0.62	0.36	−0.22	−0.14		0.70	1.88	2.50	
	Ba	−0.23	−0.15	−0.33	0.38	0.36	−0.22	−0.14		1.94	0.97	0.87	
	Ci, Cy	0.18	0.24	0.50	0.38	−0.22	−0.22	0.43		1.95	1.20	2.40	
3	Av, An, As	0.10	0.12	−0.11	−0.62	0.17	−0.02	−0.14			1.49	2.12	
	Ba	−0.23	−0.15	−0.33	0.38	0.36	−0.22	−0.14			0.97	**0.87**	
	Ci, Cy	0.18	0.24	0.50	0.38	−0.22	−0.22	0.43			1.20	2.40	
4	Av, An, As	0.10	0.12	−0.11	−0.62	0.17	−0.02	−0.14			1.49		
	Ba, Bu	−0.42	−0.29	−0.33	0.38	0.07	0.07	−0.14			**0.64**		
	Ci, Cy	0.18	0.24	0.50	0.38	−0.22	−0.22	0.43			1.20		
5	Av, An, As	0.10	0.12	−0.11	−0.62	0.17	−0.02	−0.14					
	Ba, Bu, Br	−0.21	−0.29	−0.22	0.38	−0.02	0.17	−0.14					
	Ci, Cy	0.18	0.24	0.50	0.38	−0.22	−0.22	0.43					

(6.3), this leads to the strategy of multiple runs of K-Means starting from randomly generated sets of centroids to reach as deep a minimum of (6.3) as possible. This strategy works well on illustrative small data sets but it may fail when the data set is large because in this case random settings cannot cover the space of solutions in a reasonable time. Nature inspired approach provides a well-defined framework for using random centroids in parallel, rather than in sequence, to channel them to deeper minima as an evolving population of admissible solutions. The main difference of the nature inspired optimization from the classical optimization is that the latter reaches for a single solution, provably optimal, whereas the former runs a population of solutions and does not much care for the provability.

A nature inspired algorithm mimics some natural process to set rules for the population behavior and/or evolution. Among the nature inspired approaches, the following are especially popular:

A. Genetic
B. Evolutionary
C. Particle swarm optimization

A K-Means method according to each of these will be described in this section.

A nature inspired algorithm proceeds as a sequence of steps of evolution for a population of possible solutions, that is, clusterings represented by specific data structures. A K-Means clustering comprises two items: a partition S of the entity set I in K clusters and a set of clusters' K centroids $c = \{c_1, c_2, \ldots, c_K\}$. Typically, only one of them is carried out in a nature-inspired algorithm. The other is easily recovered according to K-Means rules. Given a partition S, centroids c_k are found as vectors of within cluster means. Given a set of centroids, each cluster S_k is defined as the set of points nearest to its centroid c_k, according to the Minimum distance rule ($k = 1, 2, \ldots, K$). Respectively, the following two representations are most popular in nature inspired algorithms:

(i) Partition as a string,
(ii) Centroids as a string.

Consider them in turn.

(i) Partition as a string

Having pre-specified an order of entities, a partition S can be represented as a string of cluster labels $k = 1, 2, \ldots, K$ of the entities thus ordered. If, for instance, there are eight entities ordered as e1, e2, e3, e4, e5, e6, e7, e8, then the string 12333112 represents partition S with three classes according to the labels, $S1 = \{e1, e6, e7\}$, $S2 = \{e2, e8\}$, and $S3 = \{e3, e4, e5\}$, which can be easily seen from the diagram relating the entities and labels:

e1 e2 e3 e4 e5 e6 e7 e8
 1 2 3 3 3 1 1 2

6.2 K-Means Clustering

A string of N integers from 1 to K is considered not admissible, if some integer between 1 and K is absent from it (so that the corresponding cluster is empty). Such a not admissible string for the entity set above would be 11333111, because it lacks label 2 and, therefore, makes class S2 empty.

(ii) **Centroids as a string**

Consider the same partition as in (i) on the set of eight objects, the companies in Company data Table 6.1 in their order. Clusters S1 = {e1, e6, e7}, S2 = {e2, e8}, and S3 = {e3, e4, e5}, as well as their centroids, are presented in Table 6.8. The three centroids form a sequence of 7×3 = 21 numbers c = (−0.24, −0.09, −0.11, 0.04, −0.02, −0.02, 0.05, 0.34, 0.31, 0.33, −0.12, 0.07, −0.22, 0.14, 0.01, −0.12, −0.11, 0.04, −0.02, 0.17, −0.14), which suffices for representing the clustering: the sequence can be easily converted back in three 7-dimensional centroid vectors to recover then clusters with the Minimum distance rule. It should be pointed out that the original clusters may be somewhat weird and not recoverable in this way. For example, entity e4, which is Ba, appears to be nearer to centroid 1 rather than to centroid 3 so that the Minimum distance rule would produce clusters S1 = {e1, e4, e6, e7}, S2 = {e2, e8}, and S3 = {e3, e5} rather than those original ones, but such a loss makes no difference, because K-Means clusters necessarily satisfy the Minimum distance rule so that all the entities are nearest to their cluster's centroids.

What is important is that any 21-dimensional sequence of real values can be treated as the clustering code for its centroids.

GA for K-Means Clustering

Genetic algorithms work over a population of strings, each representing an admissible solution and referred to as a chromosome. The optimized function is referred to as the fitness function. Let us use the string represenatation for partitions $S = \{S_1, \ldots, S_K\}$ of the entity set. The minimized fitness function is the summary within-cluster distance to centroids, the function $W(S,c)$ in (6.3):

0. *Initial setting.* Specify an even integer P for the population size (no rules exist for this), and randomly generate P chromosomes, that is, strings $s_1, \ldots, s_P$ of

Table 6.8 Centroids of clusters S1 = {e1, e6, e7}, S2 = {e2, e8}, and S3 = {e3, e4, e5} according to data in Table 6.1

Av	−0.20	0.23	−0.33	−0.63	0.36	−0.22	−0.14
Bu	−0.60	−0.42	−0.33	0.38	−0.22	0.36	−0.14
Ci	0.08	−0.10	0.33	0.38	−0.22	−0.22	0.43
Centroid1	−0.24	−0.09	−0.11	0.04	−0.02	−0.02	0.05
An	0.40	0.05	0.00	−0.62	0.36	−0.22	−0.14
Cy	0.27	0.58	0.67	0.38	−0.22	−0.22	0.43
Centroid2	0.34	0.31	0.33	−0.12	0.07	−0.22	0.14
As	0.08	0.09	0.00	−0.62	−0.22	0.36	−0.14
Ba	−0.23	−0.15	−0.33	0.38	0.36	−0.22	−0.14
Br	0.10	−0.29	0.00	0.38	−0.22	0.36	−0.14
Centroid3	0.01	−0.12	−0.11	0.04	−0.02	0.17	−0.14

K integers $1, \ldots, K$ in such a way that all K integers $1, 2, \ldots, K$ are present within each chromosome.. For each of the strings, define corresponding clusters, calculate their centroids as gravity centres and the value of criterion, $W(s_1), \ldots, W(s_P)$, according to formula (6.3).

1. *Mating selection.* Choose $P/2$ pairs of strings to mate and produce two "children" strings. The mating pairs usually are selected randomly (with replacement, so that the same string may appear in several pairs and, moreover, can form both parents in a pair). To mimic Darwin's "survival of the fittest" law, the probability of selection of string s_t ($t = 1, \ldots, P$) should reflect its fitness value $W(s_t)$. Since the fitness is greater for the smaller W value, some make the probability inversely proportional to $W(s_t)$ (see Murthy and Chowdhury 1996) and some to the difference between a rather large number and $W(s_t)$ (see Yi Lu et al. 2004). This latter approach can be taken further with the probability proportional to the explained part of the data scatter – in this case "the rather large number" is the data scatter rather than an arbitrary value.
2. *Cross-over.* For each of the mating pairs, generate a random number r between 0 and 1. If r is smaller than a pre-specified probability p (typically, p is taken about 0.7–0.8), then perform a crossover; otherwise the mates themselves are considered the result. A (single-point) crossover of string chromosomes $a = a_1 a_2 \ldots a_N$ and $b = b_1 b_2 \ldots b_N$ is performed as follows. A random number n between 1 and N-1 is selected and the strings are crossed over to produce children $a_1 a_2 \ldots a_n b_{n+1} \ldots b_N$ and $b_1 b_2 \ldots b_n a_{n+1} \ldots a_N$. If a child is not admissible (like, for instance, strings $a = 11133222$ and $b = 32123311$ crossed over at $n = 4$ would produce $a' = 11133311$ and $b' = 32123222$ so that a' is inadmissible because of absent 2), then various policies can be applied. Some authors suggest the crossover operation to be repeated until an admissible pair is produced. Some say inadmissible chromosomes are ok, just they must be assigned with a smaller probability of selection.
3. *Mutation.* Mutation is a random alteration of a character in a chromosome. This provides a mechanism for jumping to different "ravines" of the minimized fitness function. Every character in every string is subject to the mutation process, with a low probability q which can be constant or inversely proportional to the distance between the corresponding entity and corresponding centroid.
4. *Elitist survival.* This strategy suggests keeping the best fitting chromosome(s) stored separately. After the crossover and mutations have been completed, find fitness values for the new generation of chromosomes. Check whether the worst of them is better than the record or not. If not, put the record chromosome instead of the worst one into the population. Then find the record for thus obtained population.
5. *Halt condition.* Check the stop condition (typically, a limit on the number of iterations). If this doesn't hold, go to 1; otherwise, halt.

Y. Lu et al. (2004) note that such a GA works much faster if after step 3. Mutation the labels are changed according to the Minimum distance rule. They apply this instead of the elitist survival.

6.2 K-Means Clustering

Thus, a GA algorithm operates with a population of chromosomes representing admissible solutions. To update the population, mates are selected, undergone a cross-over process generating offspring which then is subjected to mutation process. Elite maintenance completes the update. In the end, the elite is output as the best solution.

A computational shortcoming of the GA algorithm is that the length of the chromosomes is the size of the entity set N, which may run in millions in contemporary applications. Can this be overcome? Sure, by using centroid, not partition, strings to represent a clustering. Centroid string sizes depend on the number of features and number of clusters, not the number of entities. Another advantage of centroid strings is in the mutation process. Rather than an abrupt swap between literals, they can be changed softly, in a quantitative manner by adding or subtracting a small change. This is utilized in evolutionary and particle swarm algorithms.

Evolutionary K-Means

Here, the chromosome is represented by a set of K centroids $c = (c_1, c_2, \ldots c_K)$ which can be considered a string of KV real ("float") numbers. In contrast to the partition-as-string representation, the length of the string here does not depend on the number of entities that can be of advantage when the number of entities is massive. Each centroid in the string is analogous to a gene in the chromosome.

The crossover of two centroid strings c and c', each of the length KV, is performed at a randomly selected place n, $1 \leq n < KV$, exactly as it is in the genetic algorithm above. Chromosomes c and c' exchange the portions lying to the right of n-th component to produce two offspring. This means that, a number of centroids in c is substituted by corresponding centroids in c'. Moreover, if n cuts across a centroid, its components change in each of the offspring chromosomes.

The process of mutation, according to Bandyopadhyay and Maulik (2002), can be organized as follows. Given the fitness W values of all the chromosomes, let $minW$ and $maxW$ denote their minimum and maximum respectively. For each chromosome, its radius R is defined as a proportion of $maxW$ reached at it: $R = (W-minW)/(maxW-minW)$. When the denominator is 0, that is, if $minW = maxW$, define $R = 1$ in all chromosomes. Here, W is the fitness value of the chromosome under consideration. Then the mutation intensity δ is generated randomly in the interval between $-R$ and $+R$.

Let $minv$ and $maxv$ denote the minimum and maximum values in the data set along feature v ($v = 1, \ldots, V$). Then every v-th component xv of each centroid c_k in the chromosome changes to

$$xv + \delta*(maxv - xv) \text{ if } \delta \leq 0 \text{ (increase), or}$$
$$xv + \delta*(xv - minv), \text{ otherwise (decrease)}.$$

The perturbation leaves chromosomes within the hyper-rectangle defined by boundaries $minv$ and $maxv$. Please note that the best chromosome, at which $W = minW$, does not change in this process because its $R = 0$.

Elitism is maintained in the process as well.

The algorithm follows the scheme outlined for the genetic algorithm.
Based on little experimentation, this algorithm is said to outperform the previous one, GA, many times in terms of the speed of convergence.

The evolutionary approach can be further modified such as, for example, the so-called Differential evolution (see Paterlini and Krink 2006 who claim that this method outperforms the others in K-Means). In Differential evolution, the crossover, mutation and elite maintenance are merged together by removing the mating stage and changing those for the following. An offspring chromosome is created for every chromosome t in the population ($t = 1, \ldots, P$) as follows. Three other chromosomes, k, l and m, are taken randomly from the population. Then, for every component (gene) $x.t$ of the chromosome t, a uniformly random value r between 0 and 1 is drawn. This value is compared to the pre-specified probability p (somewhat between 0.5 and 0.8). If $r > p$ then the component goes to the offspring unchanged. Otherwise, this component is substituted by the linear combination of the same component in the three other chromosomes: $x.m + \alpha^*(x.k - x.l)$ where α is a small scaling parameter. After the offspring's fitness is evaluated, it substitutes chromosome t if it is better; otherwise, t remains as is and the process applies to the next chromosome.

Particle Swarm Optimization for K-Means

Particle swarm mimics a drift of a bee population so that the population members here are not crossbred, nor they mutate. They just move randomly by drifting in random directions having an eye on the best places visited so far, individually and socially. This can be done because they are vectors of real numbers. Because of the change, the genetic metaphor is abandoned here, and the elements are referred to as particles rather than chromosomes, and the set of them as a swarm rather than a population.

Each particle comprises:

– a position vector x that is an admissible solution to the problem in question (such as the *KV* centroid vector in K-Means),
– the evaluation of its fitness $f(x)$ (such as the summary distance W in (6.3)),
– a velocity vector z of the same dimension as x, and
– the record of the best position b reached by the particle so far.

The swarm best position bg is determined as the best among all the individual best positions b.

At iteration t ($t = 0, 1, \ldots$) the next iteration's position is defined as the current position shifted by the velocity vector:

$$x(t+1) = x(t) + z(t+1)$$

where $z(t+1)$ is computed as a change in the direction of personal and population's best positions:

$$z(t+1) = z(t) + \alpha(b - x(t)) + \beta(bg - x(t))$$

6.2 K-Means Clustering

where

- α and β are uniformly distributed random numbers (typically, within the interval between 0 and 2, so that they are around unity),
- item $\alpha(b - x(t))$ is referred to as the cognitive component and
- item $\beta(bg - x(t))$ as the social component of the process.

Initial values $x(0)$ and $z(0)$ are generated randomly within the manifold of admissible values.

In some implementations, the group best position bg is changed for that of local best position bl that is defined by the particle's neighbors only so that some pre-defined neighborhood topology makes its effect. There is a report that the local best position works especially well, in terms of the depth of the minimum reached, when it is based on just two Euclidean neighbors.

Q.6.17. Formulate a particle swarm optimization algorithm for K-Means clustering.

6.2.6 Partition Around Medoids PAM

K-Means centroids are average points rather than individual entities, which may be considered artificial in contexts in which the user may wish to involve but only genuinely occurring real world entities rather the "synthetic" averages. Estates or art objects or countries are examples of entities for which this makes sense. To implement the idea, let us change the concept of cluster prototype from centroid to *medoid* (Kaufman and Rousseeuw 1990). An entity in a cluster S, $i^* \in S$, is referred to as its medoid if it is the nearest in S to all other elements of S, that is, if i^* minimizes the sum of distances $D(i) = \Sigma_{j \in S} d(i,j)$ over all $i \in S$. The symbol $d(i,j)$ is used here to denote any dissimilarity function, which may or may not be squared Euclidean distance, between observed entities $i, j \in I$.

The method of partitioning around medoids PAM (Kaufman and Rousseeuw 1990) works exactly as Batch K-Means with the only difference that medoids, not centroids, are used as cluster prototypes. It starts, as usual, with choosing the number of clusters K and initial medoids $c = (c_1, c_2, \ldots, c_K)$ that are not just V dimensional points but individual entities. Given medoids c, clusters S_k are collected according to the Minimum distance rule – as sets of entities that are nearest to entity c_k for all $k = 1, 2, \ldots, K$. Given clusters S_k, medoids are updated according to the definition. This process reiterates again and again, and halts when no change of the clustering occurs. It obviously will never leave a cluster S_k empty. If the size of the data set is not large, all computations can be done over the entity-to-entity distance matrix without ever changing it.

Worked example 6.2. PAM applied to Company data

Let us apply PAM to the Company data displayed in Table 6.1 with $K = 3$ and entities Av, Br and Cy as initial medoids. We can operate over the distance matrix, presented in Table 6.9, because there are only eight entities.

Table 6.9 Distances between standardized company entities. For the sake of convenience, those smaller than 1, are highlighted in bold

Entities	Av	An	Ast	Ba	Br	Bu	Ci	Cy
Av	**0.00**	**0.51**	**0.88**	1.15	2.20	2.25	2.30	3.01
An	**0.51**	**0.00**	**0.77**	1.55	1.82	2.99	1.90	2.41
As	**0.88**	**0.77**	**0.00**	1.94	1.16	1.84	1.81	2.38
Ba	1.15	1.55	1.94	**0.00**	**0.97**	**0.87**	1.22	2.46
Br	2.20	1.82	1.16	**0.97**	**0.00**	**0.75**	**0.83**	1.87
Bu	2.25	2.99	1.84	**0.87**	**0.75**	**0.00**	1.68	3.43
Ci	2.30	1.90	1.81	1.22	**0.83**	1.68	**0.00**	**0.61**
Cy	3.01	2.41	2.38	2.46	1.87	3.43	**0.61**	**0.00**

With seeds Av, Br, Cy, the Minimum distance rule would obviously produce the product-based clusters A, B, and C. At the next iteration, clusters' medoids are computed: they are obviously An in A cluster, Bu in B cluster and either of the two entities in C cluster – leave it thus at the less controversial Cy. With the set of medoids changed to An, Bu and Cy, we apply the Minimum distance rule again, leading us to the product-based clusters again. This halts the process.

Note that PAM can lead to instability in results because the assignment depends on distances to just a single entity.

Q.6.18. Why Cy is less controversial than Ci in Table 6.9? **A.** Because Cy unequivocally relates to Ci only, whereas Ci is close to Br as well.

Q.6.19. Assume that the distance d(Br, Bu) in Table 6.9 is 0.85 rather than 0.75. Show that then if one chooses Ci to be medoid of C cluster, then the Minimum distance rule would assign to Ci not only Cy but also Br, because its distance to Ci, 0.83, would be less than its distance to Bu, 0.85. Show that this cluster, {Ci, Cy, Br} will remain stable over successive iterations.

6.2.7 Initialization of K-Means

To initialize K-Means, one needs to specify:

(i) the number of clusters, K, and
(ii) initial centroids, $c = (c_1, c_2, \ldots, c_K)$.

Each of these can be of an issue in practical computations. Both depend on the user's expectations related to the level of granularity and typological attitudes, which remain beyond the scope of the theory of K-Means. This is why some suggest relying on the user's view of the substantive domain to specify the number and positions of initial centroids as hypothetical prototypes. There have been however a number of approaches for specifying the number and location of the initial centroids by exploring the structure of the data, of which we describe the following three:

6.2 K-Means Clustering

(a) multiple runs of K-Means;
(b) distant representatives;
(c) anomalous patterns.

(a) Multiple runs of K-Means

According to this approach, at a given K, a number of K-Means' runs R is pre-specified; each run starts with K randomly selected entities as the initial seeds (randomly generated points within the feature ranges have proven to give inferior results in experiments reported by several authors). Then the best result in terms of the square-error criterion $W(S,c)$ (6.3) is output. This can be further extended to choosing the "right" number of clusters K. Let us denote by W_K the minimum value of $W(S,c)$ found after R runs of K-Means over random initializations. Then the series W_K found at different K, from a pre-specified range say between 2 and 20, is usually taken to see which K would lead to the best W_K over the range. Unfortunately, the best W_K is not necessarily minimum W_K, because the minimum value of the square-error criterion cannot increase when K grows, which should be reflected in the empirically found W_K's. In the literature, a number of stop criteria utilizing W_K have been suggested based on some simplified data models and intuition such as "gap" or "jump" statistics. Unfortunately, they all may fail even in the relatively simple situations of controlled computation experiments (see Chiang and Mirkin 2010 for a review).

A relatively simple heuristic rule is based on the intuition that when there are K^* well separated clusters, then for $K<K^*$ a $(K+1)$-cluster partition should be the K-cluster partition with one of its clusters split in two, which would drastically decrease W_{K+1} from W_K. On the other hand, at $K>K^*$, both K- and $(K+1)$-cluster partitions are to be the "right" K^*-cluster partition with some of the "right" clusters split randomly, so that W_K and W_{K+1} are not that different. Therefore, as "a crude rule of thumb", Hartigan (1975, p. 91) proposed calculating index

$$H_K = (W_K/W_{K+1} - 1)(N - K - 1),$$

where N is the number of entities, while increasing K, so that the very first K at which H_K becomes smaller than 10 is to be taken as the estimate of K^*. It should be noted that, in the experiments by Chiang and Mirkin (2010), this rule came as the best of a set of nine different criteria and, moreover, the threshold 10 in the rule appears to be not very sensitive to 10–20% changes.

Case Study 6.4. Hartigan's Index for Choosing the Number of Clusters

Consider values of H_K for Iris and Town datasets computed after the results of 100 runs of Batch K-Means using the mean/range standardization starting from random K entities taken as seeds (Table 6.10). Each of the computations has been

Table 6.10 Values of Hartigan's H_K index for two data sets at K ranging from 2 to 11 as based on two different sets of 100 clusterings from random K entities as initial centroids

Dataset		K = 2	3	4	5	6	7	8	9	10	11
Iris	1st set	108.3	38.8	29.6	24.1	18.6	15.0	16.1	15.4	15.4	**9.4**
	2nd set	108.3	38.8	29.6	24.1	18.7	15.4	15.6	15.7	16.0	**7.2**
Town	1st set	13.2	10.5	**9.3**	5.0	4.7	3.1	3.0	3.2	3.2	1.6
	2nd set	13.2	10.5	**9.3**	5.8	4.1	2.5	3.0	7.2	−0.2	1.8

repeated twice (see 1st and 2nd sets in Table 6.10) to illustrate typical variations of H_K values due to the fact that empirical values of W_K may be not optimal. In particular, at the 2nd set of K-Means over Town data we can see a break of the rule that H_K is positive because of the monotonic relation between K and the optimal W_K that are to decrease when K grows. The monotonic relation here is broken because the values of W_K after 100 runs are not necessarily minimal indeed.

The "natural" number of clusters in Iris data, according to Hartigan's criterion is not 3 as claimed because of substantive considerations but much greater, 11! In Town data set, the criterion would indicate 4 naturally occurring clusters. However, one should argue that the exact value of 10 in Hartigan's rule does not bear much credibility – it should be accompanied by a significant drop in H_K value. We can see such a drop at $K = 5$, which should be taken, thus, as the "natural" number of clusters in Town data. Similarly, a substantial drop of H_K on Iris data occurs at $K = 3$, which is the number of natural clusters, taxa, in this set.

Altogether, making multiple runs of K-Means seems a sensible strategy, especially when the number of entities is not that high. With the number of entities growing into thousands, the number of tries needed to reach a representative value of W_K may become prohibitively large. Deeper minima can be sought by using the evolutionary schemes described above. On the other hand, the criterion $W(S,c)$ has some intrinsic flaws and should be used only along some domain-knowledge or data-structure based strategy.

Two data-driven approaches, (b) and (c) above, to defining initial centroids are described in the next two sections. They both employ the idea that clusters should represent some anomalous yet typical tendencies.

(b) "Build" algorithm for a pre-specified K (Kaufman and Rousseeuw 1990)

This process involves only actual entities. It starts with choosing the medoid of set I, that is, the entity whose summary distance to the others is minimum, and takes it as the first medoid c_1. Assume that a subset of m initial seeds have been selected already ($K > m \geq 1$) and proceed to selecting c_{m+1}. Denote the set of already selected seeds by c and consider all remaining entities $i \in I-c$. Define distance $d(i,c)$ as the minimum of the distances $d(i,c_k)(k = 1, \ldots, m)$ and form an auxiliary

6.2 K-Means Clustering

cluster A_i consisting of such j that are closer to i than to c so that $E_{ij} = d(j,c) - d_{ij} > 0$. The summary value $E_i = \Sigma_{j \in A_i} E_{ij}$ reflects both the number of points in A_i and their remoteness from c. That $i \in I - c$ for which E_i is maximum is taken as the next seed c_{m+1}.

Worked example 6.3. Selection of initial medoids in Company data

Let us apply Build algorithm to the matrix of entity-to-entity distances for Company data displayed in Table 6.9, at $K = 3$. First, we calculate the summary distances from all the entities to the others, see Table 6.11, and notice that Br is the medoid of the entire set I, because its total distance to the others, 9.60, is the minimum of total distances in Table 6.11. Thus, we set Bre as the first initial seed.

Now we build auxiliary clusters A_i around all other entities. To form A_{Av}, we take the distance between Av and Br, 2.20, and see, in Table 6.10, that distances from Av to entities An, As, and Ba are smaller than that, which makes them Av's auxiliary cluster with $E_{Av} = 4.06$. Similarly, A_{An} is set to consist of the same entities, but it is less remote than Av because $E_{An} = 2.98$ is less than E_{Av}. Auxiliary cluster A_{As} consists of Av and An with even smaller $E_{As} = 0.67$. Auxiliary clusters for Ba, Ci and Cy consist of one entity each (Bu, Cy and Ci, respectively) and have much smaller the levels of remoteness; cluster A_{Bu} is empty because Br is its nearest. This makes the most remote entity Ave the next selected seed. Now, we can start building auxiliary clusters on the remaining six entities again. Of them, clusters A_{An} and A_{Bu} are empty and the others are singletons, of which A_{Cy} consisting of Ci is the remotest, with $E_{Cy} = 1.87 - 0.61 = 1.26$. This completes the set of initial seeds: Br, Av, and Cy. Note, these are companies producing different products. It is this set that was used to illustrate PAM in Section 6.2.6.

(c) Anomalous patterns (Mirkin 2005)

This method involves remote clusters, as Build does, too, but it does not discard them after finding, which allows for obtaining the number of clusters K as well. Besides, it is less computationally intensive. The method employs the concept of reference point. A reference point is chosen to exemplify an "average" or "normal" entity, not necessarily among the dataset. For example, when analyzing student marks over different subjects, one might choose a "normal student" point which

Table 6.11 Summary distances for entities according to Table 6.9

Entity	Av	An	As	Ba	Br	Bu	Ci	Cy
Distance to others	12.30	11.95	10.78	10.16	9.60	13.81	10.35	16.17

would indicate levels of marks in tests and work in projects that are considered normal for the contingent of students under consideration, and then see what patterns of observed behavior deviate from this. Or, a bank manager may set as his reference point, a customer having specific assets and backgrounds, to see what patterns of customers deviate from this. In engineering, a moving robotic device should be able to segment the view into homogeneous chunks according to the robot's location as its reference point, with objects that are nearer to it having finer resolution than objects that are farther away. In many cases the gravity center of the entire entity set, its "grand mean", can be taken as a reference point of choice.

Using the chosen reference point allows for the comparison of entities with it, not with each other, which drastically reduces computations: instead of mulling over all the pair-wise distances, one may focus on entity-to-reference-point distances only – a reduction to the order of N from the order of N^2.

An anomalous pattern is found by building a cluster which is most distant from the reference point. To do this, the cluster's seed is defined as the entity farthest away from the reference point. Now a version of K-Means at $K = 2$ is applied with two seeds: the reference point which is never changed in the process and the cluster's seed, which is updated according to the standard procedure. In fact, only the anomalous cluster is of interest here. Given a centroid, the cluster is defined as the set of entities that are closer to it than to the reference point. Given a cluster, its centroid is found as the gravity center, by averaging all the cluster entities. The procedure is reiterated until convergence (see Fig. 6.10).

Assuming the reference point has been shifted into the origin, the Anomalous pattern method is a version of K-Means in which:

(i) the number of clusters K is 2;
(ii) centroid of one of the clusters is 0, which is forcibly kept there through all the iterations;
(iii) the initial centroid of the anomalous cluster is taken as an entity farthest away from 0.

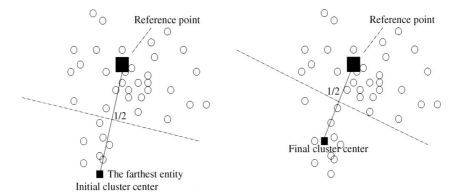

Fig. 6.10 Extracting an Anomalous pattern cluster with the reference point in the gravity center: the first iteration is on the *left* and the final one on the *right*

6.2 K-Means Clustering

Property (iii) mitigates the issue of determining appropriate initial seeds. This provides for using Anomalous pattern algorithm iteratively to obtain an initial setting for K-Means.

There is a certain similarity between selecting initial centroids in iK-Means and initial medoids with Build. But there are certain differences as well:

- K must be pre-specified in Build and not necessarily in iK-Means;
- The central point of the entire set I is taken as an initial seed in Build and is not in iK-Means;
- Addition of a new seed is based on different criteria in the two methods.

A clustering algorithm should present the user with a comfortable set of options. The iK-Means or PAM with Build/AP can be easily extended so that some entities can be removed from the data set because they are either (i) "deviant" or (ii) "intermediate" or (iii) "trivial". These can be defined as the contents of small AP or Build clusters, for the case (i), or entities that are far away from their centroids/medoids, for the case (ii), or entities that are close to the grand mean, the center of gravity of the entire data set, for the case (iii).

Worked example 6.4. Anomalous pattern in Market towns

Let us apply the Anomalous pattern method to Town data assuming the grand mean as the reference point and scaling by range. That means that after mean-range standardization the reference point is 0.

The point farthest from 0, to be taken as the initial "anomalous" centroid, appears to be entity 35 (St Austell) whose distance from zero (remember – after standardization!) is 4.33, the maximum. There are only three entities, 26, 29 and 44 (Newton Abbot, Penzance and Truro) that are closer to the seed than to 0, thus forming the cluster along with the original seed, at this stage. After one more iteration, the anomalous pattern cluster stabilizes with 8 entities 4, 9, 25, 26, 29, 35, 41, 44. Its centroid is displayed in Table 6.12.

As follows from the fact that all the standardized entries are positive and mostly fall within the range of 0.3–0.5, the anomalous cluster, according to Table 6.12, consists of better off towns – all the centroid values are larger than the grand mean by 30–50% of the feature ranges. This probably relates to the fact that they comprise eight out of the eleven towns that have a resident population greater than 10,000.

Table 6.12 Centroid of the anomalous pattern cluster of town data in real and standardized forms

Centroid	P	PS	Do	Ho	Ba	Sm	Pe	DIY	Sp	Po	CAB	FM
Real	18484	7.6	3.6	1.1	11.6	4.6	4.1	1.0	1.4	6.4	1.2	4.0
Std'zed	0.51	0.38	0.56	0.36	0.38	0.38	0.30	0.26	0.44	0.47	0.30	0.18

The other three largest towns have not made it into the cluster because of their deficiencies in services such as Hospitals and Farmers' Markets. The fact that the scale of population is by far the largest in the original table doesn't much affect the computation here as it runs with the range standardized scales at which the total contribution of this feature is not high, just about 8.5% only. It is rather the concerted action of all the features associated with a greater population which makes the cluster.

This process is illustrated on Fig. 6.11. The stars show the origin and the anomalous seed at the beginning of the iteration. Curiously, this picture does not fit well into the concept of the anomalous pattern cluster, as illustrated on the previous Fig. 6.10 – the anomalous pattern is dispersed here across the plane, which is at odds with the property that the entities in it must be closer to the seed than to the origin. The cause is not an error, but the fact that this plane represents all 12 original variables and presents them rather selectively. It is not that the plane makes too little of the data scatter – on the contrary, it makes a decent 76% of the data scatter. The issue here is the second axis in which the last feature FM expressing whether there is a Farmers market or not takes a lion share – thus stratifying the entire image over y axis.

The Fig. 6.11 has been produced with commands:

≫ subplot(1,2,1);plot(x1,x2,'k.', 0,0,'kp',x1(35),x2(35),'kp');text(x1(fir),x2(fir),ftm);
≫ subplot(1,2,2);plot(x1,x2,'k.', 0,0,'kp',x1a,x2a,'kp');text(x1(sec),x2(sec),fsm);

Here fir and sec are lists of indices of towns belonging to the pattern after the first and second iterations, respectively, while ftm and fsm refer to lists of their names.

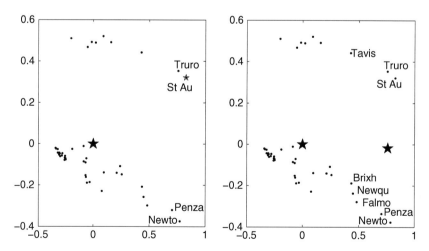

Fig. 6.11 The first and second iterations of Anomalous pattern cluster on the principal component plane; the visual separation of the pattern over y axis is due to a very high loading of the presence (*top*) or absence (*bottom*) of a farmer's market

6.2.8 Anomalous Pattern and Intelligent K-Means

P6.2.8.1 Anomalous Pattern and iK-Means: Presentation

The Anomalous pattern method can be used as a procedure to automatically determine both the number of clusters and initial seeds for K-Means. Preceded by this option, K-Means is referred to as intelligent K-Means, iK-Means for brevity, because it relieves the user from the task of specifying the initial setting.

In iK-Means method, the user is required to specify an integer, t, the threshold of resolution, to be used to discard all the Anomalous patterns consisting of t or less entities. When $t = 0$, nothing is discarded. At $t = 1$ – the default option, singleton anomalous patterns are considered a nuisance and put back to the data set. If $t = 10$, all patterns with 10 or less entities are discarded as too small to deserve any attention at all – the level of resolution which may be justified at larger datasets and coarser details needed.

In our experiments, the entities comprising singleton Anomalous pattern clusters are frequently erroneous, that is, errors are in some of their features such as, for instance, the human age of 5,000 years. That means, that Anomalous pattern clustering can be used as a device for checking against huge errors in data entries.

The iK-Means method is flexible with regard to outliers and the "swamp" of inexpressive – normal or ordinary – entities around the grand mean. For example, at its step 4, K-Means can be applied to either the entire dataset or to the set from which the smaller APs have been removed. This may depend on the domain: in some problems, such as structuring of a set of settlements for better planning or monitoring or analysis of climate changes, no entity should be dropped out of the consideration, whereas in other problems, such as developing synoptic descriptions for text corpora, some "deviant" texts could be left out of the coverage at all.

In a series of experiments with overlapping Gaussian clusters described by Chiang and Mirkin (2010), iK-Means has performed rather well and appeared superior to many other options for choosing K. These options included approaches based on post-processing of results of multiple runs of K-Means and then treating them according to either of the following:

(a) Variance based approach: using intuitive or model based functions of criterion (6.3) which should get extreme or "elbow" values at a correct K such as Hatigan's rule above;
(b) Structural approach: comparing within-cluster cohesion versus between-cluster separation at different K;
(c) Consensus distribution approach: choosing K according to the distribution of the consensus matrix for sets of K-Means clusterings at different K.

Some other approaches rely on different ideas for choosing K such as

(d) using results of a divisive or agglomerative clustering procedure or
(e) using the similarity of K-Means clustering results on randomly perturbed or sampled data.

Worked example 6.5. Iterated Anomalous patterns in Market towns

Applied to the range-standardized Market town data, AP algorithm, iterated until no unclustered entities remained, has produced 12 clusters of which 5 are singletons. These singletons have strange patterns of facilities indeed. For example, entity 19 (Liskeard, 7,044 residents) has an unusually large number of Hospitals (6) and CABs (2), which makes it a singleton cluster. Lists of seven non-singleton clusters are in Table 6.13, in the order of their extraction in the iterated AP.

This cluster structure doesn't much change when, according to the iK-Means algorithm, Batch K-Means is applied to the seven centroids (with the five singletons put back into the data). Moreover, similar results have been observed with clustering of the original all-England list of about thirteen hundred Market towns described by a wider list of eighteen characteristics of their development: the number of non-singleton clusters was the same, with very similar descriptions.

Q.6.20. Why is the contribution of AP 4, 18.6%, greater than that of the preceding AP3, 10.0%? **A.** Because of a much larger number of entities, 18 against 6 in AP 3. Even if the centroid of AP 3 is further away from 0 than centroid of AP 4, which is the cause that AP 3 is extracted first, the contribution takes into account the number of entities as well!

FC6.2.8.2 Anomalous Pattern and iK-Means: Formulation and Computation

Before substantiating AP algorithm, let us give it a more explicit formulation.

Anomalous Pattern (AP) Algorithm

1. *Pre-processing.* Specify a reference point $a = (a_1, \ldots, a_V)$ (when in doubt, take a to be the data grand mean) and standardize the original data table by shifting the origin to $a = (a_1, \ldots, a_V)$.
2. *Initial setting.* Put a tentative centroid, c, as the entity farthest away from the origin, 0.

Table 6.13 Iterated AP market town non-singleton clusters

Cluster #	Size	Contents	Contribution (%)
1	8	4, 9, 25, 26, 29, 35, 41, 44	35.1
3	6	5, 8, 12, 16, 21, 43	10.0
4	18	2, 6, 7, 10, 13, 14, 17, 22, 23, 24, 27, 30, 31, 33, 34, 37, 38, 40	18.6
5	2	3, 32	2.4
6	2	1, 11	1.6
8	2	39, 42	1.7
11	2	20, 45	1.2

6.2 K-Means Clustering

3. *Cluster update.* Determine cluster list S around c against the only other "centroid" 0, so that entity y_i is assigned to S if $d(y_i, c) < d(y_i, 0)$.
4. *Centroid update.* Calculate the within S mean c' and check whether it differs from the previous centroid c. If c' and c do differ, update the centroid by assigning $c \Leftarrow c'$ and go to Step 3. Otherwise, go to 5.
5. *Output.* Output list S and centroid c, with accompanying interpretation aids (as advised in the next section), as the anomalous pattern.

It is not difficult to prove that, like K-Means itself, the Anomalous pattern alternately minimizes a specific version of K-Means general criterion $W(S,c)$ (6.3),

$$W(S, c) = \sum_{i \in S} d(y_i, c) + \sum_{i \notin S} d(y_i, 0) \qquad (6.14)$$

where S is a subset of I rather than partition and c its centroid. Yet AP differs from 2-Means in the following aspect: there is only one centroid, c, which is updated in AP; the other centroid, 0, never changes and serves only to attract not-anomalous entities. This is why 2-Means produces two clusters whereas AP – only one, that is farthest away from the reference point, 0.

In fact, criterion (6.14) can be equivalently rephrased using Equations (6.6) and (6.7) representing the complimentary criterion $B(S,c)$. When (6.7) applies to the situation of two clusters, one with centroid in c, the other in 0, it becomes of finding a cluster S maximizing its contribution to the data scatter $T(Y)$:

$$\mu^2 = z^T Y Y^T z / z^T z = c_v^2 |S| = d(0, c) |S| \qquad (6.15)$$

This means that AP algorithm straightforwardly follows the Principal Component Analysis one-by-one extraction strategy extended to binary scoring vectors. That is, the model behind AP is a version of the PCA Equation (5.10) in which the scoring values z^*_i are but zeros or ones:

$$y_{iv} = \begin{cases} c_v + e_v, i \in S \\ 0 + e_v, i \notin S \end{cases} \qquad (6.16)$$

where S is the cluster list of the anomalous pattern to be found.

In spite of the rather simplistic assumption presented in (6.16), AP clusters fare well with real data. They can be extracted one-by-one, along with their contributions to the data scatter (6.15) showing cluster saliencies. These saliencies can be used to halt the process when the contribution of the next cluster drops decisively, thus leading to an incomplete clustering when needed.

Here are steps of iK-Means(t) where t is the cluster discharge threshold – the minimum number of entities in a pattern that can be considered a cluster on its own. In most applications dealing with moderately sized data (up to a few hundred entities) t can be put to be equal to 1.

iK-Means(*t*) Algorithm

0. *Setting.* Preprocess and standardize the data set. Take t as the threshold of resolution. Put $k = 1$ and $I_k = I$, the original entity set.
1. *Anomalous pattern.* Apply AP to I_k to find k-th anomalous pattern S_k and its centroid c_k.
2. *Test.* If Stop-condition (see below) does not hold, remove S_k from I_k to make $k \Leftarrow k + 1$ and $I_k \Leftarrow I_k - S_k$, after which step 1 is executed again. If it does, go to 3.
3. *Discarding small clusters.* Remove all of the found clusters containing t entities or less. Denote the number of remaining clusters by K and re-label them so that their centroids are $c_1, c_2, \ldots, c_K$.
4. *K-Means.* Do Batch K-Means using $c_1, c_2, \ldots, c_K$ as initial seeds.

Case Study 6.5. iK-Means Clustering of a Normally Distributed 1D Dataset

Let us generate a one dimensional set X of 280 points according to Gaussian N(0,10) distribution (see Fig. 6.12). This data set is attached in the appendix as Table A5.2. Many would say that this sample constitutes a single, Gaussian, cluster. Yet the idea of applying a clustering algorithm seems attractive as a litmus paper to capture the pattern of clustering embedded in iK-Means algorithm.

In spite of the symmetry in the generating model, the sample is slightly biased to the negative side; its mean is -0.89 rather than 0, and its median is about -1.27. Thus the maximum distance from the mean is at the maximum of 32.02 rather than at the minimum of -30.27.

The Anomalous pattern starting from the furthest away value of maximum comprises 83 entities between the maximum and 5.28. Such a stripping goes along

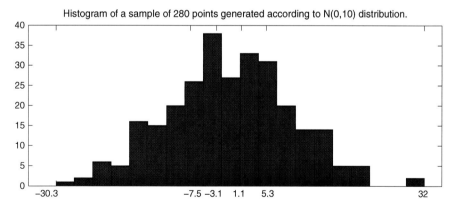

Fig. 6.12 Histogram of the sample of 280 values generated by Matlab's randn command from the Gaussian distribution N(0,10)

6.2 K-Means Clustering

Table 6.14 A summary of the iterative Anomalous pattern clustering results for the sample of Gaussian distribution in Table A5.2. Clusters are shown in the extraction order, along with their sizes, left and right boundary entity indices, means and contributions to the data scatter

Order of extraction	Size	Left index	Right index	Mean	Contrib (%)
1	83	198	280	11.35	34.28
2	70	1	70	−14.32	46.03
3	47	71	117	−5.40	4.39
4	41	157	197	2.90	1.11
5	18	118	135	−2.54	0.38
6	10	147	156	0.27	0.002
7	6	136	141	−1.42	0.039
8	2	145	146	−0.49	0.002
9	3	142	144	−0.77	0.006

real-world conventional procedures. For example, consider the heights of a sample of young males to be drafted for a military action whose histogram is known to be bell shaped like Gaussian. Those on the either side of the bell shaped height histogram are not quite fitting for action: those too short cannot accomplish many a specific task whereas those too tall may have problems in closed spaces such as submarines or aircraft.

The iterative Anomalous pattern clustering would sequentially strip the remaining margins off too. The set of fragments of the sorted sequence in Table 6.14 that have been found by the Anomalous pattern clustering algorithm in the order of their forming, including the cluster means and contributions to the data scatter.

The last extracted clusters are all around the mean and, predictably, small in size. One also can see that the contribution of a following cluster can be greater than that of the preceding cluster thus reflecting the local nature of the Anomalous pattern algorithm which intends to find the maximally contributing cluster each time. The total contribution of the nine clusters is about 86% to which the last five clusters contribute next to nothing.

Project 6.1. Using contributions to determine the number of clusters

The question of determining the number K can be addressed with the model (6.16) itself, applied to the cluster contribution values as the raw data. Assume the contributions are sorted in the descending order and denoted by h_k so that $h_1 \geq h_2 \geq \ldots$ ($k = 1, 2, \ldots$).

If one assumes that the first K values are all approximately equal to each other, whereas the rest approximate zero, then the optimal K can be derived as follows.

Denote the average of the first K contributions as $h(K)$. Then criterion (6.16) to maximize is the product $Kh^2(K)$. The optimal K obviously satisfies inequality $Kh^2(K) > (K+1)h^2(K+1)$. Since the average $h(K+1)$ can be expressed as $h(K+1) = (K*h(K) + h_{K+1})/(K+1)$, the inequality can be easily transformed to $h^2(K) -$

$2h(K)h_{K+1} + h_{K+1}^2/K > 0$ which can be further presented as $(h(K) - h_{K+1})^2 > h_{K+1}^2(1 - 1/K)$. Since $h(K) \geq h_{K+1}$, this inequality can be further simplified to $h(K) - h_{K+1} > h_{K+1}\sqrt{(1 - 1/K)}$, that is,

$$h(K) > h_{K+1}\left(1 + \sqrt{1 - 1/K}\right) \qquad (6.17)$$

which is, roughly, $h_{K+1} < h(K)/2$. This has an advantage that the threshold is not pre-specified but rather determined according to the structure of gaps between the numbers h_k in their sorted order. The value of K at which (6.17) holds can be considered as a candidate for the right number of clusters or components or, in fact, anything evaluated by contributions.

Similar inequalities can be derived at different models for the chosen contribution values. One may try, for example, the power law assumption that $h(k) = ak^{-b}$ for $k = 1, \ldots, K$ and $h(k) = 0$ for $k > K$ (see also Cangelosi and Goriely 2007).

Method iK-Means utilizes a slightly different strategy for choosing the right K. This strategy involves (i) all the anomalous patterns rather than those most contributing, thus involving the patterns close to the reference points too, and (2) a different scoring device – the intuitively clear number of entities rather than a purely geometric contribution whose intuitive value is unclear.

Project 6.2. Does PCA clean the data structure indeed: K-Means after PCA

There is a wide-spread opinion that in a situation of many features, the data structure can come less noisy if the features are first "cleaned off" by applying PCA and using a few principal components instead of the original features. Although strongly debated by specialists (see, for example, Kettenring 2006), the opinion is wide-spread among the practitioners. A voice of dissent was raised by late A. Kryshtanowski (2008) who provided an example of data structure that "becomes less pronounced in the space of principal components".

The example refers to data of two Gaussian clusters, each containing 500 of 15-dimensional entities. The first cluster can be generated by the following MatLab commands:

```
≫b(1:500,1)=10*randn(500,1);
≫b(1:500,2:15)=repmat(b(1:500,1),1,14)+20*randn(500, 14);
```

The first variable in the cluster is Gaussian with the mean 0 and standard deviation 10, whereas the other fourteen variables add to that another Gaussian variable whose mean and standard deviation are 0 and 20, respectively. That is, this set is a sample from a 15-dimensional Gaussian with a diagonal covariance matrix, whose center is in or near the origin of the space, with the standard deviations of all features at 22.36, the square root of $10^2 + 20^2$, except for the first one that has the standard deviation of 10.

6.2 K-Means Clustering

The entities in the second cluster are generated as the next 500 rows in the same matrix in a similar manner:

≫b(501:1000,1)=20+10*randn(500,1);
≫b(501:1000,2:15)=repmat(b(501:1000,1),1,14)+20*randn(500,14)+10;

The first variable now is centered at 20, and the other variables, at 30. The standard deviations follow the pattern of the first cluster.

Since the standard deviations by far exceed the distance between centroids, these clusters are not easy to distinguish: see Fig. 6.13 illustrating the data cloud, after centering, on the plane of the first principal components.

When applying iK-Means to this data, preliminarily centered and range-normalized, the algorithm finds indeed much more clusters, 13 of them, at the discarding threshold $t = 1$. However, when the discarding threshold is set to $t = 200$, to remove any less populated anomalous patterns, the method arrives at just two clusters that differ from those generated by 96 entities (see the very first resulting column in Table 6.15 presenting the results of the computation) constituting the total error of 9.6%. The same method applied to the data z-score standardized, that is, centered and normalized by the standard deviations, arrives at 99 errors; a rather modest increase, probably due to specifics of the data generation.

Since Kryshtanowsky (2008) operated with the four most contributing principal components, we also take the first four principal components, after centering by the means and normalizing data by the range:

≫n=1000; br=(b−repmat(mean(b),n,1))./ repmat(max(b)−min(b),n,1);
≫[zr,mr,cr] =svd(br);
≫ zr4=zr(:,1:4);

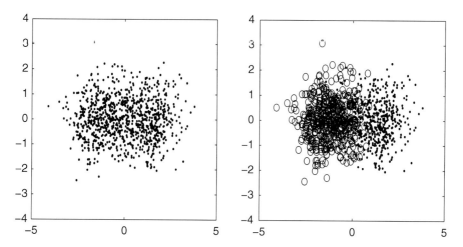

Fig. 6.13 Data of two clusters generated as described above, after centering, are represented by points on the plane of the two first principal components (on the *left*); the second cluster is represented by *circles* on the *right*

Table 6.15 The numbers of errors of iK-Means clustering at different data transformations: over the original data differently normalized and over four principal components derived at different data normalizations

Data	Original 15 features		Four principal components		
Normalized by	Range	St. deviation	No normalization	Range	Standard deviation
Cluster 1	44	43	37	51	47
Cluster 2	52	56	47	47	45
Total	96	99	84	98	92

These four first components bear 66% of the data scatter. We also derived four principal components from the data non-normalized (yet centered) and from the data normalized by the standard deviations (z-score standardized). The latter is especially important in this context, because Kryshtanowsky (2008) used the conventional form of PCA based on the correlation matrix between the variables, which is equivalent to the model-based PCA applied to the data after z-score standardization. The iK-Means method applied to each of these data sets at the discarding threshold of 200, has shown rather consistent results (see Table 6.15).

Overall, these results seem to support the idea of better structuring under principal components rather than to refute it. The negative results of using principal components by Kryshtanowsky (2008) probably could be attributed to his indiscriminate usage of Batch K-Means method with random initializations that failed to find a "right" pair of initial centroids, in contrast to iK-Means.

6.3 Cluster Interpretation Aids

P6.3.1 Cluster Interpretation Aids: Presentation

Results of K-Means clustering, as well as any other method resulting in the list of clusters $S = \{S_1, S_2, \ldots, S_K\}$ and their centroids, $c = \{c_1, c_2, \ldots, c_K\}$, can be interpreted by using

(a) cluster centroids versus grand means (feature averages on the entire data set)
(b) cluster representatives
(c) cluster-feature contributions to the data scatter
(d) conceptual descriptions of clusters

One should not forget that, under the zero-one coding system for categories, cluster-to-category cross-classification frequencies are, in fact, cluster centroids – therefore, (a) includes looking at cross-classifications between S and categorical features although this is conventionally considered a separate interpretation device.

Consider these in turn.

6.3 Cluster Interpretation Aids

(a) Cluster centroids versus grand means

These should be utilized in both, original and standardized, formats. To express a standardized centroid value c_{kv} of feature v in cluster S_k resulting from a K-Means run, in the original scale of feature v, one needs to invert the scale transformation by multiplying over rescaling factor b_v with the follow up adding the shift value a_v, so that this becomes $b_v c_{kv} + a_v$.

Worked example 6.6. Centroids of Market town clusters

Let us take a look at centroids of the seven clusters of Market towns data both in real and range standardized scales in Table 6.16.

These show some tendencies rather clearly. For instance, the first cluster appears to be a set of larger towns that score 30–50% higher than the average on almost all of the 12 features. Similarly, cluster 3 obviously relates to smaller than average towns. However, in other cases, it is not always clear what features caused a cluster to separate. For instance, both clusters 6 and 7 seem too close to the average to make any real difference at all.

(b) Cluster representative

A cluster is typically characterized by its centroid consisting of the within-cluster feature means. Sometimes, the means make no sense – like the number of suppliers 4.5 above. In such a case, it is more intuitive to characterize a cluster by its

Table 6.16 Patterns of Market towns in the cluster structure found with iK-Means (see Table 6.13). For each of the clusters, real values are on the top line and the standardized values are in the bottom; GM is the grand mean

#	Pop	PS	Do	Ho	Ba	Su	Pe	DIY	SP	PO	CAB	FM
1	18484	7.63	3.63	1.13	11.63	4.63	4.13	1.00	1.38	6.38	1.25	0.38
	0.51	0.38	0.56	0.36	0.38	0.38	0.30	0.26	0.44	0.47	0.30	0.17
2	5268.00	2.17	0.83	0.50	4.67	1.83	1.67	0.00	0.50	1.67	0.67	1.00
	−0.10	−0.07	−0.14	0.05	0.02	−0.01	−0.05	−0.07	0.01	−0.12	0.01	0.80
3	2597	1.17	0.50	0.00	1.22	0.61	0.89	0.00	0.06	1.44	0.11	0.00
	−0.22	−0.15	−0.22	−0.20	−0.16	−0.19	−0.17	0.07	−0.22	−0.15	−0.27	−0.20
4	11245	3.67	2.00	1.33	5.33	2.33	3.67	0.67	1.00	2.33	1.33	0.00
	0.18	0.05	0.16	0.47	0.05	0.06	0.23	0.15	0.26	−0.04	0.34	−0.20
5	5347	2.50	0.00	1.00	2.00	1.50	2.00	0.00	0.50	1.50	1.00	0.00
	−0.09	−0.04	−0.34	0.30	−0.12	−0.06	−0.01	−0.07	0.01	−0.14	0.18	−0.20
6	8675	3.80	2.00	0.00	3.20	2.00	2.40	0.00	0.00	2.80	0.80	0.00
	0.06	0.06	0.16	−0.20	−0.06	0.01	0.05	−0.07	−0.24	0.02	0.08	−0.20
7	5593	2.00	1.00	0.00	5.00	2.67	2.00	0.00	1.00	2.33	1.00	0.00
	−0.08	−0.09	−0.09	−0.20	0.04	0.10	−0.01	−0.07	0.26	−0.04	0.18	−0.20
GM	7351.4	3.02	1.38	0.40	4.31	1.93	2.04	0.22	0.49	2.62	0.64	0.20

"typical" representative. This is especially appealing when the representative is a well known object. Such an object can give much better intuition to a cluster than a logical description in situations in which entities are complex and the features are superficial. This is the case, for instance, in mineralogy where a class of minerals can be represented by its "stratotype" mineral, or in art studies where a general concept such as "surrealism" can be represented by an art object such as a painting by S. Dali.

A cluster representative must be the nearest to its cluster's centroid. An issue is that two different expressions for K-Means lead to two different measures. The sum of entity-to-centroid distances W(S,c) in (6.3) leads to the strategy that can be referred to as "the nearest in distance." The sum of entity-to-centroid inner products for B(S,c) in (6.8) leads to the strategy "the nearest in inner product". Intuitively, the choice according to the inner product follows tendencies represented in c_k towards the whole of the data expressed in grand mean position whereas the distance follows just c_k itself. These two principles usually lead to similar choices, though sometimes rather not.

Worked example 6.7. Representatives of Company clusters

Consider, for example, A product cluster in Company data as presented in Table 6.17: The nearest to centroid in distance is Ant and nearest in inner product is Ave.

To see why is that, let us take a closer look at the two companies. Ant and Ave are similar on all four binary features. Each is at odds with the centroid's tendency on one feature only: Ant is zero on NSup while centroid is negative, and Ave is negative on Income while centroid is positive on that. The difference, however, is in feature contributions to the cluster; that of Income is less than that of NSup, which makes Ave to win, as a follower of NSup, over the inner product expressing contributions of entities to the data scatter. With the distance measure, the cluster tendency by itself does not matter at all because it is expressed in the signs of the standardized centroid.

Q.6.21. Find representatives of Company clusters B and C.

Table 6.17 Standardized entities and centroid of cluster A in company data. The nearest to centroid are: Ant, in distance, and Ave, in inner product (both are in thousandth)

Cluster	Income	SharP	NSup	EC	Util	Indu	Retail	Distance	InnerPr
Centroid	0.10	0.12	−0.11	−0.63	0.17	−0.02	−0.14		
Ave	−0.20	0.23	−0.33	−0.63	0.36	−0.22	−0.14	222	**524**
Ant	0.40	0.05	0.00	−0.63	0.36	−0.22	−0.14	**186**	521
Ast	0.08	0.09	0.00	−0.63	−0.22	0.36	−0.14	310	386

6.3 Cluster Interpretation Aids

(c) Feature-cluster contributions to the data scatter

To see what features do matter in each of the clusters, the contributions of feature-cluster pairs to the data scatter are to be invoked. The feature-cluster contribution is equal to the product of the squared feature (standardized) centroid component and the cluster size. In fact, this is proportional to the squared difference between the feature's grand mean and its within-cluster mean: the further away the latter from the former, the greater the contribution! This is illustrated on Fig. 6.14:

Worked example 6.8. Contributions of features to Market town clusters

The cluster-specific feature contributions are presented in Table 6.18, along with their total contributions to the data scatter in row Total. The intermediate rows Exp and Unexp show the explained and unexplained parts of the totals, with Exp being the sum of all cluster-feature contributions and Unexp the difference between the Total and Exp rows.

The columns on the right show the total contributions of clusters to the data scatter, both as is and per cent. The cluster structure in total accounts for 73.3%

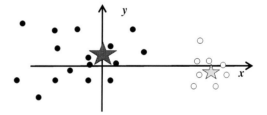

Fig. 6.14 Contributions of features *x* and *y* in the group of *blank-circled* points are proportional to the squared differences between their values at the grand mean (*large star*) and within-group centroid (*small star*)

Table 6.18 Decomposition of the data scatter over clusters and features at market town data; row Exp sums all the cluster contributions, row Total gives the feature contributions to the data scatter, and row Unexp is the difference, Total-Exp

#	P	PS	Do	Ho	Ba	Su	Pe	DIY	SP	PO	CAB	FM	Total	Total (%)
1	**2.09**	1.18	**2.53**	1.05	1.19	1.18	0.71	0.54	1.57	1.76	0.73	0.24	14.77	35.13
2	0.06	0.03	0.11	0.01	0.00	0.00	0.02	0.03	0.00	0.09	0.00	**3.84**	4.19	9.97
3	0.86	0.43	0.87	0.72	0.48	0.64	0.49	0.10	0.85	0.39	**1.28**	0.72	7.82	18.60
4	0.10	0.01	0.07	**0.65**	0.01	0.01	0.16	0.07	0.20	0.00	0.36	0.12	1.75	4.17
5	0.02	0.00	**0.24**	0.18	0.03	0.01	0.00	0.01	0.00	0.04	0.06	0.08	0.67	1.59
6	0.02	0.02	0.12	0.20	0.02	0.00	0.01	0.03	**0.30**	0.00	0.03	**0.20**	0.95	2.26
7	0.02	0.02	0.03	0.12	0.00	0.03	0.00	0.02	**0.20**	0.00	0.09	0.12	0.66	1.56
Exp	3.16	1.69	3.96	2.94	1.72	1.88	1.39	0.79	3.11	2.29	2.56	5.33	30.81	73.28
Unexp	0.40	0.59	0.70	0.76	0.62	0.79	1.02	0.96	1.20	0.79	1.52	1.88	11.23	26.72
Total	3.56	2.28	4.66	3.70	2.34	2.67	2.41	1.75	4.31	3.07	4.08	7.20	42.04	100.00

of the data scatter, a rather high proportion. Of the total contributions, three first clusters have the largest ones totaling to 26.78, or 87% of the explained part of the data scatter, 30.81. Among the variables, FM gives the maximum contribution to the data scatter, 5.33. This can be attributed to the fact that FM is a binary variable on the data set – binary variables, in general, have the largest total contributions because they are bimodal. Indeed, FM's total contribution to the data scatter is 7.20 so that its explained part amounts to $5.33/7.20 = 0.74$ which is not as much as that of, say, the Population resident feature, $3.16/3.56 = 0.89$, which means that overall Population resident better explains the clusters than FM.

Worked example 6.9. Contributions and relative contributions of features at Company clusters

Consider the clustering of Companies data according to their main product, A or B or C, to find out what features can be associated with each of the clusters. The cluster centroids as well as feature-cluster contributions are presented in Table 6.19. The summary contributions over clusters are presented in the last column of Table 6.19, and over features, in the first line of third row of Table 6.19 termed "Explain". The feature contributions to the data scatter, that is, the sums of squares of the feature's column entries, are in the second line of the third row – these allow us to express the explained feature contributions per cent, in the third line.

Now we can take a look at the most contributing feature-to-cluster pairs. This can be done by considering relative contributions within individual lines (clusters) or within individual columns (features). For example, in the third row of Table 6.19, each contribution that covers half or more of the explained contribution by the feature is highlighted in bold. Obviously, the within-line maxima do not necessarily

Table 6.19 Centroids and feature-to-cluster contributions for product clusters in company data

Item			Income	SharP	NSup	EC	Util	Indu	Retail	Total
Cluster	A		0.10	0.12	−0.11	−0.63	0.17	−0.02	−0.14	
centroids	B		−0.21	−0.29	−0.22	0.38	−0.02	0.17	−0.14	
standardized	C		0.18	0.24	0.50	0.38	−0.22	−0.22	0.43	
Cluster	A		0.03	0.05	0.04	**1.17**	0.09	0.00	0.06	1.43
contributions	B		**0.14**	**0.25**	0.15	0.42	0.00	0.09	0.06	1.10
	C		0.06	0.12	**0.50**	0.28	0.09	**0.09**	**0.38**	1.53
Total	Ex		0.23	0.41	0.69	1.88	0.18	0.18	0.50	4.06
contributions	Data		0.74	0.69	0.89	1.88	0.63	0.63	0.50	5.95
	Ex%		31.1	59.4	77.5	100.0	28.6	28.6	100.0	68.3
Relative	A		16.7	29.5	18.5	**258.1**	59.9	0.0	49.5	
contribution	B		101.2	**191.1**	90.2	120.2	0.0	77.7	64.2	
indexes, %	C		31.7	67.0	**219.5**	58.5	56.7	56.7	**297.0**	
Cluster			24.10	39.23	2.67	**0.00**	0.67	0.33	0.00	
centroids real			18.73	**22.37**	2.33	1.00	0.33	0.67	0.00	
			25.55	44.10	**4.50**	1.00	0.00	0.00	**1.00**	

6.3 Cluster Interpretation Aids

match those within columns. The relative contribution indexes in the fourth row of Table 6.19 combine these two perspectives: they are ratios of two relative contributions: the relative explained feature contribution within a cluster to the relative feature contribution to the data scatter. For example, relative contribution index of feature ShareP to cluster B, 1.911, is found by relating its relative explained contribution 0.25/1.10 to its relative contribution to the data scatter, 0.69/5.95. Those of the relative contribution indexes that are greater than 150%, so that the feature contribution to the cluster structure is at least 50% greater than its contribution to the data scatter, are highlighted.

The most contributing features are those that make the clusters different. To see this, Table 6.19 is supplemented with the real values of the within-cluster feature means, in its last row. The values corresponding to the outstanding contributions are highlighted in bold. Cluster A differs by feature EC – A-listed companies do not use e-commerce; cluster B differs by the relatively low Share Prices; and cluster C differs by either the fact that it all falls within Retail sector or the fact that its companies have relatively high numbers of suppliers, 4 or 5. It is easy to see that each of these statements not only points to a tendency but distinctly describes the cluster as a whole.

Case-Study 6.6. 2D Analysis of Most Contributing Features

Consider 2D analysis of the relationship between the Company data partition in three product classes, A, B, and C, and the most contributing of the quantitative features in Table 6.19 – the Number of suppliers (77.5%) as illustrated in Table 6.20.

To calculate the correlation ratio of the NSup feature according to formula ( ), let us first calculate the average within-class variance $\sigma_u^2 = (3*0.22 + 3*0.022 + 2*0.25/8 = 0.23$; the correlation ratio then will be equal to $\eta^2 = (\sigma^2 - \sigma_u^2)/\sigma^2 = (1.00 - 0.23)/1.00 = 0.77$. According to (6.13), this, multiplied by $N = 8$ and $\sigma^2 = 1$, must be equal to the total explained contribution of feature NSup to the data scatter in Table 6.21, 0.69 – which is clearly not! Why? Because the correlation ratio in (6.13) refers to the standardized, not original feature, and to make up for this, one needs to divide the result by the squared scaling parameter, the range r, which is 3 in this case. Now we get things right: $\eta^2*N/r^2 = 0.77*8/9 = 0.69$ indeed!

Table 6.20 Tabular regression of NSup feature over the product-based classes in the company dataset in Table 5.2

Classes	#	NSup mean	NSup variance
A	3	2.67	0.22
B	3	2.33	0.22
C	2	4.50	0.25
Total	8	3.00	1.00

Table 6.21 Contingency table between the product-based classes and nominal feature Sector in the company dataset according to Table 5.1

Category Class	Utility	Industrial	Retail	Total
A	2	1	0	3
B	1	2	0	3
C	0	0	2	2
Total	3	3	2	8

Table 6.22 Relative frequencies together with absolute and relative Quetélet indexes for contingency Table 6.21

Cat. class	Utility	Indust	Retail	Total	Utility	Indust	Retail	Utility	Indust	Retail
	Relative frequencies				Absolute Quetélet ind.			Relative Quetélet ind.		
A	0.25	0.12	0.00	0.37	0.29	−0.04	−0.25	0.78	−0.11	−1.00
B	0.12	0.25	0.00	0.37	−0.04	0.29	−0.25	−0.11	0.78	−1.00
C	0.00	0.00	0.25	0.25	−0.38	−0.38	0.75	−1.00	−1.00	3.00
Total	0.37	0.37	0.25	1.00						

The case of a nominal feature can be analyzed similarly. Consider contingency table between the product based partition S and feature Sector in Company data (Table 6.19).

In contrast to the classical statistics perspective, the small and even zero values are not of an issue here. Table 6.22 presents, on the left, the same data in the relative format; the other two parts present absolute and relative Quetelet indexes as described in Section 3.4.

These indexes have something to do with the cluster-feature contributions in Table 6.19. Given that the categories have been normalized by unities as well as the other features, the absolute Quetélet indexes are involved. [To use the relative Quetélet indexes, the categories have to be normalized by the square roots of their frequencies, as explained in the Formulation part.] Their squares multiplied by the cluster cardinalities and additionally divided by the squared rescaling parameter, 3 in this case, are the contributions, according to formula (6.14), as presented in the Table 6.23.

Table 6.23 Absolute Quetélet indexes from Table 6.22 and their squares factored according to formula (6.14)

Cat. class	Size	Utility	Industrial	Retail	Utility	Industrial	Retail	Total
		Absolute Quetélet indexes			Contributions			
A	3	0.29	−0.04	−0.25	**0.085**	0.002	0.062	0.149
B	3	−0.04	0.29	−0.25	0.002	**0.085**	0.062	0.149
C	2	−0.38	−0.38	0.75	0.094	0.094	**0.375**	0.563
Total	8				0.181	0.181	0.500	0.862

6.3 Cluster Interpretation Aids

Obviously, all entries in the right part of Table 6.23 are items in the total Proportional prediction index (6.16) divided by 3 – because of the specifics of the data normalization with the additional normalization by the square root of the number of categories.

(d) Conceptual description of clusters

If a contribution is high, then, as can be seen on Fig. 6.14, it is likely that the corresponding feature can be utilized for conceptual description of the corresponding class.

Worked example 6.10. Describing Market town clusters conceptually

Consider, for example, Table 6.18 of contributions of the clusters found at Market towns data. Several entries in Table 6.18 are highlighted in bold as those most contributing to the data scatter parts explained by clusters, the columns on the right. Take a look at them, cluster-wise.

Cluster 1 is indeed characterized by its two most contributing features, Population resident (P, contribution 2.09) and the number of doctor surgeries (Do, contribution 2.53). It can be described as a "set of towns with the population resident P not less than 10,200 and number of doctor surgeries Do not less than 3" – this description perfectly fits the cluster with no errors, be it false positive or false negative. Cluster 2 is blessed with an unusually high relative contribution of FM, 3.84 of the total 4.19; this may be seen as the driving force of the cluster's separation: it comprises all the towns with a Farmers market that have not been included in cluster 1! Other clusters can be described similarly. Let us note the difference between clusters 6 and 7, underlined by the high contributions of swimming pools (SW) to both, though by different reasons: every town in cluster 7 has a swimming pool whereas any town in cluster 6 has none.

Worked example 6.11. Describing Company clusters conceptually

Conceptual descriptions can be drawn for the product clusters in Company data according to Table 6.19. This table shows that feature EC is the most contributing to the Product A cluster, feature ShaP to the Product B cluster, and features SupN and Retail to the Product C cluster. The relatively high contribution of ShaP to B cluster is not that obvious because that of EC, 0.42, is even higher. It becomes clear only on the level of relative contributions when the contributions are related to their respective Total counterparts, 0.25/0.69 and 0.42/1.88 – the former prevails indeed. Clusters A, B, and C can be distinctively described by the statements "EC==0", "ShaP < 28", and "SupN >3" (or "Sector is Retail"), respectively.

Unfortunately, high feature contributions not always lead to clear-cut conceptual descriptions. The former are based on the averages whereas the latter on clear-cut divisions, and division boundaries can be at odds with the averages.

F6.3.2 Cluster Interpretation Aids: Formulation

According to (6.4) and (6.5), clustering (S,c) decomposes the scatter $T(Y) = \Sigma_{i,v} y_{iv}^2$ of data matrix Y in the explained and unexplained parts, $B(S,c)$ and $W(S,c)$, respectively. The latter is the square-error K-Means criterion, whereas the explained part $B(S,c)$ is clustering's contribution to the data scatter, which is equal, according to (6.6), to

$$B(S,c) = \sum_{k=1}^{K} \sum_{v \in V} c_{kv}^2 N_k$$

This is the sum of additive items $B_{kv} = N_k c_{kv}^2$, each accounting for the contribution of feature-cluster pair, $v \in V$ and S_k ($k = 1, 2, \ldots, K$).

Since the total contribution of feature v to the data scatter is $T_v = \sum_{i \in I} y_{iv}^2$, its unexplained part can be expressed as $W_v = T_v - B_{+v}$ where $B_{+v} = \sum_{k=1}^{K} B_{kv}$ is feature's v explained part, the total contribution of v to the cluster structure. This can be displayed as a Scatter Decomposition (ScaD) table whose rows correspond to clusters, columns to variables and entries to the contributions B_{kv} (see Table 6.24).

The summary rows, Explained, Unexplained and Total, as well as column Total can be expressed as percentages of the data scatter $T(Y)$. The contributions highlight relative roles of features both at individual clusters and in total.

The explained part $B(S,c)$ is, ccording to (6.6), the sum of contributions of individual feature-to-cluster pairs $B_{kv} = c_{kv}^2 N_k$ which can be used for interpretation of the clustering results. The sums of B_{kv}'s over features or clusters

Table 6.24 ScaD: Data scatter decomposed over clusters and features using notation introduced above

Cluster	Feature				
	f_1	f_2		f_M	Total
S_1	B_{11}	B_{12}		B_{1V}	B_{1+}
S_2	B_{21}	B_{22}		B_{2V}	B_{2+}
S_K	B_{K1}	B_{K2}		B_{KV}	B_{K+}
Explained	B_{+1}	B_{+2}		B_{+V}	$B(S,c)$
Unexplained	W_{+1}	W_{+2}		W_{+V}	$W(S,c)$
Total	T_1	T_2		T_V	$T(Y)$

6.3 Cluster Interpretation Aids

express total contributions of individual clusters or features into the explanations of clusters.

As has been shown in Section 4.5.3, summary contributions of individual data features to clustering (S,c) have something to do with statistical measures of association in bivariate data, such as correlation ratio η^2 (3.15) in Section 3.3 and chi-squared X^2 (3.13) in Section 3.4 (Mirkin 2005). In fact, the analysis in Section 4.5.3.2 applies in full to the case when target features are those used for building clustering S.

Specifically, for a quantitative feature v represented by the standardized column y_v, its summary contribution B_{+v} to the data scatter is equal to

$$B_{+v} = N\sigma_v^2 \eta_v^2 \qquad (6.18)$$

Note that the correlation ratio in (6.18) has been computed over the normalized feature y_v. The correlation ratio of the original non-standardized feature x_v differs from that by factor equal to the squared rescaling parameter b_v^2.

Consider now a nominal feature v represented by a set of binary columns, dummies, corresponding to individual categories $l \in v$. The grand mean of binary column for $v \in F$ is obviously the proportion of this category in the set, p_{+v}. To standardize the column, one needs to subtract the mean, p_{+v}, from all its entries and divide them by the scaling parameter, b_v. After the standardization, the centroid of cluster S_k can be expressed through co-occurrence proportions too, as expressed by the formula on p. 149:

$$c_{kv} = \left(\frac{p_{kv}}{p_k} - p_{+v}\right)/b_v$$

where p_{kv} is the proportion of entities falling in both category v and cluster S_k; the other symbols: p_{+v} is the frequency of v, p_k the proportion of entities in S_k, and b_v the normalizing scale parameter.

According to Equations (4.18) and (4.19), the summary contribution of all pairs category-cluster (l,k) is equal to

$$B(v/S) = N \sum_{l \in v} \sum_{k=1}^{K} \frac{(p_{kl} - p_k p_{+l})^2}{p_k b_l^2} \qquad (6.19)$$

This is akin to several contingency table association measures considered in the literature including Pearson chi-squared X^2 in (3.15) and Gini impurity function, or summary absolute Quetelet index, in (3.22). To make $B(v/S)$ equal to the chi-squared coefficient, the scaling of binary features must be done by using $b_l = \sqrt{p_l}$, which is the standard deviation of the Poisson probabilistic distribution that randomly throws $p_l N$ unities into an N-dimensional binary vector. To make $B(v/S)$ equal to Gini impurity function, no normalization of the dummies is to be done, or rather the recommended option of normalization by ranges applies since the range of a dummy is 1.

One should not forget the additional normalization of the binary columns by the square root of the number of categories in a nominal feature v, $\sqrt{|v|}$ leading to both the individual contributions B_{kv} in (6.18) and the total contribution $B(v/S)$ in (6.19) divided by the number of categories $|v|$. When applied to Pearson chi-squared, the division by $|v|$ can be considered as another normalization of the coefficient. As mentioned in Section 3.4, the maximum of Pearson chi-squared (related to N) is $\min(|v|, K) - 1$. Therefore, when $|v| \leq K$, the division would lead to a normalized index whose values are between 0 and $1 - 1/|v|$. If, however, the number of categories is larger so that $K < |v|$, then the normalized index could be very near 0 indeed. In this regard, it should be of interest to mention that in the literature some other normalizations have been considered. Specifically, Pearson chi-squared is referred to as Cramer coefficient if related to $\min(|v|, K) - 1$, and as Tchouproff coefficient if related to $\sqrt{(|v|-1)(K-1)}$ (Kendall and Stewart 1973).

Q.6.22. Prove that, for any cluster k in K-Means clustering, $\sum_{i \in S_k} y_{iv}^2 = |S_k|(c_{kv}^2 + \sigma_{kv}^2)$.

Q.6.23. How one should interpret the normalization of a category by the $\sqrt{p_v}$? What category gets a greater contribution: that more frequent or that less frequent?

Comment 6.1. When the chi-squared contingency coefficient or related indexes are applied in the traditional statistics context, the presence of zeros in a contingency table becomes an issue because it contradicts the hypothesis of statistical independence. In the context of data recovery clustering, zeros are treated as any other numbers and create no problems at all because the coefficients are measures of contributions and bear no other statistical meaning in this context.

Comment 6.2. K-Means advantages: The method

(i) Models typology building activity
(ii) Computationally effective both in memory and time
(iii) Can be utilized incrementally, "on-line"
(iv) Straightforwardly associates feature salience weights with feature scales
(v) Applicable to both quantitative and categorical data and mixed data provided that care has been taken of the relative feature scaling
(vi) Provides a number of interpretation aids including cluster prototypes and features and entities most contributing to cluster specificity.

K-Means issues:

(vii) Simple convex spherical shape of clusters.
(viii) Choosing the number of clusters and initial seeds.
(ix) Instability of results with respect to initial seeds.

Although conventionally considered as shortcomings, issues vii-ix can be beneficial too. To cope with issue vii, the feature set should be chosen carefully. Then the simple shape of a cluster will provide for a simpler conceptual description of it. To cope with issue viii, the initial seeds should be selected not randomly but rather based

6.4 Extension of K-Means to Different Cluster Structures

Table 6.25 Decomposition of the data scatter over product clusters in Company data; notations are similar to those in Table 6.18

Product	Income	ShaP	SupN	EC	Util	Indu	Retail	Total	Total %
A	0.03	0.05	0.04	**1.17**	0.00	0.09	0.06	1.43	24.08
B	0.14	**0.25**	0.15	0.42	0.09	0.00	0.06	1.10	18.56
C	0.06	0.12	**0.50**	0.28	0.09	0.09	**0.38**	1.53	25.66
Exp	0.23	0.41	0.69	1.88	0.18	0.18	0.50	4.06	68.30
Unexp	0.51	0.28	0.20	0.00	0.44	0.44	0.00	1.88	31.70
Total	0.74	0.69	0.89	1.88	0.63	0.63	0.50	5.95	100.00

on preliminary analysis of the substantive domain or using anomalous approaches Build or AP. Another side of issue ix is that solutions are close to pre-specified centroids, which is good when the centroids have been chosen carefully.

Q.6.24. Find SCAD decomposition for the product clusters in Company data.
A. This is in Table 6.25. Table 6.25 shows feature EC as the one most contributing to the Product A cluster, feature ShaP to the Product B cluster, and features SupN and Retail to the Product C cluster. The relatively high contribution of ShaP to B cluster is not that obvious because that of EC, 0.42, is higher. It becomes clear only on the level of relative contributions relating the absolute values to their respective Exp counterparts, 0.25/0.41 and 0.42/1.88 – the former prevails indeed. Clusters A, B, and C can be distinctively described by statements "EC==0", "ShaP < 28", and "SupN >3" (or "Sector is Retail"), respectively.

6.4 Extension of K-Means to Different Cluster Structures

So far the clustering was to encode a data set with a number of clusters forming a partition. Yet there can be differing partition-like clustering structures of which, arguably, the most popular are:

I *Fuzzy*: Cluster membership of entities may be not necessarily confined to one cluster only but shared among several clusters;
II *Probabilistic*: Clusters can be represented by probabilistic distributions rather than manifolds;
III *Self-Organizing Map (SOM)*: Capturing clusters within cells of a plane grid along with the grid's neighborhood structure.

Further on in this section extensions of K-Means to these structures are presented.

6.4.1 Fuzzy K-Means Clustering

A fuzzy cluster is represented by its membership function $z = (z_i)$, $i \in I$, in which z_i ($0 \leq z_i \leq 1$) is interpreted as the degree of membership of entity i to the cluster.

This extends the concept of conventional, hard (crisp) cluster, which can be considered a special case of the fuzzy cluster corresponding to membership z_i restricted to only 1 or 0 values.

A conventional (crisp) cluster k ($k = 1, \ldots, K$) can be thought of as a pair consisting of centroid $c_k = (c_{k1}, \ldots, c_{kv}, \ldots, c_{kV})$ in the V feature space and membership vector $z_k = (z_{1k}, \ldots, c_{ik}, \ldots, c_{Nk})$ over N entities so that $z_{ik} = 1$ means that i belongs to cluster k, and $z_{ik} = 0$ means that i does not. Moreover, clusters form a partition of the entity set so that every i belongs to one and only one cluster if and only if $\Sigma_k z_{ik} = 1$ for every $i \in I$.

These are extended to the case of fuzzy clusters, so that fuzzy cluster k ($k = 1, \ldots, K$) is a pair comprising centroid $c_k = (c_{k1}, \ldots, c_{kv}, \ldots, c_{kV})$, a point in the feature space, and membership vector $z_k = (z_{1k}, \ldots, z_{ik}, \ldots, z_{Nk})$ such that all its components are between 0 and 1, $0 \leq z_{ik} \leq 1$, expressing the extent of belongingness of i to each of the clusters k. Fuzzy clusters form what is referred to as a fuzzy partition of the entity set, if the summary membership value of every entity $i \in I$ is unity, that is, $\Sigma_k z_{ik} = 1$ for each $i \in I$. One may think of the total membership of any entity i as a substance that can be differently distributed among the centroids.

These concepts are especially easy to grasp if membership value z_{ik} is considered as the probability of belongingness. However, in many cases fuzzy partitions have nothing to do with probabilities. For instance, dividing all people by their height may involve fuzzy categories "short," "medium" and "tall" with fuzzy meanings such as those shown in Fig. 6.15.

Fuzzy clustering can be of interest in applications related with natural fuzziness of cluster boundaries such as in image analysis, robot planning, geography, etc.

If fuzzy cluster membership values are put into the PCA model, as K-Means crisp memberships have been (see formula (6.12) in Section 6.2.2.2), they make a rather weird structure in which centroids are not average but rather extreme points in their clusters, which can be relaxed in a certain way and make clusters appealing, if somewhat unusual (Nascimento 2005).

An empirically convenient criterion (6.20) below differently extends that of (6.3) by using the Euclidean squared distance d and factoring in an exponent of the membership, z^α. The value α affects the fuzziness of the optimal solution: at $\alpha = 1$,

$$F(\{c_k, z_k\}) = \sum_{k=1}^{K} \sum_{i=1}^{N} z_{ik}^\alpha d(y_i, c_k) \tag{6.20}$$

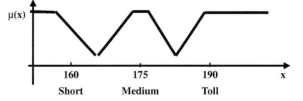

Fig. 6.15 Possible trapezoid fuzzy sets corresponding to fuzzy concept of man's height: short, regular, and toll

6.4 Extension of K-Means to Different Cluster Structures

the optimal membership values are proven to be crisp, the larger the α the "smoother" the membership values. Usually α is taken to be $\alpha = 2$.

Globally minimizing criterion (6.20) is a difficult task. Yet the alternating minimization of it appears rather easy. As usual, this works in iterations starting, from somehow initialized centroids. Each iteration proceeds in two steps: (1) given cluster centroids, cluster membership values are updated; (2) given membership values, centroids are updated – after which everything is ready for the next iteration. The process stops when the updated centroids are close enough to the previous ones. Updating formulas are derived from the first-order optimality conditions. They require the partial derivatives of the criterion over the optimized variables to be set to 0.

Membership values update formula:

$$z_{ik} = 1/\sum_{k'=1}^{K} [d(y_i, c_k)/d(y_i, c_{k'})]^{\frac{1}{\alpha-1}} \qquad (6.21)$$

Centroids update formula:

$$c_{kv} = \sum_{i=1}^{N} z_{ik}^{\alpha} y_i / \sum_{i'=1}^{N} z_{i'k}^{\alpha} \qquad (6.22)$$

Since Equations (6.21) and (6.22) are the first-order optimality conditions for criterion (6.20) leading to unique solutions, convergence of the method, usually referred to as fuzzy K-Means (c-means, too, assuming c is the number of clusters, see Bezdek et al. 1999), is guaranteed.

Yet the meaning of criterion (6.20) has not been paid much attention to until recently. It appears, criterion F in (6.20) can be presented as $F = \Sigma_i F(i)$, the sum of weighted distances $F(i)$ between points $i \in I$ and cluster centroids, so that $F(i)$ is equal to the harmonic average of the individual memberships at $\alpha = 2$ (see Stanforth, Mirkin and Kolossov 2007, where this fact is used for the analysis of domain of applicability for predicting toxicity of chemical compounds). Figure 6.16 presents the indifference contours of the averaged F values versus those of the nearest centroids. The former look much smoother.

The Anomalous pattern method is applicable as a tool for initializing Fuzzy K-Means as well as crisp K-Means, leading to reasonable results as reported by Stanforth, Mirkin and Kolossov 2007. Nascimento and Franco (2009) applied this method for segmentation of sea surface temperature maps; found fuzzy clusters closely follow the expert-identified regions of the so-called coastal upwelling, that are relatively cold, and nutrient rich, water masses. In contrast, the conventional fuzzy K-Means, with user defined K, under- or over-segments the images.

Q.6.25. Regression-wise clustering. In general, centroids c_k can be defined in a space which is different from that of the entity points y_i ($i \in I$). Such is the case of regression-wise clustering. Recall that a regression function $x_V =$

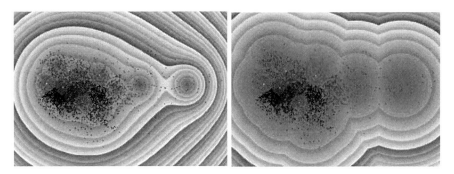

Fig. 6.16 Maps of the indifference levels for the membership function $F(i)$ at about 14,000 chemical compounds clustered with iK-Means in 41 clusters (**a**); (**b**) scores membership using only the nearest cluster's centroid

$f(x_1, x_2, \ldots, x_{V-1})$ may relate a target feature, x_V, to (some of the) other features $x_1, x_2, \ldots, x_{V-1}$ as, for example, the price of a product to its consumer value and production cost attributes. In regression-wise clustering, entities are grouped together according to the degree of their correspondence to a regression function rather than according to their closeness to the gravity center. That means that regression functions play the role of centroids in regression-wise clustering (see Fig. 6.17).

Consider a version of Straight K-Means for regression-wise clustering to involve linear regression functions relating standardized variable y_V to other standardized variables, $y_1, y_2, \ldots, y_{V-1}$, in each cluster. Such a function is defined by the equation $y_V = a_1 y_1 + a_2 y_2 + \ldots + a_{V-1} y_{V-1} + a_0$ for some coefficients $a_0, a_1, \ldots, a_{V-1}$. These coefficients form a vector, $a = (a_0, a_1, \ldots, a_{V-1})$, which can be referred to as a regression-wise centroid.

When a regression-wise centroid is given, its distance to an entity point $y_i = (y_{i1}, \ldots, y_{iV})$ is defined as $r(i, a) = (y_{iV} - a_1 y_{i1} - a_2 y_{i2} - \ldots - a_{V-1} y_{i,V-1} - a_0)^2$, the squared difference between the observed value of y_V and that calculated from the regression equation. To determine the regression-wise centroid $a(S)$, given a

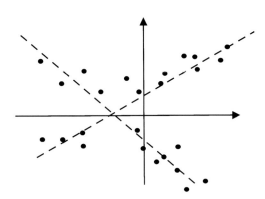

Fig. 6.17 Two regression-wise clusters with their regression *lines* as centroids

cluster list $S \subseteq I$, the standard technique of multivariate linear regression analysis is applied, which is but minimizing the within cluster summary residual $\sum_{i \in S} r(i, a)$ over all possible a.

Formulate a version of the Straight K-Means for this situation.
Hint: Same as Batch K-Means, except that:

(1) centroids must be regression-wise centroids and
(2) the entity-to-centroid distance must be $r(i,a)$.

6.4.2 Mixture of Distributions and EM Algorithm

Data of financial transactions or astronomic observations can be considered as a random sample from a (potentially) infinite population. In such cases, the data structure can be analyzed with probabilistic approaches of which arguably the most radical is the mixture of distributions approach.

According to this approach, each of the yet unknown clusters k is modeled by a density function $f(x, \alpha_k)$ which represents a family of density functions over x defined up to a parameter vector α_k. Consider a one-dimensional density function $f(x)$, that, for any x and very small change dx, assigns its probability $f(x)dx$ to the interval between x and $x+dx$, so that the probability of any interval (a,b) is integral $\int_a^b f(x)dx$, which is the area between x-axis and $f(x)$ within (a,b) as illustrated on Fig. 6.18 for interval (6,8). Multidimensional density functions have a similar nterpretation.

Usually, the cluster density $f(x, \alpha_k)$ is considered uni-modal with the mode corresponding to the cluster standard point. Such is the normal, or Gaussian, density function defined by α_k consisting of its mean vector m_k and covariance matrix Σ_k:

$$f(x, m_k, \Sigma_k) = \exp(-(x - m_k)^T \Sigma_k^{-1} (x - m_k)/2) \Big/ \sqrt{(2\pi)^V |\Sigma_k|} \qquad (6.23)$$

The shape of Gaussian clusters is ellipsoidal because any surface at which $f(x, \alpha_k)$ is constant satisfies equation $(x - m_k)^T \Sigma_k^{-1} (x - m_k) = c$, where c is any constant, that defines an ellipsoid. This is why the PCA representation is highly compatible with the assumption of the underlying distribution being Gaussian. The mean vector m_k specifies the k-th cluster's location.

The mixture of distributions clustering model can be set as follows. The row points $y_1, y_2, \ldots, y_N$ are considered a random sample of $|V|$-dimensional observations from a population with density function $f(x)$ which is a mixture of individual cluster density functions $f(x, \alpha_k)(k = 1, 2, \ldots, K)$ so that $f(x) = \sum_{k=1}^{K} p_k f(x, \alpha_k)$, where $p_k \geq 0$ are the mixture probabilities such that $\sum_{k=1}^{K} p_k = 1$.

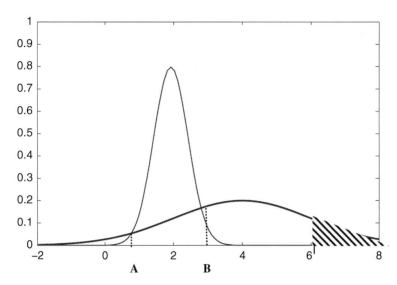

Fig. 6.18 Two Gaussian clusters represented by their density functions drawn with a thin and *bold lines*, respectively. The probability of interval (6,8) in the *bold line* cluster is shown by the area with diagonal filling. The interval (A,B) is the only place in which the thin *line* cluster is more likely than the *bold line* cluster

To estimate the individual cluster parameters, the principle of maximum likelihood, one of the main approaches in mathematical statistics, applies. The approach is based on the postulate that the events that have really occurred are those that are most likely. In general, this is not correct – everybody can recall a situation in which a less likely event has occurred. But the principle, applied for parameter estimation, is as much effective as a similarly wrong principle of the maximum parsimony, and even more. In its simplest version, the approach requires to find the mixture probabilities p_k and cluster parameters a_k, $k = 1, 2, \ldots, K$ by maximizing the likelihood of the observed data under the assumption that the observations come independently from a mixture of distributions. It is not difficult to show, under the assumption that the observations come independenly of each other, that the likelihood is the product of the density values, $P = \prod_{i=1}^{N} \sum_{k=1}^{K} p_k f(y_i, a_k)$. To computationally handle the maximization problem for P with respect to the unknown parameter values, its logarithm, $L = log(P)$, is maximized in the form of the following expression:

$$L = \sum_{i=1}^{N} \sum_{k=1}^{K} g_{ik}[\log(p_k) + \log(f(y_i, \alpha_k)) - \log(g_{ik})], \qquad (6.24)$$

where g_{ik} is the posterior density of cluster k defined as $g_{ik} = p_k f(y_i, \alpha_k) / \Sigma_k p_k f(y_i, \alpha_k)$.

6.4 Extension of K-Means to Different Cluster Structures

Criterion L can be considered a function of two groups of variables:

(1) the mixture probabilities p_k and cluster parameters α_k, and
(2) posterior densities g_{ik},

to apply the method of alternating optimization. The alternating maximization algorithm for this criterion is referred to as EM-algorithm since computations are performed as a sequence of Expectation (E) and Maximization (M) steps. As usual, to start the process, the variables must be initialized. Then E-step is executed: Given p_k and α_k, optimal g_{ik} are found. Given g_{ik}, M-step finds the optimal p_k and α_k. This brings the process to an E-step again to follow by an M-step. And so forth. The computation stops when the current parameter values approximately coincide with the previous ones. This algorithm has been developed, in various versions, for Gaussian density functions as well as for some other parametric families of probability distributions. It should be noted that developing a fitting algorithm is not that simple, and not only because there are too many parameters here to estimate. One should take into consideration that there is a tradeoff between the complexity of the probabilistic model and the number of clusters: a more complex model may fit to a smaller number of clusters. To select a better model one can utilize the likelihood criterion penalized for the complexity of the model. A popular penalized log-likelihood criterion is referred to as Bayesian Information Criterion (BIC) and is defined, in this case, as

$$\text{BIC} = 2\ log\ p(X/p_k, \alpha_k) - \lambda log(N), \qquad (6.25)$$

where X is the observed data matrix, λ the number of parameters to be fitted, and N the number of observations, that is, rows in X. The greater the value, the better. BIC analysis has been shown to be useful, for example, in assessing the number of clusters K for the mixture of Gaussians model.

The goal of EM algorithm is determining the density functions rather than assigning entities to clusters. If the user needs to see the "actual clusters", the posterior probabilities g_{ik} can be utilized: i is assigned to that k for which g_{ik} is the maximum. Since this "optimal assignment" rule deviates from the distribution of g_{ik}, the proportions of entities in clusters obtained in this way will deviate from the mixture probabilities p_k. This is why it is advisable to consider the relative values of g_{ik} as fuzzy membership values.

The situation, in which all Gaussian clusters have their covariance matrices constant diagonal and equal to each other, so that $\Sigma_k = \sigma^2 E$, where E is identity matrix and σ^2 the variance, is of a theoretical interest. In this case, all clusters have uniformly spherical distributions of the same radius. The maximum likelihood criterion P in this case is equivalent to the criterion of K-Means and, moreover, there is a certain homology between the EM and Batch K-Means algorithms in this case.

To see what is going on here, consider feature vectors corresponding to entities x_i, $i \in I$, as randomly and independently sampled from the population, with an unknown assignment of the entities to clusters S_k. The likelihood of this sample is determined by the following equation:

$$P = C \prod_{k=1}^{K} \prod_{i \in S_k} \sigma^{-V} \exp\{-(x_i - m_k)^T \sigma^{-2}(x_i - m_k)/2\},$$

because in this case the determinant in (6.23) is equal to $|\Sigma_k| = \sigma^{2V}$. and the inverse covariance matrix is σ^{-2}E. The logarithm of the likelihood is proportional to

$$L = -2V\log(\sigma) - \sum_{k=1}^{K} \sum_{i \in S_k} (x_i - m_k)^T (x_i - m_k)/\sigma^2$$

It is not difficult to see from the first-order optimality conditions for L that, given partition $S = \{S_1, S_2, \ldots, S_K\}$, the optimal values of m_k and σ are determined according to the usual formulas for the mean and the standard deviation. Moreover, given m_k and σ, the partition $S = \{S_1, S_2, \ldots, S_K\}$ maximizing L will simultaneously minimize the double sum in the right part of its expression above, which is exactly the summary squared Euclidean distance from all entities to their centroids, that is, criterion $W(S,m)$ for K-Means in (6.3) except for notation: the cluster gravity centers are denoted here by m_k rather than by c_k, which is not a big deal after all.

Thus the mixture model leads to the conventional K-Means method as a method for fitting the model, under the condition that all clusters have spherical Gaussian distribution of the same variance. This leads some authors to conclude that K-Means is applicable only under the assumption of such a model. However, this conclusion is wrong because it involves a logic trap: it is well known that the fact that A implies B does not necessarily mean that B implies A – there are plenty of examples to the opposite. Note however that the K-Means data recovery model, also leading to K-Means, assumes no restricting hypotheses on the mechanism of data generation. It also implies, through the data scatter decomposition, that useful data standardization options should involve dividing by range or similar range-related indexes rather than by the standard deviation, associated with the spherical Gaussian model. In general, the situation here is similar to that of the linear regression, which is a good method to apply when there is a Gaussian distribution of all variables involved, but it can and should be applied under any other distribution of observations if they tend to lie around a straight line.

6.4.3 Kohonen's Self-Organizing Maps SOM

Kohonen's Self-Organizing Map is an approach to visualize the data cluster structure by explicitly mapping it onto a plane grid (Kohonen 1995). Typically, the grid is rectangular and its size is determined by the user-specified numbers of its rows and columns, r and c, respectively, so that there are $r \times c$ nodes on the grid. Each of the grid nodes, $g_k(k = 1, 2, \ldots, rc)$, is one-to-one associated with the so-called

6.4 Extension of K-Means to Different Cluster Structures

model, or reference, vector m_k which is of the same dimension as the entity points y_i, $i \in I$.

The grid has a neighborhood structure which is to be set by the user. In a typical case, the neighborhood G_k of node g_k is defined as the set of all the grid nodes whose path distance from g_k is less than a pre-selected threshold value (see Fig. 6.19).

Then each m_k is associated with some data points – a process that can be reiterated. In the end, data points associated at each m_k are visualized at the grid point ($k = 1, \ldots, rc$) (see Fig. 6.20). Historically, all SOM algorithms have been set in an incremental manner as neuron networks do, but later, after some theoretical investigation, straight/ batch versions appeared, such as the following.

Initially, vectors m_k are initialized in the data space either randomly or according to an assumption of the data structure such as, for instance, centroids of K-Means clusters found at $K = rc$. Given vectors m_k, entity points y_i are partitioned into "neighborhood" sets I_k. For each $k = 1, 2, \ldots, rc$, the neighborhood set I_k is defined as consisting of those y_i that are assigned to m_k according to the Minimum distance rule. Given sets I_k, model vectors m_k are updated as centers of gravity of all entities y_i assigned to grid nodes in the neighborhood of g_k, that is, such y_i that $i \in I_t$ for some $g_t \in G_k$. Then a new iteration of building I_k with the follow-up updating

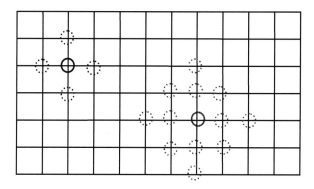

Fig. 6.19 A 7×12 SOM grid on which nodes $g1$ and $g2$ are shown along with their neighborhoods defined by thresholds 1 and 2, respectively

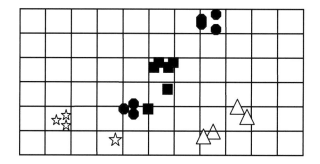

Fig. 6.20 A pattern of final SOM structure using entity labels of geometrical shapes

m_k's, is run. The computation stops when new m_k are close enough to the previous ones or after a pre-specified number of iterations.

As one can see, SOM in this version is much similar to Straight/Batch K-Means except for the following:

(a) number $K = rc$ of model vectors is large and has nothing to do with the number of final clusters – this comes visually as the number of grid clusters;
(b) data points are averaged over the grid neighbourhood, not the feature space neighborhood;
(c) there are no interpretation rules except according to positioning of points on the grid.

Item (a) results in the fact that many of final I_k's are empty, so that relatively very few of grid nodes are populated, which may create a powerful image of a cluster structure that may go to a deeper – or more interesting – minimum than K-Means, because of (b).

6.5 Summary

This Chapter is devoted to K-Means, arguably the most popular clustering method. The method partitions the entity set into clusters along with centroids representing them. It is very intuitive and usually does not require that much space to get presented, except of course its various versions such as incremental or nature inspired or medoid based algorithms. This text also includes less popular subjects that are important when using K-Means for real-world data analysis:

- Presentation and analysis of examples of its failures
- Innate tools for interpretation of clusters
- Reformulations of the criterion that could yield different algorithms for K-Means
- Initialization – the choice of K and location of centroids

Three modifications of K-Means onto different cluster structures are presented as well. These are: Fuzzy K-Means for finding fuzzy clusters, Expectation-Maximization (EM) for finding probabilistic clusters as items of a mixture of distributions, and Kohonen self-organizing maps (SOM) that tie up the sought clusters to a visually comfortable two-dimensional grid.

References

Bandyopadhyay, S., Maulik, U.: An evolutionary technique based on K-means algorithm for optimal clustering in R^N. Inf. Sci. **146**, 221–237 (2002).
Bezdek, J., Keller, J., Krisnapuram, R., Pal, M.: Fuzzy Models and Algorithms for Pattern Recognition and Image Processing. Kluwer Academic Publishers, Dordrecht (1999).

References

Cangelosi, R., Goriely, A.: Component retention in principal component analysis with application to cDNA microarray data. Biol. Direct. **2**, 2 (2007). http://www.biolgy-direct.com/con-tent/2/1/2.

Green, S.B., Salkind, N.J.: Using SPSS for the Windows and Mackintosh: Analyzing and Understanding Data. Prentice Hall, Upper Saddle River, NJ (2003).

Hartigan, J.A.: Clustering Algorithms. Wiley, New York (1975).

Kaufman. L., Rousseeuw, P.: Finding Groups in Data: An Introduction to Cluster Analysis. Wiley, New York (1990).

Kendall, M.G., Stewart, A.: Advanced Statistics: Inference and Relationship (3d edition). Griffin, London (1973). ISBN: 0852642156.

Kettenring, J.: The practice of cluster analysis. J. Classific. **23**, 3–30 (2006).

Kohonen, T.: Self-Organizing Maps. Springer, Berlin (1995).

Kryshtanowski, A.: Analysis of Sociology Data with SPSS. Higher School of Economics Publishers, Moscow (in Russian) (2008).

Lu, Y., Lu, S., Fotouhi, F., Deng, Y., Brown, S.: Incremental genetic algorithm and its application in gene expression data analysis. BMC Bioinform. **5**,172 (2004).

Ming-Tso Chiang, M., Mirkin, B.: Intelligent choice of the number of clusters in K-Means clustering: an experimental study with different cluster spreads. J. Classif. **27**(1), 3–40 (2010).

Mirkin, B.: Clustering for Data Mining: A Data Recovery Approach. Chapman & Hall/CRC, Roca Baton, FL (2005). ISBN 1-58488-534-3.

Mirkin, B.: Mathematical Classification and Clustering. Kluwer Academic Press, Boston-Dordrecht (1996).

Murthy, C.A., Chowdhury, N.: In search of optimal clusters using genetic algorithms. Pattern Recognit. Lett. **17**, 825–832 (1996).

Nascimento, S., Franco, P.: Unsupervised fuzzy clustering for the segmentation and annotation of upwelling regions in sea surface temperature images. In: Gama, J. (ed.) Discovery Science, LNCS 5808, pp. 212–226. Springer (2009).

Nascimento, S.: Fuzzy Clustering via Proportional Membership Model. ISO Press, Amsterdam (2005).

Paterlini, S., Krink, T.: Differential evolution and PSO in partitional clustering. Comput. Stat. Data Anal. **50**, 1220–1247 (2006).

Stanforth, R., Mirkin, B., Kolossov, E.: A measure of domain of applicability for QSAR modelling based on intelligent K-Means clustering. QSAR Comb. Sci. **26**(7), 837–844 (2007).

Chapter 7
Hierarchical Clustering

7.1 General

Term hierarchy can mean different things in different contexts. Here it is a decision tree nested structure drawn like that on Fig. 7.1 below (see also figures in Section 4.5). Such a hierarchy may relate to mental or real processes such as

(a) conceptual structures (taxonomy, ontology);
(b) genealogy; or
(c) evolutionary tree.

The top node, referred to as the root, represents all the entity set I under consideration. Every interior node of the hierarchy has a number of children nodes representing division of the subset – or cluster – represented by the node into smaller clusters. The terminal nodes that have no children are referred to as leaves and usually correspond to singletons. A hierarchical structure should be annotated to reflect the correspondence between the nodes and entity sets. Such an annotation, according to bases of division was utilized in classification trees of Section 4.5. In clustering, another annotation is frequently used – that imposed by the leaf contents.

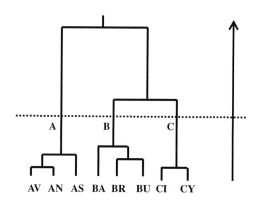

Fig. 7.1 A cluster hierarchy of Company data entities: nested node clusters, each comprising a set of leaves. Cutting the tree at a certain height leads to a partition of the three product clusters here

B. Mirkin, *Core Concepts in Data Analysis: Summarization, Correlation and Visualization*, Undergraduate Topics in Computer Science,
DOI 10.1007/978-0-85729-287-2_7, © Springer-Verlag London Limited 2011

Every node of the tree corresponds to cluster of those entities that annotate the leaves descending from the node.

On the right of the hierarchy on Fig. 7.1, there is a y-axis to represent the node heights. The node height is a useful device for positioning nodes in layers. Typically, all leaves have zero heights whereas the root is assigned with the maximum height, usually taken as unity or 100%. Some hierarchies are naturally assigned with node heights, e.g., the molecular clock in evolutionary trees, some not, e.g. the decimal classification of library subjects. But to draw a hierarchy as a figure, one needs to define positions for each node, thus its height as well, even if implicitly.

Nodes may be linked by using what is called edges. Only one edge ascends from each node – this is a defining property of nested hierarchies that each node, except for the root, has one and only one parent. Each hierarchy node, or its parental edge, represents cluster of all leaves descending from the node; such are the edges labeled by product names A, B, and C on Fig. 7.1 – they represent the corresponding clusters. These clusters have a very special pattern of overlapping: for any two clusters of a hierarchy, their intersection is either empty or coincides with one of them – this is one more characteristic property of a nested hierarchy.

The tree on Fig. 7.1 has one more specific property – it is binary: each interior node in the tree has exactly two children, that is, split in two parts. Most clustering algorithms, including those presented below, do produce binary trees, along with node heights.

Q.7.1. Given a binary hierarchy H with leaf set I, prove that the number of edges in the hierarchy is $2(|I|-1)$.

Q.7.2. Consider a binary hierarchy H with node set J and height function $h(j)$, $j \in J$, such that $h(j) = 0$ at each leaf j. Assume that $h(j)$ is monotone, that is, the closer the node to the root the greater the value of $h(j)$. Define the distance $u(i1,i2)$ between each pair of leaves $i1,i2 \in I$ as the height of the least cluster node $j(i1,i2)$ such that both $i1$ and $i2$ are among its descendants, $u(i1,i2) = h(j(i1,i2))$. Prove that the distance u is an ultrametric, that is, it is not only symmetric, $u(i1,i2) = u(i2,i1)$, and reflexive, $u(i1,i1) = 0$, but also satisfies ultrametric inequality

$$u(i1,i2) \leq max[u(i1, i3), u(i2, i3)] \tag{7.1}$$

for every triplet of leaves $i1, i2$, and $i3$.

Q.7.3. Prove that if distance u is ultrametric then, for each three entities, the three distances between them satisfy the following property: those two larger ones are equal to each other. This can be rephrased as follows: under an ultrametric, every triangle is isosceles.

Q.7.4. Define Baire distance $b(x,y)$ between non-coinciding real numbers x and y, both located in interval [0,1], as follows. Consider their decimal digits, $x = 0.x_1x_2\ldots$ and $y = 0.y_1y_2\ldots$, and set $b(x,y) = 2^{-n}$ where n is the very first digit at which $x_n \neq y_n$. If, for example $x = 0.125$, $y = 0.128$ and $z = 0.250$, then

7.2 Agglomerative Clustering and Ward's Criterion

$b(x, y) = 2^{-3}$ and $b(x, z) = 2^{-1}$ (Murtagh et al. 2008). Prove that Baire distance is ultrametric and, moreover, every finite ultrametric can be represented as Baire metric.

Methods for hierarchic clustering are divided in two classes:

- *Divisive* methods: they build a cluster hierarchy by proceeding top-to-bottom, starting from the entire data set and recursively splitting clusters into parts; and
- *Agglomerative* methods: they build a cluster hierarchy by proceeding bottom-up, starting from the least clusters available, usually singletons, and merging those nearest to each other at each step.

7.2 Agglomerative Clustering and Ward's Criterion

P7.2.1 Agglomerative Clustering: Presentation

At each step of an agglomerative clustering algorithm a set of already formed clusters is considered along with the matrix of distances between maximal clusters $S_1, S_2, \ldots, S_K$. These maximal clusters form a partition of the entity set I. At the step, two nearest maximal clusters are merged and the newly formed cluster is supplied with its height and distances to other clusters. The process ends, typically, when all clusters have been merged into the universal root cluster consisting of the entire entity set.

Worked example 7.1. Agglomerative clustering of Company dataset

Consider the Company dataset. The starting point of the algorithm is the set of singletons – eight clusters consisting of one entity each. Squared Euclidean distances between the corresponding rows of the standardized data matrix are presented in Table 6.9 in Section 6.2.6, which is reproduced here as Table 7.1.

Table 7.1 Distances between standardized Company entities from Table 6.9. For the sake of convenience, row-wise non-diagonal minima are highlighted in bold

Entities	Ave	Ant	Ast	Bay	Bre	Bum	Civ	Cyb
Ave	0.00	**0.51**	0.88	1.15	2.20	2.25	2.30	3.01
Ant	**0.51**	0.00	0.77	1.55	1.82	2.99	1.90	2.41
Ast	0.88	**0.77**	0.00	1.94	1.16	1.84	1.81	2.38
Bay	1.15	1.55	1.94	0.00	0.97	**0.87**	1.22	2.46
Bre	2.20	1.82	1.16	0.97	0.00	**0.75**	0.83	1.87
Bum	2.25	2.99	1.84	0.87	**0.75**	0.00	1.68	3.43
Civ	2.30	1.90	1.81	1.22	0.83	1.68	0.00	**0.61**
Cyb	3.01	2.41	2.38	2.46	1.87	3.43	**0.61**	0.00

Table 7.2 Distances between the merged cluster and the others according to different rules

Initial clusters		Av	An	As	Ba	Br	Bu	Ci	Cy
{Av}		0.00	0.51	0.88	1.15	2.20	2.25	2.30	3.01
{An}		0.51	0.00	0.77	1.55	1.82	2.99	1.90	2.41
Merged	Method								
	NN	*	*	0.77	1.15	1.82	2.25	1.90	2.41
{Av, An}	FN	*	*	0.88	1.55	2.20	2.90	2.30	3.01
	AN	*	*	0.82	1.35	2.01	2.68	2.10	2.71

The minimum distance is $d(\text{Ave, Ant}) = 0.51$, which leads us to merging these singletons into a doubleton {Ave, Ant}. Now we have 7 clusters of which only one, that merged, is new. To do further agglomeration steps, we need to define distances between the merged cluster and the others. This can be done in many ways including those based only on the distances in Table 6.9, such as the Nearest Neighbor (NN), also termed Single Linkage, or Farthest Neighbor (FN), also termed Complete Linkage, or the Average neighbor (AN), also termed Average Linkage. These utilize the minimum distance or maximum distance or the average distance, respectively (see Table 7.2).

In each of these, the height of the merged cluster can be accepted to be equal to the distance between the clusters being merged, $h = 0.51$. Since other distances cannot be less than that, the rule guarantees the monotonicity of the height over further mergers.

Q.7.5. Complete the process of building cluster hierarchies according to the Nearest Neighbor rule, Farthest Neighbor rule, and Average Neighbor rule (Table 7.2).

7.2.1.1 Ward's Criterion

Consider a partition $S = \{S_1, S_2, \ldots, S_K\}$ arrived at on an agglomeration step. According to Ward's rule the distance between two clusters, S_k, S_l, is defined as the increase in the value of K-Means criterion $W(S,c)$ at the partition obtained from S by merging them into $S_k \cup S_l$. As shown in Eqs. (7.2) and (7.3) further on, the increase can be computed as the so-called Ward distance between centroids of the two clusters: the usual squared Euclidean distance scaled by a factor whose numerator is the product of cardinalities of the clusters and denominator is the sum of them. Note that Ward distance between singletons is just half the squared Euclidean distance between the corresponding entities.

Ward's agglomeration starts with singletons whose variance is zero and proceeds by merging those clusters that effect as small increase in the square-error criterion as possible, at each agglomeration step. This justifies the use of Ward agglomeration results to get a reasonable initial setting for K-Means when K is preset. The two methods, K-Means and Ward, supplement each other in that clusters are carefully built with Ward agglomeration, whereas K-Means allows overcoming the inflexibility of the agglomeration process over individual entities by reshuffling

7.2 Agglomerative Clustering and Ward's Criterion

them. There is an issue with this strategy though: Ward agglomeration, unlike K-Means, is a computationally intensive method, not applicable to large sets of entities.

Worked example 7.2. Ward algorithm with distances only

Let us apply Ward agglomerative algorithm to Company data. In spite of the fact that Ward distance is defined as the weighted distance between centroids in (7.3), it can also be computed, within each recursive agglomeration step by using only the distance matrix – which is provided by formula (7.4). Moreover, the cluster heights defined as within-cluster square errors, that, is deviations from the centroid, can be computed by using distances only using formula (7.5). That means we can run the entire agglomeration process by using only the distance matrix in Table 7.1. The only thing to be taken into account that it is a matrix of squared Euclidean distances which is to be halved to become a matrix of Ward distances.

The minimum value in the matrix is 0.51/2 so that the first merger is to be {Av, An}. To compute the distance between that and, say, entity Ba, according to formula (7.4), Ward distances between the merger's parts and Ba are weighted by the summary cardinalities of the corresponding clusters, which are both 2 in this case, and summed up: $2 * (1.15/2) + 2 * (1.55/2) = 2.70$, after which the Ward distance between the merger's parts, 0.51/2, multiplied by the singleton Ba's cardinality, is subtracted: 2.70–0.26 = 2.44. The result is related then to the summary cardinality of the merged cluster and singleton Ba, that is, 3, to obtain 2.44/3 = 0.81. The Ward distances from the merged clusters to the rest, computed in this way, are presented in Table 7.3. Of course the distance matrix changes from an agglomeration step to another by dynamically recomputing the distances as described above.

The height of the merged cluster is taken to be its squared error, which coincides with the original distance 0.51 between Av and An divided by 2.

After the first merger, the minimum distance is between Ci and Cy, 0.31, followed by the distance between Br and Bu, 0.38. The distance matrix, after these mergers, will be as presented in Table 7.4. The minimum values highlighted in bold indicate mergers to do at further agglomeration steps.

The hierarchy on Fig. 7.2 reflects these agglomeration steps and values of the within-cluster error height function. The height of the root, under this definition, is equal to the data scatter so that all the heights can be expressed as proportions of the data scatter. The total data scatter is the sum of all distances in Table 7.1 divided

Table 7.3 Ward distances between the merged cluster and the others according to Ward's rule

Initial clusters		Av	An	As	Ba	Br	Bu	Ci	Cy
{Av}		0.00	0.26	0.44	0.58	1.10	1.12	1.15	1.56
{An}		0.26	0.00	0.38	0.78	0.91	1.54	0.95	1.20
Merged	Method								
{Av, An}	Ward's	*	*	**0.46**	0.82	1.26	1.66	1.31	1.72

Table 7.4 Ward distances between clusters after three mergers

Clusters	Av+An	As	Ba	Br+Bu	Ci+Cy
Ave+Ant	0.00	**0.46**	0.82	2.00	2.12
Ast	**0.46**	0.00	0.97	0.87	1.27
Bay	0.82	0.97	0.00	**0.49**	1.10
Bre+Bum	2.00	0.86	**0.49**	0.00	1.62
Civ+Cyb	2.12	1.27	1.10	1.62	0.00

by their number, 8, that is 11.89. Note, this time the original distances are taken rather than Ward distances – according to formula (7.5). The product based clusters have much smaller within cluster errors, 1.44 for A, 1.73 for B, and 0.61 for C, thus constituting 12.1, 14.5, and 5.1% of the data scatter, respectively. This is reflected in the cluster heights on Fig. 7.2. The merged A+B cluster's error is 7.22 making its height 60.7%. Such a drastic rise is due to the super-additive property (7.2') of the cluster error: it not only sums the heights of the merged clusters but also adds Ward distance between them.

The hierarchy may drastically change if a different feature scaling system is applied. For example, with the standard deviation based standardization (z-scoring), the two product C companies do not constitute a single cluster but are separately merged within the product A and B clusters.

Q.7.6. Complete the agglomeration steps according to Ward distance matrix in Table 7.4.

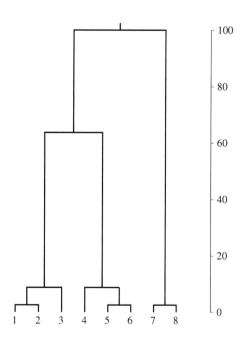

Fig. 7.2 Hierarchy produced by applying Ward agglomerative clustering algorithm to Company data. The node heights are cluster squared errors that are scaled as percentages of the pre-processed data scatter

F7.2.2 Square-Error Criterion and Ward Distance: Formulation

Consider a partition $S = \{S_1, S_2, \ldots, S_k\}$ on set I, together with centroids $c=\{c_1, c_2, \ldots, c_K\}$, and the square error criterion $W(S,c) = \sum_{k=1}^{K} \sum_{i \in S_k} d(i, c_k)$ of K-Means. Let two of the clusters, S_f, S_g, be merged so that the resulting partition is $S(f,g)$ coinciding with S except for the merged cluster $S_f \cup S_g$; the new centroid obviously being $c_{f \cup g} = (N_f c_f + N_g c_g)/(N_f + N_g)$, where N_f and N_g are cardinalities of clusters S_f and S_g, respectively. As proven previously – and rather evident indeed (see Fig. 7.3) – the value of square error criterion on partition $S(f,g)$ is greater then $W(S,c)$. But how much greater? The answer is

$$W(S_{f \cup g}, c_{f \cup g}) - W(S, c) = \frac{N_f N_g}{N_f + N_g} \sum_{v \in V} (c_{fv} - c_{gv})^2 = \frac{N_f N_g}{N_f + N_g} d(c_f, c_g), \quad (7.2)$$

the squared Euclidean distance between centroids of the merged clusters S_f and S_g weighted by a factor proportional to the product of cardinalities of the merged clusters (Ward 1963).

To prove this, let us follow the definition and do some elementary transformations. First, we notice that the distances within unchanged clusters do not change in the partition $S(f,g)$ so that the difference between the values of criterion W is $\sum_{i \in S_f \cup S_g} \sum_{v \in V} (y_{iv} - c_{f \cup g, v})^2 - \sum_{i \in S_f} \sum_{v \in V} (y_{iv} - c_{fv})^2 - \sum_{i \in S_g} \sum_{v \in V} (y_{iv} - c_{gv})^2 =$
$\sum_{i \in S_f} \sum_{v \in V} (y_{iv} - c_{f \cup g, v})^2 - \sum_{i \in S_f} \sum_{v \in V} (y_{iv} - c_{fv})^2 + \sum_{i \in S_g} \sum_{v \in V} (y_{iv} - c_{f \cup g, v})^2 - \sum_{i \in S_g} \sum_{v \in V} (y_{iv} - c_{gv})^2.$

Since $c_{f \cup g, v} = c_{fv} + N_g(c_{gv} - c_{fv})/(N_f + N_g) = c_{gv} + N_f(c_{fv} - c_{gv})/(N_f + N_g)$ and the binomial rule $(a+b)^2 = a^2 + b^2 + 2ab$, the sum $\sum_{i \in S_f} \sum_{v \in V} (y_{iv} - c_{f \cup g, v})^2$ can be presented as

$$\sum_{i \in S_f} \sum_{v \in V} (y_{iv} - c_{fv})^2 + \sum_{i \in S_f} \sum_{v \in V} \left(\frac{N_g}{N_f + N_g}\right)^2 (c_{fv} - c_{gv})^2 + 2 \sum_{i \in S_f} \sum_{v \in V} \frac{N_g}{N_f + N_g} (y_{iv} - c_{fv})(c_{fv} - c_{gv})$$

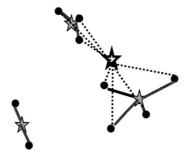

Fig. 7.3 The distances in criterion $W(S,c)$ before (*solid lines*) and after the merger (*dashed lines*) of two clusters on the *upper right*. The numbers of *dashed* and *solid* lines are the same, but the *dashed-line* distances are longer overall

where the right hand item is equal to zero, because $\sum_{i \in S_f}(y_{iv} - c_{fv}) = 0$. A similar decomposition holds for the sum $\sum_{i \in S_g} \sum_{v \in V}(y_{iv} - c_{f \cup g,v})^2$. These two combined make the difference $W(S(f,g), c(f,g))$-$W(S,c)$ equal to

$$\sum_{v \in V} N_f \left(\frac{N_g}{N_f+N_g}\right)^2 (c_{fv} - c_{gv})^2 + \sum_{v \in V} N_g \left(\frac{N_f}{N_f+N_g}\right)^2 (c_{gv} - c_{fv})^2 = \frac{N_f N_g}{N_f+N_g} \sum_{v \in V}(c_{fv} - c_{gv})^2$$

which completes the proof of Equation (6.2).
The weighted distance

$$dw(S_f, S_g) = \frac{N_f N_g}{N_f + N_g} d(c_f, c_g) \qquad (7.3)$$

is referred to as Ward distance between clusters. Its weight coefficient highly depends on the distribution of entities between clusters being merged. This may affect the results of agglomerative or divisive algorithms that utilize Ward distance. Indeed, in an agglomerative process, the Ward distance between clusters to be merged must be as small as possible – which favors merging big and small clusters. On the other hand, in divisive clustering, when splitting, the Ward distance between split parts must be as large as possible, which favors splitting large clusters into relatively equal-sized parts. It is the effect of this weighting that underlies the odd behavior of the square error K-Means criterion noted in the discussion of Fig. 6.8 at Section 6.2.1.

Given a cluster S with its centroid c, let us denote the square error within S by $W(S) = \sum_{i \in S} d(y_i, c)$. Using this, Equation (7.2) can be rewritten as

$$W(S_f \cup S_g) = W(S_f) + W(S_g) + dw(S_f, S_g). \qquad (7.2')$$

This explains the additive properties of the square error $W(S)$ when used as the height index in drawing the clustering tree. According to this equation, the height of the parent is equal to the sum of heights of its children plus Ward distance between them. This warrants a specific heights distribution over the tree: the closer to the root, the longer the edges!

Q.7.7. Prove that Ward distance after a merger can be recursively calculated from the distances before the merger according to the following formula:

$$wd(S_{f \cup g}, S_k) = [(N_f + N_k)wd(S_f, S_k) + (N_g + N_k)wd(S_g, S_k) - N_k wd(S_f, S_k)$$
$$- N_k wd(S_f, S_g)]/(N_f + N_g + N_k). \qquad (7.4)$$

Q.7.8. Prove that the square error of cluster (S_k, c_k), $W(S_k) = \sum_{i \in S_k} d(y_i, c_k)$, can be expressed in terms of within cluster distances only:

7.2 Agglomerative Clustering and Ward's Criterion

$$W(S_k) = \sum_{i,j \in S_k} d(y_i, y_j)/N_k. \qquad (7.5)$$

where d is the squared Euclidean distance and N_k is the number of entities in S_k. Hint: Use Equation (6.9) in Section 6.2.2.

Equations (7.4) and (7.5) allow carrying Ward's agglomeration process by using only the distances, so that the cluster centroids are involved neither for calculating Ward distances nor for the cluster's square errors (see also Hartigan 1975, Jain and Dubes 1988).

C7.2.3 Agglomerative Clustering: Computation

All agglomerative clustering algorithms follow the same scheme. They transform the original matrix of dissimilarity indexes between them into a binary cluster hierarchy. The dissimilarities can be virtual, that is, computed on the fly from other data such as entity-to-feature data. Also, they can be expressed as similarities or proximities – then a similarity maximum should be taken rather than a minimum.

7.2.3.1 Agglomerative Clustering

1. *Initial setting.* Make all entities $k \in I$ to form singleton clusters $S_k = \{k\}$, with their cardinalities set to unity and heights to zero; form a matrix of dissimilarities $D = (d(k,l))$ between them and form a list of maximal clusters including all the singletons.
2. *Finding minimum.* Find the minimum $d(f,g)$ in D.
3. *Clusters update.* Two maximal clusters, S_f and S_g, that are closest to each other, are merged together to form their parent, a new maximal cluster $S_{f \cup g} = S_f \cup S_g$. The new cluster's cardinality is defined as $N_{f \cup g} = N_f + N_g$, with the height computed accordingly. (Usually, the height is taken to be equal to $d(f,g)$. In Ward clustering, $h_{f \cup g} = h_f + h_g + wd(f,g)$.) Clusters S_f, S_g are removed from the list of maximal clusters.
4. *Distance update.* Remove rows and columns f and g from D; put in a new row and column of distances between new cluster $S_{f \cup g}$ and the remaining maximal clusters.
5. *Stop condition.* If the number of maximal clusters is larger than 1, go to step 2.
6. *Output*: the set of all clusters along with their heights. Draw a cluster tree.

This algorithm remains a scheme unless a method for computing distances between the merged cluster and the other maximal clusters is defined. A very general formula covering many a method has been proposed by Lance and Williams (1967):

$$d(k, f \cup g) = \alpha_f d(k,f) + \alpha_g d(k,g) + \beta d(f,g) + \gamma |d(k,f) - d(k,g)|. \qquad (7.6)$$

Table 7.5 Lance-Williams coefficients for some popular agglomerative clustering methods

Method	α_f	α_g	β	γ
Single linkage	½	½	0	−½
Complete linkage	½	½	0	½
UPGMA	$N_f/(N_f+N_g)$	$N_g/(N_f+N_g)$	0	0
Ward	$(N_f+N_k)/(N_f+N_g+N_k)$	$(N_f+N_k)/(N_f+N_g+N_k)$	$(N_f+N_k)/(N_f+N_g+N_k)$	0

Values of coefficients for some popular methods are presented in Table 7.5. One of the methods, popular in bioinformatics, is referred to as UPGMA (Unweighted Pair Group Method with Arithmetic Means): the dissimilarity between two clusters is defined as the average distance between all entities of the two.

From the computational point of view, there are two weak points in the algorithm. One of them concerns the operation of finding the minimum on Step 2, which is computationally intensive. This, however, can be softened a bit by some prior computations such as finding the nearest neighbor for each entity or when the dissimilarity measure satisfies some conditions that allow to limit the search span or by imposing some neighborhood structure so that the search is constrained within the neighborhoods only (Murtagh 1985). The other point concerns the storage room for dissimilarity matrix D. Its size is quadratic on the number of entities so that a thousand strong sample would relate to half a million dissimilarities and a ten thousand strong set would require memory for 50 million dissimilarities. One of the approaches to tackle the problem would be keeping the data in the original format, such as entity-to-feature table, and computing dissimilarities on the fly, only when needed. This could be at odds with the need to find the minimum dissimilarities by comparing them. Another tackle, using a part of the data only with a follow-up extending the results is in a very early stage of development currently.

Q.7.9. Formulate a version of agglomerative clustering for Ward criterion using the definition of Ward distance with centroids.

7.3 Divisive and Conceptual Clustering

P7.3.1 Divisive Clustering: Presentation

A divisive method works in a top-down manner, starting from the entire data set and splitting each cluster in two, which is reflected in drawing the split cluster as a parental node with two children corresponding to split parts. The splitting process goes on in such a way that each time a leaf cluster is split, two children nodes are sprung from the leaf which thus becomes an internal node.

7.3 Divisive and Conceptual Clustering

To specify a method of divisive clustering, one should define the following:

(i) splitting criterion – how one decides which split is better;
(ii) splitting method – how the splitting is actually done;
(iii) choice of cluster – which of the current leaf clusters is to be split;
(iv) stopping criterion – at what point one decides to stop the splitting.

Let us cover some options that can be recommended based on some theoretical and/or experimental evidence, in this sequence.

(i) Splitting criterion

The only splitting criterion that is considered here is K-Means criterion of the summary square error which is implemented as Ward-like criterion, that is the maximum possible reduction in the total squared error caused by the split: the greater the better.

When applied to categorical features represented by their categories enveloped into the corresponding binary features, this criterion can be reinterpreted in terms of what is referred to a goodness-of-split criterion, which usually measures the improvements in the predictability of the categories, from the split partition. The "predictability" can be measured differently, most frequently by involving the general concept of "uncertainty" in the distribution of possible feature values that is captured by the concepts like Gini index, entropy, variance, etc. (see detail in Sections 2.2, 2.3, 4.5, 7.3.2).

Three popular goodness-of-split criteria that are compatible with the least-squares data recovery framework are: (a) impurity function (Breiman et al. 1984), (b) category utility function (Fisher 1987), and (c) the summary Pearson chi-squared coefficient. The category utility function, in fact, is the sum of impurity functions over all categories in the data, related to the number of clusters in the partition being built. All the three can be expressed in terms of the cluster-category contributions to the data scatter and, thus, amount to be special cases of the square-error clustering criterion at the conventional zero-one coding of categories along with different normalizations of the data, as explained in Sections 4.5.3 and 7.3.2.

(ii) Splitting method

For Ward's criterion, we consider two splitting approaches:

A. K-means at $K = 2$, or Two-splitting – this is a popular option, frequently referred to as Bisecting K-Means; this leads, typically, to good results if care is taken to find good initial centroids. Since the criterion of bisecting K-Means is equivalent to the criterion of maximizing Ward distance between split parts, the divisive algorithm utilizing Two-splitting is referred to as Ward-like divisive algorithm here.
B. Conceptual clustering – in this, just one of the features is involved in each of the splits, which leads to a straightforward conceptual interpretation of all the clusters. Conceptual clustering builds a cluster hierarchy by

sequentially splitting clusters, as all divisive algorithms do, yet here each of the splits uses a single attribute in contrast to the classic clustering that utilizes distances involving all of them. The criteria such as summary impurity function or summary Pearson chi-squared are part of the Ward-like divisive clustering algorithm under a corresponding normalization option (see Section 4.5). In this aspect, conceptual clustering should be equated to building classification trees over a multiple target feature set – the only difference is that the very same target features are simultaneously input features!

(iii) **Choice of cluster to split**

The order of splitting conventionally is not considered important: if the set is divided all the way down to singletons, then the order does not much matter indeed. If, however, the goal is to produce a partition by finishing after just a few splits, then Ward's criterion gives the following guiding principle: after each split, all leaf clusters S_k are supplied with their square errors $W(S_k)$. The square-error is the contribution to the unexplained data scatter, that is, the sum of Euclidean squared distances between cluster's entities and its centroid, which is proportional to the cluster summary variance weighted by its size. To minimize the unexplained part, that cluster whose square-error is maximum is to be split first.

(iv) **Stopping criterion**

Conventionally, the divisions stop when there remains nothing that can be split, that is, when all the leaves are singletons. Yet for the Ward's criterion one can specify a threshold on the value of the square-error at a cluster, the level of "noise" reached by $W(S_k)$ at which a cluster is considered next to noise and not split anymore because of that. This threshold can be set as a proportion of the data scatter, say 5%. Another criterion of course can be just the cluster size – say, clusters whose cardinality is less than 1% of the original data size are not to be split anymore. Tasoulis et al. (2010) propose a cluster to stop splitting when the density of the cluster points projections to the first principal component has no minima inside the range.

Case Study 7.1. Divisive Clustering of Companies with Two-Splitting

Consider the Ward-like divisive clustering method for the Company data, range standardized with the follow-up rescaling the dummy variables corresponding to the three Sector categories in Table 6.1. Using Two-splitting algorithm may produce a rather poorly resolved picture if the most distant entities, 6 and 8 according to the distance matrix in Table 7.1, are taken as the initial seeds. Then step 2 would produce tentative clusters {1,3,4,5,6} and {2,7,8} because 2 is nearer to 8 than to 6 as easily seen in Table 7.1. This partition is at odds with the product clusters.

7.3 Divisive and Conceptual Clustering

Unfortunately, no further iterations can change that. This shows that the choice of the initial seeds at the two farthest entities can be not that good an option that it may seem to be. Usage of Build or Anomalous pattern algorithms, explained in Section 6.2.7, could lead to better results. Because of our reluctance in using the original Company dataset in this Chapter, let us use Build algorithm which relies on distances only.

According to distance data in Table 7.1, entity 5 is a medoid there since the sum of its distances to the rest, 9.6, is the minimum. Now we form a cluster around each entity to consist of those that are nearer to the entity than to medoid 5: these will be 2 and 3 around 1, 1 and 3 around 2, 1 and 2 around 3, 7 and 8 around each other, and clusters for entities 4,5,6 are singleton themselves. This would give an edge to entity 1 as the next seed, because it is further away from 5 and surrounded by two. Indeed the summary E value for 1, 2.20+(1.82−0.51)+(1.16−0.88) = 3.79 is by far the greatest. Using the entities 5 and 1 as seeds, indeed brings the bisecting K-Means to a desired split {1, 2, 3} versus {4,5,6,7,8}. The hierarchy on Fig. 7.2 then will be found with further splits. The node heights are the same – within cluster squared errors that are scaled as percentages of the pre-processed data scatter.

Case Study 7.2. Anomalous Cluster Versus Two-Split Cluster

Consider 280 values generated according to the one dimensional Gaussian distribution $N(0,10)$ with zero mean and standard deviation equal to 10 (see Table A4 on p. 384), presented on Fig. 7.4 and try divide it in two clusters. When it is done with a splitting criterion, the division goes just over the middle, cutting the bell-shaped curve in two equal halves. When one takes Anomalous clusters, though, the divisions are much different: first goes a quarter of the entities on the right (to the

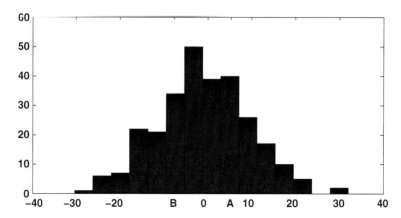

Fig. 7.4 Histogram of the one dimensional sample of 280 entities from $N(0,10)$ distribution. Points *A* and *B* denote the boundaries of the *right* and *left* anomalous fragments found with the Anomalous pattern algorithm

Table 7.6 Two-class partitions found using different strategies. A better result by the incremental two-splitting should be attributed to its one-by-one entity moving procedure

	iK-Means with t = 60			Incremental two-splitting		
Cluster	Size	Mean	Explained (%)	Size	Mean	Explained (%)
1	139	7.63	32.6	136	7.82	26.8
2	141	−9.30	32.1	144	−9.12	38.6

right of point A on Fig. 7.4), because the right end in this individual sample is a bit farther from the mean than the left one; then a similar chunk to the left of point B, etc. (see case study 6.4 in Section 6.2.8). Yet if one uses the anomalous clusters in iK-Means, just as an initial centers generator, things differ. With the discarding threshold of 60, only two major Anomalous patterns found in the beginning remain. Further 2-Means iterations bring a rather symmetric solution reflected in the leftmost part of Table 7.6. The right hand part of the table shows results found with an incremental version of Two-splitting method.

This leads to a slightly different partition, with three entities swapping their membership, which is slightly better, achieving 65.4% of the explained scatter versus 64.8% at iK-Means. This once again demonstrates that the incrementally taking into account entities is a more precise option than the all-as-one switching in iK-Means.

Let us now turn to conceptual clustering.

Case Study 7.3. Conceptual Clustering of Digit Data as Related to Ward Clustering

Let us take the set of 10 styled digits presented on Fig. 1.2 and turn the figure into a dataset by considering each of the seven rectangle edges a feature with two categories, "Present" and "Absent" (see Fig. 7.5 and Table 7.7).

To produce a classification tree leading to a partition S, we use summary Gini index (impurity function) $G(v1/S) + G(v2/S) + \ldots + G(v7/S)$ as the criterion to maximize. Start by trying each of the features as the split base to select the best of them. Consider, for example, partition $S = \{S_1, S_2\}$ of the Digit set according to

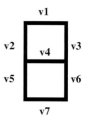

Fig. 7.5 Digit rectangle edges as features

7.3 Divisive and Conceptual Clustering

Table 7.7 Digit dataset

	v1	v2	v3	v4	v5	v6	v7
1	0	0	1	0	0	1	0
2	1	0	1	1	1	0	1
3	1	0	1	1	0	1	1
4	0	1	1	1	0	1	0
5	1	1	0	1	0	1	1
6	1	1	0	1	1	1	1
7	1	0	1	0	0	1	0
8	1	1	1	1	1	1	1
9	1	1	1	1	0	1	1
0	1	1	1	0	1	1	1
Total	8	6	8	7	4	9	7

attribute v2 which is present at $S_1 = \{4, 5, 6, 8, 9, 0\}$ and is absent at $S_2 = \{1, 2, 3, 7\}$. Cross-classification of S and v7 (see Table 7.8) yields G(v7/S) = 0.053.

To see what this has to do with the setting in which Ward's criterion applies, let us pre-process the Digit data matrix by subtracting the column averages without rescaling them (because the scaling coefficients must be all unity to make Ward's criterion equivalent to the summary Gini index, see Sections 4.5.3 and 6.2.7). However, the data in Table 7.9 is not exactly the data matrix Y considered theoretically in Section 6.2.7. Indeed, the theoretical data matrix in 6.2.7 and Equation (6.4) comprises columns corresponding to all of the categories, whereas the data matrices in Tables 7.7 and 7.9 reflect only half of the categories – those of the presence of edges v1–v7, never an absence. Indeed, a column corresponding to an "Absent" category is a mirror of the column corresponding to the "Presence" category, with all ones made zero and, vice versa, all zeros made ones. After the centering, the lacking half of the data table would be the Table 7.9 negated, that is, multiplied by –1. The data scatter is formed by squares of the entries which are the same. That means that this lacking part can be taken into account by just doubling the contributions accounted for with Table 7.9.

The data scatter of matrix in Table 7.9 is the summary column variance times $N = 10$, which is 13.1. However, to get the data scatter in the left hand side of (6.4), this must be doubled to 26.2 to reflect the "missing half" of the virtual data matrix Y.

Let us now calculate the within class averages c_{kv} of each of the variables, $v = v1,\ldots, v7$, in clusters $k = 1,2$ and take contributions $B_{kv} = N_{kv} c_{kv}^2$ summed up over clusters S_1 and S_2. This is done in Table 7.10, the last line of which contains contributions of all features to the explained part of the data scatter.

Table 7.8 Cross-classification of S = v2 and v7 on Digit dataset

	S_1	S_2	Total
v7 = 1	5	2	7
v7 = 0	1	2	3
Total	6	4	10

Table 7.9 Digit dataset pre-processed by centering its columns

	v1	v2	v3	v4	v5	v6	v7
1	−0.8	−0.6	0.2	−0.7	−0.4	0.1	−0.7
2	0.2	−0.6	0.2	0.3	0.6	−0.9	0.3
3	0.2	−0.6	0.2	0.3	−0.4	0.1	0.3
4	−0.8	0.4	0.2	0.3	−0.4	0.1	−0.7
5	0.2	0.4	−0.8	0.3	−0.4	0.1	0.3
6	0.2	0.4	−0.8	0.3	0.6	0.1	0.3
7	0.2	−0.6	0.2	−0.7	−0.4	0.1	−0.7
8	0.2	0.4	0.2	0.3	0.6	0.1	0.3
9	0.2	0.4	0.2	0.3	−0.4	0.1	0.3
0	0.2	0.4	0.2	−0.7	0.6	0.1	0.3

Table 7.10 Feature contributions to Digit clusters according to v2

	v1	v2	v3	v4	v5	v6	v7
v2 = 1	0.007	0.960	0.107	0.107	0.060	0.060	0.107
v2 = 0	0.010	1.440	0.160	0.160	0.090	0.090	0.160
Total	0.017	2.400	0.267	0.267	0.150	0.150	0.267

The last item, 0.267, is the contribution of $v7$. Has it anything to do with the reported value of impurity function $G(v7/S) = 0.053$? Yes, it does. There are two reasons to make these two quantities different. First, to get to the contribution from $G(v7/S)$, it must be multiplied by $N = 10$, which would make it 0.533. Second, the 0.267 value is the contribution to the data scatter of matrix Y obtained after enveloping of all 14 categories – not just 7 present in Table 7.8. After the contribution 0.267 is properly doubled, the quantities do coincide. Similar calculations made for the other six attributes, $v1, v2, v3, v4, v5$, and $v6$, would lead to the total contribution of S to the data scatter equal $10\Sigma_j G(vf/S) = 7.03$ which is 26.8% of the scatter 26.2.

To find out which of the features is to be used for the first split, all pair-wise Gini index values have been computed and presented in Table 7.11. According to these,

Table 7.11 Pairwise Gini indexes for all 7 features in Digit dataset

	v1	v2	v3	v4	v5	v6	v7
v1	0.320	0.003	0.020	0.015	0.053	0.009	0.187
v2	0.005	0.480	0.080	0.061	0.030	0.080	0.061
v3	0.020	0.053	0.320	0.034	0.003	0.009	0.034
v4	0.020	0.053	0.045	0.420	0.003	0.020	0.115
v5	0.080	0.030	0.005	0.004	0.480	0.080	0.137
v6	0.005	0.030	0.005	0.009	0.030	0.180	0.009
v7	0.245	0.053	0.045	0.115	0.120	0.020	0.420
Total	0.695	0.703	0.520	0.658	0.720	0.398	0.963

7.3 Divisive and Conceptual Clustering

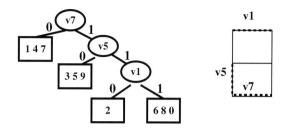

Fig. 7.6 Conceptual clustering of Digit dataset and features involved in the splits

feature $v7$ supplies the maximum summary contribution $10*0.963 = 9.63$ which is 36.8% of the total data scatter.

Therefore, the first split must be done according to $v7$. Two more splits are due $v5$, contributing 3.90, and $v1$, contributing 3.33, resulting in a four-cluster partition $S = \{1-4-7, 3-5-9, 2, 6-8-0\}$. This partition contributes $9.63+3.90+3.33 = 16.87 = 64.4\%$ to the total data scatter. The next partition step would contribute less than 10% of the data scatter, which is less than the contribution of one entity on average – a signal to stop the splitting. The classification tree, or conceptual tree, produced with the splits is presented on Fig. 7.6 along with a visualization of the set of tree-making features on the rectangle base of the Digit data.

What is nice about the tree is that the clusters are well matching those found by using the data on Confusion between the digits in a psychological experiment (see Chapter 8). This should lead to further analysis of possible importance of features $v7$, $v5$, $v1$ for the human perception of similarity between the digits.

F7.3.2 Divisive and Conceptual Clustering: Formulation

In this section, two aspects will be covered following Mirkin (2005): the appropriateness of using Bisecting K-Means as a splitting device in Ward-like divisive clustering and the relation between the square-error criterion and the summary Gini index.

On the first glance, the Ward criterion for dividing an entity set in two clusters – maximize Ward distance between the split parts – has nothing to do with that of K-Means. The K-Means criterion, in this case, given a parental cluster $J \subseteq I$, is to minimize $W(J,c) = \sum_{i \in S_1} d(i,c_1) + \sum_{i \in S_2} d(i,c_2)$ where S_1 and S_2 are the split parts of J, c_1 and c_2 their respective centroids and d squared Euclidean distance. The complementary part of the data scatter forms another criterion presented in Equation (6.6). This complementary criterion is to maximize $B(J,c) = N_1 <c_1,c_1> + N_2 <c_2,c_2>$ where N_1 and N_2 are respective cardinalities of the clusters.

Let us prove that that Ward distance between the two clusters, $dw(S_1, S_2) = \frac{N_1 N_2}{N_1 + N_2} d(c_1,c_2)$, is just that. To proceed, we need two equations. The first just expresses the squared Euclidean distance through inner products, $d(c_1,c_2) = (<c_1,c_1> - <c_1,c_2>) + (<c_2,c_2> - <c_1,c_2>)$. The second is relation between the cluster centroids and the parental cluster centroid c: $N_1 c_1 + N_2 c_2 = (N_1 + N_2)c$. Since c is not involved in $dw(S_1, S_2)$ and, in fact, is irrelevant to it, we may take it to be $c = 0$. Then the latter equation implies that $c_2 = -(N_1/N_2)c_1$. This can be put into $<c_1, c_1> - <c_1, c_2> = <c_1, c_1> + N_1/N_2 <c_1, c_1> = .= (N_1+N_2)/N_2 <c_1, c_1>$. Similarly, equation $<c_2, c_2> - <c_1, c_2> = (N_1+N_2)/N_1 <c_2, c_2>$ is obtained. Substituting these through the first equation in Ward distance, we find $dw(S_1, S_2) = \frac{N_1 N_2}{N_1+N_2} d(c_1, c_2) = \frac{N_1 N_2}{N_1+N_2}((N_1+N_2)/N_2 <c_1, c_1> + (N_1+N_2)/N_1 <c_2, c_2>) = B(J, c)$, which proves the statement.

That means that Ward-like divisive clustering is adequately served with Two-splitting, or Bisecting K-Means.

Now we will show that, in the situation in which all the features are categorical, maximizing the summary Gini index $\sum_{v \in V} G(v/J)$ is adequate as well. Assume that data matrix Y in this case is drawn by putting a dummy variable for each of the categories with a follow up centering it with the mean which is the category frequency. Then, according to Statement 4.4.2.1(c) in Section 4.5.3, Gini index $G(v/S)$, multiplied by the number of entities, is the contribution of the partition of J to the summary scatter of the dummies corresponding to categories of feature v – this, in fact, easily follows from Eqs. (3.13) and (4.19). This implies that the summary Gini index, multiplied by the number of entities, is the contribution of the partition to the summary scatter of all the dummy variables, that is, the data scatter of matrix Y. That means that maximum of the summary Gini index is reached at a partition minimizing the total unexplained contribution which is exactly the square error criterion. The statement is proved.

Q.7.10. What data standardization should be applied if one wants to build a conceptual clustering tree maximizing the summary Pearson chi-squared by using Ward distance maximization? **A.** Each category is to be represented by a dummy variable which then should be centered by subtracting its frequency and normalized by the square root of the frequency (see Statement 4.4.2.2(c)).

C7.3.3 Divisive and Conceptual Clustering: Computation

The process of divisive clustering is much like that of building a classification tree – the only difference is the criterion. It is correlation to a target variable in the latter case and it is a summary correlation with all the features forming the clustering space, in the former case, even if it is expressed as maximizing Ward distance between split parts. The equivalence between K-Means criterion at $K = 2$ and the criterion of maximization of Ward distance justifies the following algorithm.

7.3.3.1 Ward-Like Divisive Clustering

1. *Start*
   Put $J \Leftarrow I$ and draw tree root as a node corresponding to I at the height of $W(I)$ which is the data scatter, by itself or 100%.

7.3 Divisive and Conceptual Clustering

2. *Split*
   Split J in two parts, S_1 and S_2, to maximize Ward distance $wd(S_1, S_2)$.
3. *Draw*
   In the drawing, add two children nodes corresponding to S_1 and S_2 at the parent node corresponding to J, their heights being their square errors.
4. *Update*
   Find the cluster of maximum height among the leaves of the current cluster hierarchy and make it J.
5. *Stop-Condition*
   Check the stopping condition as described below. If it holds, halt and output the hierarchy and possible interpretation aids; otherwise, go to Step 2.

Developing a good splitting algorithm at Step 2 in divisive clustering can be an issue. Here are two versions: Two-splitting and C-splitting.

7.3.3.2 Two-Splitting (2-Means Splitting, Bisecting K-Means)

1. *Initialization*
   Given J, specify initial seeds of its split parts, c_1 and c_2.
2. *Batch 2-Means.*
   Apply Bisecting K-Means with initial seeds specified at step 1 and the squared Euclidean distance.
3. *Output*
   Output: (a) final split parts S_1 and S_2;
   (b) their centroids c_1 and c_2;
   (c) their heights, $h(S_1)$ and $h(S_2)$;
   (d) contribution of the split which is Ward distance $dw(S_1, S_2)$.

To specify two initial seeds in Two-splitting, either option can be applied:

(1a) random selection;
(1b) maximally distant entities;
(1c) centroids of two Anomalous pattern clusters derived on J as described in Section 6.2.7.
(1d) two centroids derived with algorithm Build in Section 6.2.7.

Random selection must be repeated many times to get a reasonable solution for any sizeable dataset. Maximally distant entities not necessarily reflect the structure of a good split (see case study 7.1). Therefore, two latter options should be preferred.

7.3.3.3 C-Splitting (Conceptual Clustering with Binary Splits)

1. *Initial setting*
   Set J to consist of the universal cluster, the entire entity set I.

2. *Evaluation*

   In a loop over all leaf clusters J and variables $v \in V$, for each J and v, consider all possible splits y_{Jv} of J over v in two parts. If v is quantitative or ordinal, J-splits are defined by splits of its range in two intervals: one part consists of entities at which feature v is less than or equal to y_{iv} and the other of those at which feature v is greater than y_{Jv}. If v is nominal, J is split over each of v's categories l in "yes" or "no" parts. This amounts to using quantitative dummy variables for each category.

3. *Split*

   Select that triplet (J, v, y_{Jv}) which received the highest score and perform the binary split of J, thus generating two its offspring nodes S_1 and S_2.

4. *Output*

   This is the same as in the previous versions plus the variables v and split values y_{iv} for each of the splits.

Comment 7.1. Some may argue that the framework of divisive clustering is deliberately set as a greedy optimization procedure: at each local splitting step the best solution is taken which is not necessarily the best if one considers summary results of several sequential steps. The greedy-wise nature of the setting is true. Yet it is not easy to formulate a holistic optimization problem for divisive clustering. If for example, the process of splitting goes all the way down to singleton clusters, then perhaps the greedy-wise setting is most natural (Mirkin 1996). When there is a stopping condition such as a pre-specified number K of terminal clusters, then the problem becomes of globally minimizing K-Means square-error criterion. One should remember that the K-Means criterion has some innate drawbacks related to its rigidity in putting the goal of getting split parts as uniform as possible (see, for example, case study 6.2 in Section 6.2.1). This means that achieving the global minimum is not necessarily beneficial from the point of view of data analysis.

7.4 Single Linkage Clustering, Connected Components and Maximum Spanning Tree

P7.4.1 Maximum Spanning Tree and Clusters: Presentation

Consider a similarity, rather than dissimilarity, matrix, for a change. All the contents of this section applies to dissimilarity data as well with the only change – of taking maximum for taking minimum.

Weighted graphs, or networks, is a natural way for representing similarity matrices such as those in Tables 1.7 and 1.8. Single link clustering method applies to symmetric matrices, such as that presented in Table 7.12 – a symmetric version of the Confusion data Table 1.7 obtained by a most conventional way: given a possibly non-symmetric matrix A, take its transpose A^T and define $\tilde{A} = (A + A^T)/2$. This

7.4 Single Linkage Clustering and MST

Table 7.12 A symmetric version of confusion data in Table 1.7

Stimulus	Response									
	1	2	3	4	5	6	7	8	9	0
1	877	11	18	86	9	20	165	6	15	11
2	11	782	38	13	31	31	9	29	18	11
3	18	38	681	6	31	4	31	29	132	11
4	86	13	6	732	9	11	26	13	44	6
5	9	31	31	9	669	88	7	13	104	11
6	20	31	4	11	88	633	2	113	11	31
7	165	9	31	26	7	2	667	6	13	16
8	6	29	29	13	13	113	6	577	75	122
9	15	18	132	44	104	11	13	75	550	32
0	11	11	11	6	11	31	16	122	32	818

is a technical way to express the idea that every symmetric pair of non-coinciding entries such as 7 in position (1,3) and 29 in position (3,1) should be substituted by their half-sum: $36/2 = 18$. To obtain data in Table 7.12, the result was rounded up to the nearest larger integer.

The similarity matrix in Table 7.12 can be represented by a graph whose nodes correspond to the entities $i \in I$ and edge weights to the similarity values. Frequently, a threshold t applies so that only those edges $\{i,j\}$ are put in the graph for which the similarity values are greater than the threshold.

For the threshold $t = 20$, this graph is presented on Fig. 7.7.

In graph theory, a number of concepts have been developed to reflect the structure of weighted graphs of which one of the most popular is the concept of Maximum Spanning Tree (MST), see Johnsonbaugh and Schaefer (2004). A tree is a graph with no cycles, and a spanning tree is a tree over all the entities under consideration

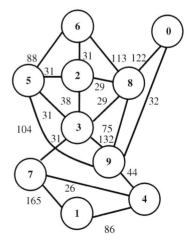

Fig. 7.7 Network of connections corresponding to similarity weights of 21 or greater in matrix of Table 7.12

as its nodes. The length of a spanning tree is defined as the sum of weights of all its edges. An MST is a spanning tree whose length is maximum.

Worked example 7.3. Concept of MST

Consider graph of Fig. 7.7. Figure 7.8 highlights two of its spanning trees. The length of that on the left is $165\{1-7\} + 31\{7-3\} + 44\{4-9\} + 132\{9-3\} + 31\{3-5\} + 38\{3-2\} + 29\{3-8\} + 31\{2-6\} + 122\{8-0\} = 623$; here, curly braces' contents correspond to the edges in the tree. The length of that on the right is $86\{1-4\} + 165\{1-7\} + 31\{7-3\} + 132\{3-9\} + 104\{9-5\} + 88\{5-6\} + 113\{6-8\} + 38\{3-2\} + 122\{8-0\} = 879$, which is much greater. In fact the latter is a Maximum Spanning Tree.

Given a weighted graph, or similarity matrix, an MST T can be built by using Prim's algorithm which collects T step by step starting from a singleton node, in fact any of the nodes, and then adding a maximum outside link to T one by one.

Worked example 7.4. Building an MST on Confusion data

Let us build a Maximum Spanning Tree for the network on Fig. 7.7 by using Prim's algorithm. Start, for example, with $T = \{0\}$ and add to T that link which is maximum, that is, obviously $122\{0-8\}$. Since T has two nodes now, we check external links of each of them to find a maximum external link from T to the rest, $113\{8-6\}$, thus getting three nodes, 0, 8, 6 and two links, $\{0-8\}$ and $\{8-6\}$, in T. The maximum external link now is $88\{6-5\}$ bringing 5 and link $\{6-5\}$ into T. Next maximum links are $104\{5-9\}$, $132\{9-3\}$ and $31\{3-7\}$ bringing 9, 3 and 7, respectively, into T. Of the three remaining nodes outside T, 2, 4 and 1, the maximum link

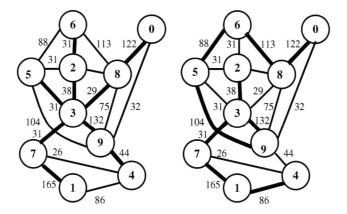

Fig. 7.8 Two spanning trees on the graph of Fig. 7.7 are highlighted by *bold* edges. The length of the tree on the *left* is 623, and that on the *right*, 879

7.4 Single Linkage Clustering and MST

is 165{7−1} followed by 86{1−4}. Node 2's maximum connection is 38{2−3} thus completing the MST drawn on the right hand side of Fig. 7.8.

Prim's algorithm is what is called greedy – it works node-by-node and picks up the best solution at the given step, paying no attention to what happens next. The MST problem is one of a very few combinatorial problems that can be solved indeed by a greedy algorithm. On the other hand, one should not be overly optimistic about performances of the algorithm because it finds, at each step a maximum of a number of elements, on average – half the number of entities, and one should not forget that finding a maximum is a rather expensive operation.

Another potential drawback, related to the data size, which is quadratic over the number of entities, is not that bad. Specifically, if the similarities are computed from data in the entity-to-feature format, the difference between the data sizes can grow fast indeed: say 500 entities over 5 features take about 2,500 numbers, whereas the corresponding similarity matrix will have 250,000 numbers – a hundred times greater. Yet if the number of entities grows 20 times to 10,000 – a rather modest size nowadays – the raw data table will take 50,000 numbers whereas the similarity matrix, of the order of 100,000,000, a 100 millions, which is two thousand greater! Yet it is possible to organize the computation of an MST by storing and updating lists of nearest neughbors in the network in such a manner that the quadratic size increase is not necessary, because almost all necessary similarities can be calculated from the raw data when needed.

Two cluster-analysis concepts are related to MST: connected components of threshold graphs and single link clustering.

A connected component of a graph is a maximum subset of nodes such that each pair of its nodes can be connected by a path in the graph. Given a similarity matrix or weighted graph on its entities, a threshold graph is defined as an ordinary, unweighted, graph with the same set of nodes and set of edges *{i,j}* such that their weights in the original graph are greater than threshold *t*, for some real *t*. This gives a most natural concept of cluster: a connected component in a threshold graph. On the first glance, there can be myriads of different threshold graphs, but in fact all of them are defined by an MST.

The components of a threshold graph are fragments of any MST found by cutting its weakest edges – those at which weights are less than the threshold. It should be clear from the definition that all connections within the MST fragments are weaker than the threshold.

Worked example 7.5. MST and connected components

Let us sort all the edges in MST found at the graph on Fig. 7.8 in the ascending order: 31{3−7}, 38{3−2}, 86{1−4}, 88{5−6}, 104{9−5}, 113{6−8}, 122{8−0}, 132{9−3}, 165{1−7}. Given a threshold *t*, say *t* = 50, cut the 2 edges in the tree that are less than the threshold, 3−2 and 3−7 – the tree will be partitioned in 2+1 = 3 fragments corresponding to connected components of the corresponding threshold graph.

306　　7 Hierarchical Clustering

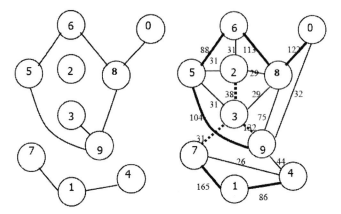

Fig. 7.9 Threshold graph at $t = 50$ for the graph of Fig. 7.6 is on the *left*, and MST with 2 weakest links shown using *dashed lines* is on the *right*

Figure 7.9 presents, on the left, a threshold graph at $t = 50$, along with clearly seen components consisting of subsets {1,4,7}, {2}, and {3,5,6,8,9,0}. The same subsets are seen on the right where the two weakest links are cut out of the MST. The fact that the threshold graph components are MST fragments is not a coincidence but rather a mathematically proven property of the MSTs.

The single linkage clustering is a hierarchical clustering approach, either agglomerative or divisive, that is based on the following definition of similarity between clusters: given a similarity matrix $A = (a_{ij})$ between entities $i, j \in I$, the similarity between subsets S_1 and S_2 in I is defined by the maximum similarity between their elements, $a(S_1, S_2) = \max_{i \in S_1, j \in S_2} a_{ij}$. This is why this approach sometimes is referred to as the Nearest Neighbor (NN) approach. The single linkage clusters are but continuous fragments of an MST.

Specifically, given an MST, the entire hierarchy of single link clustering can be recovered by one-by-one cutting the weakest links (divisive clustering) or, starting from the trivial singletons, one by one merging the strongest links. The similarity values over the MST can be used as the height function for drawing over the process of mergers/divisions.

Worked example 7.6. Single link hierarchy corresponding to an MST

Let us build a binary classification tree according to the MST on Fig. 7.8. This tree is presented on Fig. 7.10. Note that in contrast to the heights defined in Section 7.2 over dissimilarities, the direction of the height axis goes down here to reflect the principle that the smaller the similarity the further away are the clusters. Specifically, 7 and 1 merge together first because of the maximum similarity, 165, followed by merging 3 and 9 (link 132) and 0 and 8 (link 122). After that 6 joins into {8,0}

7.4 Single Linkage Clustering and MST

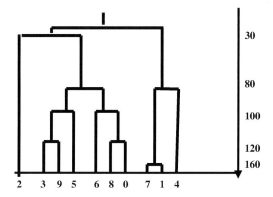

Fig. 7.10 A binary tree of the single link clustering for the MST of Fig. 7.8; the heights of the branches reflect the similarities between corresponding nodes

(similarity 113) and 5 joins to {3, 9} (similarity 104). Then these two clusters merge together (similarity 88). Then 4 joins into cluster {1,7} (similarity 86). At last, 2 joins the larger cluster at similarity 38. The process is complete when two clusters, {1,4,7} and the rest, merge together at similarity 31.

Case-study 7.4. Difference Between K-Means and Single Link Clustering

Consider a set of 2D points presented on Fig. 7.11. Those on the left have been clustered by using the single link approach, whereas those on the right, by using the square error criterion of K-Means.

Overall, this example demonstrates the major difference between conventional clustering and single link clustering: the latter finds elongated structures whereas the former cuts out convex parts. Sometimes, especially in the analysis of results of physical processes or experiments over real-world particles, the elongated structures

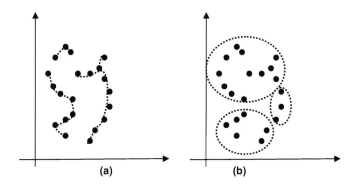

Fig. 7.11 A 2D point set: clustered with the single link method (**a**) and K-Means (**b**)

do capture the essence of the data and are of great interest. In other cases, especially when entities/features have no intuitive geometric meaning – think of bank customers or internet users, for example, convex clusters make much more sense as groupings around their centroids.

Comment 7.2. An interesting property of the single linkage method is that it involves just *N-1* similarity entries occurring in an MST rather than all entries in *A*. This results in a threefold effect:

(1) a nice mathematical theory,
(2) fast computations, and
(3) poor application capability.

Worked example 7.7. MST and single linkage clusters for Company dataset

To illustrate point (3) in Comment 7.2 above, let us consider a Minimum Spanning Tree built on distances between Companies in Table 7.1. Starting from Av, we add minimum link An(0.51)Av to tree *T* being built, then we add to *T* the minimum distance link As(0.77)An. Then the minimum distance is 1.15 between Ba and Av, which brings next links Bu(0.87)Ba, Br(0.75)Bu, followed by Ci(0.83)Br and Cy(0.61)Ci. (Note that all row-wise minimum distances highlighted on Table 7.1 have been brought in the tree *T*. These minimum distances can be used, in fact, in a different method for building an MST – Boruvka's algorithm (1926), arguably the very first clustering method!). This MST *T*, which is in fact a path, is presented on the left side of Fig. 7.12.

This path goes along the product clusters so that B companies are all between A and C companies. Yet the single link clusters shown on the right side of Fig. 7.12 do not reflect the structure of the set but separate a distant company Ba and mix

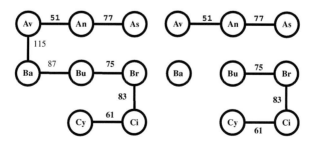

Fig. 7.12 Minimum spanning tree for company dataset in Table 7.1 (on the *left*; two weakest links, 115 and 87, are shown using ordinary font), and three single link clusters according to it (on the *right*): the structure of company products is reflected on the tree and lost on the clusters because of wrong cuts. (For the sake of convenience, the distances are multiplied by 100)

together products B and C – all this just because the right-to-remove link Br-Ci (0.83) appears a wee smaller than wrong-to-remove link Ba-Bu (0.87).

Q.7.11. Explain the sequence of splits in a divisive algorithm according to tree of Fig. 7.10.

Q.7.12. How many edges in an MST are to be cut if the user wants to find 5 clusters?

Q.7.13. Prove that the number of edges in an MST is always the number of entities short one.

Q.7.14. Apply Prim's algorithm to Amino acid similarity data in Table 1.8.

F7.4.2 MST, Connected Components and Single Link Clustering: Formulation

Here we present some mathematical properties relating the concepts of Maximum Spanning Tree, connected component of a graph and Nearest Neighbor clustering.

F7.4.2.1 MST and Connected Components

Let us first recall some definitions from graph theory.

A weighted (similarity) graph $\Gamma = (I, G, A)$ is defined as a triplet of: (i) an N-element set of nodes I; (ii) set of edges, that is, two-element subsets of I, G; and (iii) edge weight function represented by a symmetric matrix $A = (a_{ij})$ so that $a_{ij} = 0$ if $\{i,j\} \notin G$. A graph is referred to as an ordinary graph if its nonzero weights are all unities.

A path between nodes i and j in Γ is a sequence of nodes $i_1, i_2, \ldots, i_n$ such that $\{i_m, i_{m+1}\} \in G$ for each $m = 1, 2, \ldots, n-1$ and $i_1 = i$, $i_n = j$. A path is referred to as a cycle if $i_1 = i_n$. A subset of nodes S is referred to as a connected component if there is a path within S between each pair of nodes in S, and S is maximal in this sense so that addition of any supplementary node to S breaks the property. Graph Γ is called connected if it consists of just one connected component.

Given a connected weighted graph $\Gamma = (I, G, A)$, a connected weighted graph $T = (J, H, B)$, with no cycles, is referred to as its spanning tree if $J = I$, $H \subset G$, and B is A restricted to H, so that $b_{ij} = a_{ij}$ for $\{i,j\} \in H$ and $b_{ij} = 0$ for $\{i,j\} \notin H$. A characteristic property of a spanning tree T is that it has exactly $N-1$ edges: if there are more edges than that, T must contain a cycle, and if there are less edges than that, T cannot span the entire set I and, therefore, it would consist of several connected components.

The weight of a spanning tree $T = (I, H, B)$ is defined as the total weight of its edges, that is, the sum of all elements of weight matrix B. A spanning tree of maximum weight is referred to as a Maximum Spanning Tree, MST.

Given a weighted graph $\Gamma = (I, G, A)$ and a real t, an ordinary graph $\Gamma_t = (I, G_t, A_t)$ is referred to as a threshold graph if $G_t = \{\{i,j\}: a_{ij} > t\}$. Given a spanning tree T and a threshold t, its threshold graph can be found by cutting out those edges whose weights are smaller than or equal to t. It appears T bears a lot of structural information of the corresponding graph $\Gamma = (I, G, A)$.

In particular, connected components of an MST found by cutting those links from MST that are less than t, one-to-one correspond to connected components of the threshold graph Γ_t.

Indeed, consider a component S of MST T obtained by cutting all edges of the T whose weights are less than t. We need to prove that for each pair $i, j \in S$ there is a path between i and j such that it all belongs to S and the weight of each edge in the path is greater than t, and for all i,k such that $i \in S$ and $k \notin S$, if $\{i, k\} \in G$, then $a_{ik} \leq t$. But the former obviously follows from the fact that S is a connected component of T in which all weights are greater than t since the others have been cut out. The latter is not difficult to prove either: assumption that $a_{ik} > t$ for some $i \in S$ and $k \notin S$ such that $\{i, k\} \in G$ would contradict the assumption that T is an MST, that is, that the weight of T is maximal, because by substituting the edge connecting S and the component containing k by edge {i,k}, one would obtain a spanning tree of a greater weight. Assume now that an S is a connected component of the threshold graph (at threshold t), and prove that S is a component of the threshold graph, at the same threshold t, for any MST. Indeed, if S overlaps two components, $S1$ and $S2$, of the threshold graph of some MST T, then there must be a pair i,j in S such that $i \in S1$ and $j \in S2$ and $a_{ij} > t$, which again contradicts the fact that $S1$ and $S2$ are not connected in the threshold graph of T. This completes the proof.

F7.4.2.2 MST and Single Link Clustering

Single link clustering is, primarily, an agglomerative clustering method in which the similarity between two clusters, S_1 and S_2, is defined according to the nearest neighbor rule as the maximum similarity between elements of these clusters, $a(S_1, S_2) = \max_{i \in S_1, j \in S_2} a_{ij}$ – the fact that the between-cluster similarity is defined by just one link underlies the name of the method.

There is no need to revise all the maximum similarities after every merger step. New similarities can be revised dynamically in the agglomeration process according to the following rule:

$$a(S, S_1 \cup S_2) = max[a(S, S_1), a(S, S_2)],$$

where $S_1 \cup S_2$ is the result of the agglomeration step.

Thus, the only intensive computation is finding the maximum in the newly formed column $a(S, S_1 \cup S_2)$ of the similarity matrix over all current clusters S.

Yet there is another way to proceed – by building an MST first. All the merger steps can be made according to the MST topology. First, the $N-1$ edges of the tree are to be sorted in the descending order. Then the following recursive steps apply. On the first step, take any maximum similarity edge {i,j}, combine its nodes into a cluster and merge i and j by removing the edge. On the general step, take any remaining similarity edge {i,j} of the maximum similarity value (among those left)

7.4 Single Linkage Clustering and MST 311

and combine clusters containing i and j nodes into a merged cluster. Halt, when no edges remain in the sorted order.

This operation is legitimate because the following property holds: clusters found in the process of mergers according to the sorted list of MST edges are clusters obtained in the agglomerative single link clustering procedure.

There is no straightforward divisive version of the Single Link method as originally defined. However, it is rather easy to do if an MST is built first. Then cutting the tree over any of its weakest, that is, minimum, links produces the first single link division. Each of the split parts is divided in the same way – by cutting out one of the weakest links.

Q.7.15. Let us refer to a similarity matrix A as ultrametric if it satisfies the following property: for any triplet $i, j, k \in I$, $a_{ij} \geq min(a_{ik}, a_{kj})$, that is, two of the values a_{ij}, a_{ik}, a_{kj} are equal, whereas the third one may be greater than that. Given an MST T, define a new similarity measure between any nodes i and j by using the unique path $T(i,j)$ connecting them in T: $at(i,j) = \min_{k,l \in T(i,j)} a_{kl}$. Prove that:

(i) Similarity $at(i,j)$ coincides with that defined by the agglomerative hierarchy built according to the Single Link algorithm;
(ii) Similarity $at(i,j)$ is an ultrametric;
(iii) Similarity $at(i,j)$ is the maximum ultrametric satisfying condition $at(i,j) \leq a_{ij}$ for all $i, j \in I$.

C7.4.3 Building a Maximum Spanning Tree: Computation

To find an MST, several "greedy" approaches can be undertaken. One of them, by J. Kruskal (1956), finds an MST by picking up edges; the other, by R.C. Prim (1957), picks up nodes. Prim's algorithm builds an MST T from an arbitrary node by finding the weakest link to the tree from outside and adding it to tree at each step. An exact formulation is this.

7.4.3.1 Prim's Algorithm

1. *Initialization.* Start with tree T consisting of an arbitary node $i \in I$ with no edges.
2. *Tree update.* Find $j \in I - T$ maxiimizing a_{ij} over all $i \in T$ and $j \in I - T$. Add j and edge $\{i,j\}$ with the maximal a_{ij} to T.
3. *Stop-condition.* If $I - T = \emptyset$, halt and output tree T. Otherwise, go to 2.

To build a computationally effective procedure for the algorithm may be a cumbersome issue, depending on how maxima are found, to which a lot of work has been devoted. A simple pre-processing step can be quite useful: in the beginning, find a nearest neighbor for each of the entities; only they may go to MST. At each step, update the neighbors of all elements in $I-T$ so that they lead to elements of T (Murtagh 1985). The claim that the algorithm builds an MST indeed can be proven using inductive statement that T at each step is part of an MST.

Q.7.16. Prove that the MST would not change if the similarities are transformed with a monotone transformation, that is, a function $\varphi(x)$ such that $\varphi(x1) > \varphi(x2)$ if $x1 > x2$. Hint: Because the sequence of events in Prim's algorithm does not change.

Q.7.17. Prove that an agglomerative version of the single linkage method can work recursively by modifying the similarities, after every merger according to formula

$$a(S, S_1 \cup S_2) = max[a(S, S_1), a(S, S_2)],$$

and each time merging the nearest neighbors in the similarity matrix.

7.5 Summary

Hierarchical clustering builds a binary hierarchy. Currently, this is usually taken as a prerequisite to partitioning the entity set rather than anything else. Yet with the current surge of research on hierarchical ontologies as practical tools for knowledge handling, it should not take long to see hierarchical clustering as serving, and of course modified by, the challenges of ontology building.

The chapter's material explains a most popular algorithm for agglomerative clustering and two different algorithms for divisive clustering. Divisive clustering splits clusters in parts and should be a more interesting approach computationally because it can utilize fast splitting algorithms and stop splitting whenever it seems right. Much of the material relates to the so-called Ward distance – an implementation of K-Means clustering criterion, the summary square error. In particular, both presented divisive clustering algorithms use this criterion, rearranged in an appropriate format. One algorithm proceeds with conventional K-Means at $K = 2$, utilized for splitting a cluster. The other maximizes summary Gini coefficient to make splits conceptual, that is based on one feature at a time. The last section explains relation between the single link clustering, a popular method to extract elongated structures from the data, and graph-theoretic structures in data: the Minimum Spanning Tree (MST) and connected components.

References

Boruvka, O.: Příspěvek k řešení otázky ekonomické stavby elektrovodních sítí (Contribution to the solution of a problem of economical construction of electrical networks)" (in Czech), Elektronický Obzor. **15**, 153–154 (1926).

Breiman, L., Friedman, J.H., Olshen, R.A., Stone, C.J.: Classification and Regression Trees. Wadswarth, Belmont, CA (1984).

References

Fisher, D.H.: Knowledge acquisition via incremental conceptual clustering. Mach. Learn. **2**, 139–172 (1987).

Hartigan, J.A.: Clustering Algorithms. Wiley, New York (1975).

Jain, A.K., Dubes, R.C.: Algorithms for Clustering Data. Prentice Hall, Upper Saddle River, NJ (1988).

Johnsonbaugh, R., Schaefer, M.: Algorithms. Pearson Prentice Hall, Upper Saddle River, NJ (2004).

Kruskal, J.B.: On the shortest spanning subtree of a graph and the traveling salesman problem. Proc. Am. Math. Soc. **7**(1), 48–50 (1956).

Lance, G.N., Williams, W.T.: A general theory of classificatory sorting strategies: 1. Hierarchical Systems. Comput. J. **9**, 373–380 (1967).

Mirkin, B.: Mathematical Classification and Clustering. Kluwer Academic Press, Boston-Dordrecht (1996).

Mirkin, B.: Clustering for Data Mining: A Data Recovery Approach. Chapman & Hall/CRC, Boca Raton, FL (2005). ISBN 1-58488-534-3.

Murtagh, F.: Multidimensional Clustering Algorithms. Physica-Verlag, Vienna (1985).

Murtagh, F., Downs, G., Contreras, P.: Hierarchical clustering of massive, high dimensional data sets by exploiting ultrametric embedding. SIAM J. Scientif. Comput. **30**, 707–730 (2008).

Prim, R.C.: Shortest connection networks and some generalizations. Bell Syst. Technic. J. **36**, 1389–1401 (1957).

Tasoulis, S.K., Tasoulis, D.K., Plagianakos, V.P.: Enhancing principal direction divisive clustering. Pattern Recognit. **43**, 3391–3411 (2010).

Ward, J.H. Jr.: Hierarchical grouping to optimize an objective function. J. Am. Stat. Assoc. **58**, 236–244 (1963).

Chapter 8
Approximate and Spectral Clustering for Network and Affinity Data

This chapter is devoted to clustering similarity, graph and network data: these are represented by square matrices rather than rectangular ones. Methods for finding a cluster or two-cluster split are described by combining three types of approaches from both early and recent developments:

(a) combinatorial clustering approach that is oriented at optimization of some reasonable indexes of cluster homogeneity,
(b) additive clustering approach that is based on a data recovery model at which the data is decoded from a cluster structure to be found by minimizing the discrepancy between the structure and observed similarities, and
(c) spectral clustering approach exploiting the machinery of matrix eigenvalues and eigenvectors as a relaxation of combinatorial clustering problems.

These methods are easily extended to partitioning or hierarchical clustering. In addition to the partitioning or hierarchical clustering structures, covered in Chapters 6 and 7, respectively, this chapter involves a somewhat more conservative approach of finding one cluster at a time. The one-cluster approach exhibits some flexibility by making it possible that some entities are left unclustered.

There are three different ideas within the approaches (a), (b), (c), that are interwoven in the text: (i) an analogy between the spectral and additive cluster decompositions of similarity matrices, (ii) the subtraction of background, or noise, similarities, (iii) the Laplacian transformation of similarities. When combining different approaches for the first time, as is the case, one or two combinations can be rather unusual but not necessarily unsound. Such is, for example, a combination of the one-cluster approach (Mirkin 1987, 1996) and modularity transformation (Newman and Girvan 2004, Newman 2006) described in Section 8.1. A more clear portrayal of the subject is yet to be developed.

8.1 One Cluster Summary Similarity with Background Subtracted

P8.1.1 Summary Similarity and Two Types of Background: Presentation

The sum of within cluster similarities seems a perfect criterion for clustering – it is simple and intuitive (see Fig. 8.1). The greater the within cluster total similarity, the tighter is the cluster. Maximizing this criterion should lead to a cluster of highest internal similarity.

Unfortunately, when all the between-entity similarities are non-negative – a quite typical situation, the criterion is of no use because it reaches its maximum at the largest possible cluster of all, the universal "cluster" $S = I$ consisting of the entire data set. A reasonable alternative, maximizing the average within cluster similarity, will not work here either: the average steadily declines when the number of entities in cluster S increases. Leaving the average aside for now (to return to it in Section 8.3), let us concentrate on modifying the summary similarity criterion by subtracting some "background" similarity pattern from data.

There can be different background similarity patterns that should be removed to sharpen up the clusters hidden in data, of which two are considered here:

(a) constant "noise" level (see Fig. 8.2), and
(b) random interactions.

Whereas the former seems rather obvious (see also some elaborated versions of that in Section 8.3), the latter involves a probabilistic interpretation of the similarities as emerging from some interactions between the entities. According to this interpretation, each entity i is assigned with a probability of interaction, equal to the proportion of the summary similarity in i-th row in the whole summary volume of the similarities. Then random interactions between two entities will occur with the probability equal to the product of their respective probabilities, which therefore should be subtracted from the similarity coefficients to clear up the nonrandom part of the similarity.

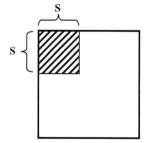

Fig. 8.1 The structure of similarity matrix regarding a cluster S, under the assumption that elements of S stand first. The checked part relates to the within cluster similarities

8.1 One Cluster Summary Similarity with Background Subtracted

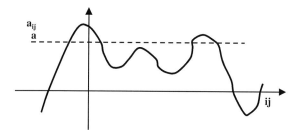

Fig. 8.2 Illustration of the effect of subtraction of a constant background "noise" from the similarity values. The graph shows similarity values (axis *y*) against some ordering of entity pairs *(i,j)* over *x*-axis. At zero noise level, the area of positive similarity values is much larger than that above the *dashed line* at which the area narrows down to two small high similarity islands

The summary criterion with the uniform noise subtracted is referred to as uniform clustering criterion (Mirkin 1996), and that with the random interaction noise subtracted is referred to as modularity function (Newman 2006). In this section, examples of similarity data are given in three different formats:

(i) genuine similarity,
(ii) networks or graphs,
(iii) affinity data derived from distances according to an entity-to-feature table.

Each of these formats has its specifics: unpredictable quirks in similarities as raw data (format i), many zeros and flat – the same – positive similarity values (format ii), and geometric nature of affinities (format iii).

Yet when presented as an entity-to-entity similarity matrix and subjected to a standardization step by subtracting background similarities, method AddRem described later in the section can be applied to maximize the total within cluster similarity. AddRem works its way by sequentially adding or removing one entity at a time. The method stops at a cluster when no change of one entity state can increase the criterion. The resulting cluster is provably tight (see Section 8.1.2).

Let us consider now application of this method, under each of the two background removal options – modularity and uniform, to instances of the three data types.

Worked example 8.1. Summary similarity clusters at a genuine similarity dataset

Consider a similarity data set such as Confusion between numerals in Table 1.7, already analyzed in Section 7.4.

A symmetric version of the Confusion data is presented in Table 8.1: the sum of $A+A^T$ without further dividing it by 2, for the sake of wholeness of the entries. In this table, care has been taken of the main diagonal. The diagonal entries are by far the

Table 8.1 Confusion data from Table 1.7 summed up with the transpose after the diagonal elements removed

	1	2	3	4	5	6	7	8	9	0
1	0	21	36	171	18	40	329	11	29	22
2	21	0	76	26	62	61	18	57	36	22
3	36	76	0	11	61	7	61	57	263	22
4	171	26	11	0	18	22	51	25	87	11
5	18	62	61	18	0	176	14	25	208	21
6	40	61	7	22	176	0	4	225	22	61
7	329	18	61	51	14	4	0	11	25	32
8	11	57	57	25	25	225	11	0	149	243
9	29	36	263	87	208	22	25	149	0	64
0	22	22	22	11	21	61	32	243	64	0
Total	677	379	594	422	603	618	545	803	883	498

largest and considerably differ among themselves, which may highly affect further computations. Since we are interested in patterns of confusion between different numerals, this would be an unwanted effect so that the diagonal entries should be made to bear no effect on the clustering process. They are changed for zeros in Table 8.1.

This matrix contains a lot of information that seems unnecessary to the human eye. Usually a symmetric similarity matrix with no main diagonal is represented by its upper triangle – that part which is over the main diagonal (see Table 8.2). One can notice that the last row, as well as first column, are absent from Table 8.2.

The results of the clustering algorithm AddRem(i) applied, at each i, in the two different settings – modularity and uniform – are presented in Table 8.3. The arbitrariness of choosing an entity to start has no effect in this case. Not too many clusters have been found anyway. The modularity criterion is capable of separating the cluster {1,4,7} from the rest, albeit with a somewhat lesser criterion value, but the rest also appears to be a cluster, in fact a tighter one. The uniform criterion at the threshold subtracted at the average level of 66.91 with no diagonal entries taken into

Table 8.2 Symmetric confusion data from Table 8.1 in the upper triangle format

	2	3	4	5	6	7	8	9	0
1	21	36	171	18	40	329	11	29	22
2		76	26	62	61	18	57	36	22
3			11	61	7	61	57	263	22
4				18	22	51	25	87	11
5					176	14	25	208	21
6						4	225	22	61
7							11	25	32
8								149	243
9									64

8.1 One Cluster Summary Similarity with Background Subtracted

Table 8.3 One-cluster structures found with the summary criterion at symmetric Confusion data in Table 8.1

Modularity		Uniform, π = Mean = 66.91		Uniform, π = 100	
Cluster	Criterion	Cluster	Criterion	Cluster	Criterion
2 3 5 6 8 9 10	1,137.2	3 5 6 8 9 10	1,200.7	1 4 7	502
1 4 7	808.2	1 4 7	700.5	3 5 9	464
				6 8 10	458
				2	

account, achieves a similar fit, though it loses digit 2 from the "rest" cluster – which is good because this digit keeps a company of its own being very rarely confused for anything else. Yet at a larger threshold value of $a = 100$, the uniform criterion leads to four high density clusters – exactly those produced in Section 7.3 by the conceptual clustering applied to the styled numerals' images. This would be a success story provided that the user knew beforehand the right threshold value, which is a rather bold hypothesis.

Results reported in Table 8.3 lead to the following idea. Would the structure be revealed in a more uniform way if clusters are taken sequentially, so that once clustered entities are removed from the set, the remainder is considered as a new set to cluster. That is, a new random interaction or average similarity data on the remaining set is compiled and AddRem is applied after that – doing the removals again and again. There may be a problem with this approach, which can be clearly seen in Table 8.3: the cluster to remove should be the set of seven numerals rather than the remainder consisting of three numerals, 1, 4 and 7. To address the issue, each part, both the remainder and cluster, should be clustered again (see case study 8.1).

Case Study 8.1. Repeated One-Cluster Clustering with Repeated Removal of Background

Let us, after each clustering step, consider the unclustered part as a fresh data set, a ground set, to perform the background similarity removal again. The results of this approach are presented in Table 8.4 in such a way that each cluster that has appeared on the right, in its column, has been clustered again. All the three modularity clusters have produced themselves as their modularity subclusters. On the contrary, at the uniform criterion, each of the three-element clusters has produced a proper subcluster as shown in the further rows of the right-hand part of the table.

To explain this phenomenon, let us take a closer look, say, at cluster {1,4,7}. Table 8.5 presents the original within cluster similarities as well as those found

Table 8.4 Partitions found at the symmetric Confusion data by sequentially extracting clusters one by one, recomputing the background similarities at each subset to be analyzed

Modularity, set adjusted		Uniform, mean set adjusted	
Ground set	Cluster	Ground set	Cluster
0–9	1 4 7	0–9	1 4 7
0 2 3 5 6 8 9	2 3 5 9	0 2 3 5 6 8 9	3 5 9
0 6 8	0 6 8	0 2 6 8	0 6 8
		1 4 7	1 7
		3 5 9	3 9
		0 6 8	0 8

Table 8.5 Similarities between numerals 1, 4 and 7 according to Table 8.2 and, also, after subtraction of the background according to each, uniform and modularity, criterion

	Raw similarities		Mean subtracted similarities		Random interactions subtracted similarities	
	4	7	4	7	4	7
1	171	329	−12.7	145.3	70.3	156.6
4		51		−132.7		−25.6

by subtracting the average similarity, for the uniform clustering, or the random interactions, for the modularity clustering.

The total sum of similarities in set $\{1,4,7\}$, the volume, according to the left part of Table 8.5 is $1{,}102 = 2{*}551$ (factor 2 applies to make up for the absent lower triangle of the similarity matrix), of which entity 1 takes 45.4%, entity 4, 20.1%, and entity 7, 34.5%. The volume of entity 4, 20.1%, is by far the smallest of the three, which straightforwardly translates to the level of its random interactions: they are smaller than those of the others so that the subtracted part of entity 4's similarities is relatively small. This is why the summary similarity of 4 in the right-hand part of the table is positive, $c_{41} + c_{47} = 70.3 - 25.6 = 44.7 > 0$, making 4 a welcome member of the cluster according to the modularity criterion. This is not so according to the uniform criterion: the summary similarity of 4 with two others is negative, $b_{41} + b_{47} = -12.7 - 132.7 = -145.4 < 0$, setting 4 apart from the rest. A similar effect is at work with entity 2: 2 is rather remote from anything else so that its similarities become negative when the average similarity is subtracted, which is not the case with the random interactions because the latter are by far smaller at 2 than those at other entities.

The analysis reported in case study 8.1 shows that the two criteria – or, better to say, the same criterion at the two different data pre-processing formulas – should be applied in different contexts: the uniform criterion is better when the meaning of similarity is uniform across the table, whereas the modularity criterion works better when the similarities should be scaled depending on the individual entities.

8.1 One Cluster Summary Similarity with Background Subtracted

Case Study 8.2. Summary Clusters at Ordinary Network Data

Consider two network graphs on Fig. 8.3a, b. The former's cluster structure is rather simple – it consists of two connected components. There is no visible cluster structure in the graph (b). The latter graph consists of just one component – but can the cluster structure hidden in it be discovered using a less rigid instrument than the concept of connected component?

An even less structured is a "cockroach" graph on Fig. 8.4, taken from Guattery and Miller (1998) as an example of a structure that is difficult for clustering (Luxburg 2007).

The results of AddRem clustering algorithm runs starting from every node for the modularity criterion at the cockroach network of Fig. 8.4 are given in the left part of Table 8.6.

There are three highly overlapping clusters, two of them reflecting the topology of the graph with the winning cluster embracing four nodes in the right-hand side of graph in Fig. 8.4. The second column reflects an attempt at finding a partition using one-by-one clustering: after first cluster is found, its entities are removed, and the method is applied to the remaining part of the data matrix, with the random interactions readjusted to the topology of the ground set to be analyzed.

Similar attempts, for the uniform criterion with the noise threshold set at the average similarity value, are presented in the right part of the table. The clusters demonstrate five patterns of which the lead, 4-5-6-10-11-12, embracing the right-hand half of the graph, concurs with the human view of the topology (see Luxburg 2007). After removal of this cluster, the algorithm finds remaining connected components – see the clusters presented in the right-hand column of Table 8.4. In contrast to the

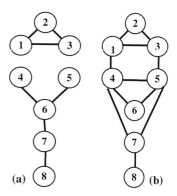

Fig. 8.3 Two graphs on a set of eight entities; that on the *left* consists of two components whereas that on the *right* has a few additional edges to make it a component

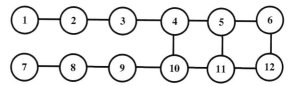

Fig. 8.4 Less than an obvious cluster structure – Cockroach network graph

Table 8.6 One cluster multiple solutions using the summary criteria at Cockroach network data in Fig. 8.4

Modularity as is and sequentially adjusted			Uniform with mean subtracted once			
Cluster	Criterion	Ground set	Cluster	Cluster	Criterion	Ground set	Cluster
5 6 11 12	5.15	1–12	5 6 11 12	4 5 6 10 11 12	8.09	1–12	4 5 6 10 11 12
1 2 3 4 5 6	4.69	1–4 7–10	1 2 3 4 10	2 3 4 5 10 11	6.09	1–3 7–9	1 2 3
3 4 7 8 9 10	4.69	7 8 9	7 8 9	4 5 8 9 10 11	6.09	7 8 9	7 8 9
				3 4 5 9 10 11	6.09		
				1 2 3 4 5 6	4.09		

8.1 One Cluster Summary Similarity with Background Subtracted

modularity criterion, the value of threshold subtracted from the data is kept the same through all the iterations because of both flat values of similarities and the thrust of the uniform criterion towards to a unified scale across the entire network.

Good clustering results found here with the uniform criterion are not easy to match with other clustering methods, which supports the view that the ordinary graphs, that is, flat networks, could be a natural niche at which the uniform criterion, with a flat value subtracted, can produce good results.

Let us turn now to clustering affinity data that are similarities between entities in an entity-to-feature table. They are usually defined by a kernel function depending on entity-to-entity distances such as a Gaussian kernel function $G(x, y) = e^{-d(x,y)/2\sigma^2}$ where $d(x,y)$ is the squared Euclidean distance between x and y if $x \neq y$ (see formula (4.12) on p. 140). The denominator $2\sigma^2$ may greatly affect results and is subject to the user's choice. In our experiences, consistent results are obtained with $2\sigma^2$ corresponding to $\sigma = 1/2$ after each feature has been normalized by its range.

One more parameter at defining the affinity data is the distance threshold, R, such that the similarity between entities is defined as 0 if the distance between them is greater than R. The usage of this parameter appears highly successful in such areas as image analysis (Shi and Malik 2000).

Worked example 8.2. Similarity clusters at affinity data

The affinity data for eight entities in Company data table (range normalized with the last three columns further divided by 3, see Section 5.1) are presented in Table 8.7. Similarity values that are greater than 0.15 are highlighted in bold – in fact, they lead to a threshold graph presented on Fig. 8.3a. The two affinity values that are at odds with the three-product cluster structure $S = \{\{1, 2, 3\}, \{4, 5, 6\}, \{7, 8\}\}$ are underlined: the absent within cluster link (1,5) and the unwanted between cluster link (6,7).

This similarity matrix after subtraction of the random interactions background is presented in Table 8.8; the positive entries are highlighted in bold and those at odds with the three-product cluster structure are underlined.

Table 8.7 Affinity similarities between eight companies in the Company dataset in Table 6.1. Those greater than 0.15 are highlighted by bold

	2	3	4	5	6	7	8
1	**0.3623**	0.1730	0.1005	0.0123	0.0111	0.0101	0.0024
2		**0.2143**	0.0447	0.0261	0.0025	0.0224	0.0080
3			0.0207	0.0989	0.0252	0.0266	0.0085
4				<u>0.1424</u>	**0.1752**	0.0880	0.0073
5					**0.2248**	<u>**0.1918**</u>	0.0236
6						0.0347	0.0011
7							**0.2982**

8 Approximate and Spectral Clustering for Network and Affinity Data

Table 8.8 Table 8.7 data after subtraction of the background of the random interactions

	2	3	4	5	6	7	8
1	0.2654	0.0922	**0.0180**	−0.0903	−0.0565	−0.0857	−0.0473
2		0.1325	−0.0389	−0.0779	−0.0660	−0.0745	−0.0424
3			−0.0490	**0.0123**	−0.0319	−0.0543	−0.0335
4				0.0540	0.1169	**0.0055**	−0.0356
5					0.1523	**0.0892**	−0.0297
6						−0.0330	−0.0341
7							0.2484

Table 8.9 One cluster and one-by-one partition structures found at the Company affinity data in Table 8.7

Modularity					Uniform, π = Mean				
One cluster		One-by-one clustering			One cluster		One-by-one clustering	
Cluster	Criterion	Ground set	Cluster		Cluster	Criterion	Ground set	Cluster
4 5 6 7 8	1.068	1−8	1 2 3		1 2 3	1.057	1−8	1 2 3
1 2 3	0.980	4−8	4 5 6		4 5 6 7 8	0.901	4−8	4 5 6 7
		7 8	7 8				4−7	4 5 6

The results of AddRem clustering for the affinity data are presented in Table 8.9. This time, both – modularity and uniform – criteria give similar results: two clusters only, with cluster of product A separated from the rest. The only difference is that the modularity criterion assigns a larger value to the combined cluster of B and C products, whereas the uniform criterion with the subtracted average affinity value gives a larger value to the cluster of A product.

Here both the uniform and modularity criteria apply to the background adjusted at the set at which the method applies. In contrast to the previous case, it is the modularity function that finds a good solution, whereas the uniform criterion cannot find the cluster of two product C companies, making each of them a singleton.

F8.1.2 One Cluster Summary Criterion and Its Properties: Formulation

Given a cluster S, its within-cluster similarities can be characterized by the summary value

$$a(S, S) = \sum_{i,j \in S} a_{ij}. \quad (8.1)$$

Obviously, the greater the sum (8.1), the better the cluster S.

Given a non-negative matrix A, the maximum of $a(S,S)$ is obviously reached at the universal cluster $S=I$, because then the sum (8.1) is the greatest possible. Provided that all rows/columns have at least one positive entry, $S=I$ is the only maximizer of (8.1). Does it mean that the summary criterion should be discarded as leading to no nontrivial clusters as is conventionally suggested?

Not at all! Just some background interrelations should be removed to help sharpening the portrait of a cluster structure hidden in the data, as illustrated in Fig. 8.2.

Two types of background data are:

(i) a constant similarity level π that has meaning of a "soft" similarity threshold (Mirkin 1987, 1996)
(ii) a "random" assignment of similarity based on the relative "strength" of entities involved (Newman and Girvan 2004, Newman 2006).

The maximum of the summary similarity criterion (8.1) applied to matrix A after a similarity shift π is subtracted will be referred to as the uniform criterion:

$$u(S, \pi) = \sum_{i,j \in S} (a_{ij} - \pi) \qquad (8.2)$$

Obviously, this criterion is the same as $b(S, S)$ (8.1) applied to matrix $B = (b_{ij})$ with $b_{ij} = a_{ij} - \pi$.

The meaning of the shift π can be derived from the criterion $b(S,S)$: pair $\{i, j\}$ should be put in cluster S if $b_{ij} > 0$, that is, $a_{ij} > \pi$, and should be not if $a_{ij} < \pi$. That means that π is a "soft" similarity threshold encouraging strong similarity in S and weak similarities out of S. The value of threshold can be defined using external information (Mirkin et al. 2010).

The background similarity in the case (ii) needs no external information. In this approach, matrix A is treated as a contingency table (Section 2.3). Consider the summary values $a_{i+} = \sum_{j \in I} a_{ij}$, and $a_{++} = \sum_{i,j \in I} a_{ij}$. Under the assumption that there is a random interaction between entities i and j, which is proportional to these summary values, the background similarity is defined as the product $k_{ij} = a_{i+}a_{j+}/a_{++}$; the denominator is added to return the product to the original scaling of similarities in A. The within-cluster summary similarity criterion (8.1) applied to matrix A after the "background" similarity is subtracted is referred to as the modularity criterion:

$$m(S) = \sum_{i,j \in S} (a_{ij} - k_{ij}) = \sum_{i,j \in S} (a_{ij} - a_{i+}a_{j+}/a_{++}) \qquad (8.3)$$

Obviously, this criterion is the same as $c(S, S)$ (8.1) applied to matrix $C = (c_{ij})$ where $c_{ij} = a_{ij} - a_{i+}a_{j+}/a_{++}$.

Let us briefly analyze some properties of these versions of the summary criterion. For the case of $u(S, \pi)$ in (8.2), let us focus on the case when the diagonal entries are not considered so that $i \neq j$ in (8.1):

$$u(S, \pi) = \sum_{\substack{i,j \in S \\ i \neq j}} a_{ij} - \pi = \sum_{\substack{i,j \in S \\ i \neq j}} a_{ij} - \pi |S|(|S| - 1) \quad (8.4)$$

where $|S|$ denotes the number of elements in S. When the diagonal elements are present, the right-hand item in (8.4) would be $\pi |S|^2$ rather than $\pi |S|(|S| - 1)$.

An irregular structure of similarities may prevent the threshold π to be a separator between all within-cluster and out-of-cluster similarities, but it certainly is on average. Indeed, let us denote the average similarity of an $i \in I$ and $S \subseteq I$ by $a(i,S)$ - this may be referred to as the uniform attraction of i to S. Obviously, $a(i, S) = \Sigma_{j \in S} a_{ij}/(|S| - 1)$ if $i \in S$, and $a(i, S) = \Sigma_{j \in S} a_{ij}/ |S|$ if $i \notin S$, because of the assumption that the diagonal similarities a_{ii} are not considered.

Let us refer to a cluster S as uniformly π-tight if, for any entity $i \in I$, its uniform attraction to S is greater than or equal to π if $i \in S$ and it is less than π, otherwise.

Then the following statement, in support of the claim that an optimal cluster S should be tight, holds.

If S maximizes criterion $u(S, \pi)$ in (8.4) then S is uniformly π-tight, that is, $a(i, S) \geq \pi$ for all $i \in S$, and $a(i, S) \leq \pi$ for all $i \notin S$.

To prove it, let us change the state of an entity i^* with respect to cluster S, that is, add i^* to S if it does not belong to S or remove it from S if it does. Now take the difference between $u(S, \pi)$ and the result of the state change, that is, $u(S - i^*, \pi)$ if $i^* \in S$, or $u(S + i^*, \pi)$ if $i^* \notin S$ where $S - i^*$ and $S + i^*$ denote S with i^* removed or added, respectively:

$$\begin{aligned} u(S, \pi) - u(S - i^*, \pi) &= 2(\Sigma_{j \in S} a_{i^*j} - \pi(|S| - 1)), \\ u(S, \pi) - u(S + i^*, \pi) &= 2(-\Sigma_{j \in S} a_{i^*j} + \pi |S|) \end{aligned} \quad (8.5)$$

Equations (8.5) are quite obvious if one consults Fig. 8.5: all the differences between $u(S, \pi)$ and its value after the change of state of i^* come from the boxed fragments of i^*th row and i^*th column. Since S is assumed to be optimal, both of the differences in (8.5) are non-negative. Take, for example, the latter: $-\Sigma_{j \in S} a_{i^*j} + \pi |S| \geq 0$ for $i^* \notin S$. Then $\pi \geq \Sigma_{j \in S} a_{i^*j}/ |S| = a(i^*, S)$ - the statement is proven for

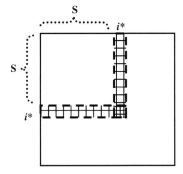

Fig. 8.5 A schematic representation of the similarities with respect to cluster S under the assumption that entities are sorted so that elements of S are followed by i^* followed by the rest; then entries related to entity i^* are in the *boxed* row and column on S's margin

8.1 One Cluster Summary Similarity with Background Subtracted

$i^* \notin S$. In the case of $i^* \in S$, take the first difference in (8.5); its being non-negative implies that $a(i^*, S) \geq \pi$ in this case, which completes the proof. In fact, a wider statement is proven. Let us refer to S as being locally optimal if, for any entity $i \in I$, $u(S, \pi)$ does not decrease under the change of i's state with respect to S. The proof warrants that any locally optimal cluster is uniformly π-tight.

A similar statement can be proven for the modularity criterion $m(S)$ in (8.3). For the sake of simplicity, assume that the diagonal entries are all zeros. Let us consider the summary similarity of S within and outside,

$$a(S) = \sum_{i \in S} \sum_{j \in I} a_{ij} = \sum_{i \in S} a_{i+}$$

and refer to it as the volume of S. In particular, an entity i's volume will be $a(i) = a_{i+}$ and the universal cluster I's volume, $a(I) = a(I, I) = a_{++}$. Then criterion (8.3) can be rewritten as

$$m(S) = a(S, S) - a(S)^2/a(I). \tag{8.3'}$$

Let us introduce the modularity attraction of an entity $i \in I$ to S, $m(i, S) = \Sigma_{i \in S} a_{ij}/a_{i+}$, and the relative volume of S in I, $v(S) = a(S)/a(I)$. Then the relative volume of an entity i would be $v(i) = a_{i+}/a_{++}$. Let us refer to a cluster S as being modularity tight if, for any entity $i \in I$, its modularity attraction to S is greater than or equal to the relative volume of S, up to a half of $v(i)$, if $i \in S$, and it is less than that, otherwise.

That is, S is modularity tight if $m(i, S) \geq a(S)/a(I) - v(i)/2$ for all $i \in S$, and $m(i, S) \leq a(S)/a(I) + v(i)/2$ for all $i \notin S$. Then the following statement is true. If S is a local maximizer of criterion $m(S)$ in (8.3) then S is modularity tight.

To prove the statement, let us take i^* and change its state with respect to S. Then the increment of criterion $m(S)$ expressed in terms of $c_{ij} = a_{ij} - a_{i+}a_{j+}/a_{++}$ will be equal to

$$m(S \pm i^*) - m(S) = \pm 2 \sum_{j \in S} c_{i^*j} + c_{i^*i^*}$$

$$= \pm 2a_{i^*+} \left(\sum_{j \in S} a_{i^*j} / a_{i^*+} - a(S)/a(I) \mp a_{i^*+}/2a(I) \right).$$

That is,

$$m(S \pm i^*) - m(S) = \pm 2a_{i^*+}(m(i^*, S) - v(S) \mp v(i^*)/2). \tag{8.6}$$

The proof follows from the fact that the increment must be non-positive at a locally optimal S.

C8.1.3 Local Algorithms for One Cluster Summary Criterion: Computation

At a preprocessed, by subtracting background similarities, similarity matrix $A = (a_{ij})$ the summary criterion is rather easy to (locally) optimize by adding entities one-by-one starting, say, from a most linked couple i and j, and at each step adding just one entity i^* – that one which is most similar to S. The computation stops when the summary similarity stops increasing, which will be the case if many of A entries are negative. There are two issues about this algorithm:

- Starting configuration – the pair of entities of the maximum similarity. This may not work in some cases such as the case of an ordinary graph matrix in which all nonzero entries are the same. Also, this choice is not flexible and may lead to a clearly suboptimal cluster and missing larger subsets whose elements are well connected but with similarity levels slightly smaller than the maximum.
- Addition with no removals. This can be of an issue because at a later stage of collecting a cluster some entities, picked up in the very beginning, can be far away from the later arrivals and should be removed at later stages.

The following algorithm AddRem tackles both of these as follows. To not get stuck in a wrong place, it runs as many times as there are entities, each time starting from another singleton $S = \{i\}$. To have an opportunity to remove a wrong element, at each step the algorithm considers the increment of the criterion caused by the change of state of every entity with respect to the current cluster. To do so, N-dimensional $1/-1$ vector $z = (z_i)$ is maintained such that $z_i = 1$ if i belongs to the current cluster and $z_i = -1$ if not. Then the change of state of $i \in I$ with respect to the cluster is equivalent to changing the sign of z_i. The change of the criterion value because of this can be expressed as follows. Denote by z the vector at current cluster S and by $z(i)$ the result of change of sign of z_i in it, so that $S(i) = S - i$ if $z_i = 1$ and $S(i) = S + i$ if $z_i = -1$. This makes the operations of addition or removal of an entity to or from the current cluster computationally similar. The increment of the summary criterion after the change is equal to

$$\Delta(i) = -2z_i \sum_{j \in S} a_{ij} + \delta a_{ii} \qquad (8.7)$$

where $\delta = 1$ if the diagonal entries are taken into account and $\delta = 0$, otherwise.

8.1.3.1 AddRem(i) Algorithm

Input: matrix $A = (a_{ij})$; output: cluster S related to entity i and value of the summary criterion.

1. *Initialization.* Set N-dimensional z to have all its entries equal to -1 except for $z_i = 1$, the summary similarity equal to δa_{ii}.

2. *General step.* For each entity $i \in I$, compute the value $\Delta(i)$ according to (8.7) and find i^* maximizing it.
3. *Test.* If $\Delta(i^*) > 0$, change the sign of z_i^* in vector z, $z_i^* \Leftarrow -z_i^*$ after which recalculate the sum by adding $\Delta(i)$ to it. (In the case of large data, computing the summary values in (8.7) can be costly. Therefore, a vector of these values should be maintained and dynamically changed after each addition/removal step.), and go to 2. Otherwise, go to 4.
4. *Output S* and the summary criterion value.

A tightness property of the resulting cluster S depending on the pre-processing step, at any starting $i \in I$, holds as established in Section 8.1.2 because S is locally optimal.

Algorithm AddRem(i) utilizes no ad hoc parameters, except for the i of course, so the cluster sizes are determined by the process of clustering itself. Multiple runs of AddRem(i) at different starting points i allow to (a) find a better cluster S maximizing the summary similarity criterion over the runs, and (b) explore the cluster structure of the dataset by analyzing both differing and overlapping clusters.

8.2 Two Cluster Case: Cut, Normalized Cut and Spectral Clustering

8.2.1 *Minimum Cut and Spectral Clustering*

P8.2.1.1 Minimum Cut and Spectral Clustering: Presentation

In this section, we turn to the issue of dividing an entity set supplied with a similarity matrix in two most isolated, or minimally connected, parts. The connection, if measured by the sum of similarities between the parts, is referred to as the cut. An illustration of such a partition is presented on Fig. 8.6 that displays the division of the similarity matrix in four blocks caused by partition of the entity set in two parts, clusters S_1 and S_2. The diagonally lined blocks pertain to the within cluster similarities and those lined vertically and horizontally, to the between cluster

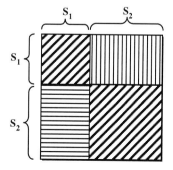

Fig. 8.6 The structure of the similarity matrix regarding partition $\{S_1, S_2\}$ of the entity set, which is assumed sorted so that elements of S_1 stand first. The blocks out of the main diagonal show similarities between S_1 and S_2, whereas those on the main diagonal refer to similarities within the parts

similarities. Since the total sum of similarities is constant, the minimum cut corresponds to the maximum sum of within-cluster similarities. This means that the minimum cut problem is akin to the problem of maximization of the within cluster summary similarity considered in the previous section. The difference is that now we are looking at splitting the set into most separated parts rather than finding just one tightly related one. A similar difference is between the concepts of a tight Anomalous Pattern cluster in Section 6.2.7 and the maximum split clusters in divisive clustering, Section 7.3.

If a similarity matrix A is conventionally non-negative, the criterion of minimum cut does not work: the optimum cut will always produce a most unbalanced partition: a singleton, that one which is least summarily related to the others, and the rest. Yet the criterion is workable if a background "noise" has been subtracted from the similarities, which brings us back to the uniform and modularity criteria. These criteria maximize the summary within cluster similarities. The former applies when a constant, typically the average similarity, is subtracted from all the similarities. The latter applies after the random interactions proportional to the products of the entity volumes, have been subtracted.

The same AddRem algorithm can be applied to the summary criterion, the only difference being that this time within cluster similarities for both clusters are summed up. Yet in this section we turn to the eigenvector, or spectral, perspective that is implied by a reformulation of the problem in terms of the Rayleigh quotient. As proven in Section 8.2.1.2 below, the minimum cut problem is equivalent to finding such $(1,-1)$ vector z that maximizes the quotient $\lambda = z^T A z / z^T z$. When relaxed to arbitrary z's, the problem is known to be of finding the maximum eigenvalue λ and corresponding eigenvector z of matrix A; this eigenvector will be referred to as the first eigenvector. Therefore, it is only natural to cluster entities according to the signs of the eigenvector: those i's with positive components go to S_1 while i's with negative components go to S_2. Although not necessarily an optimal partition, this is a practical and, in most cases, good solution.

Worked example 8.3. Spectral clusters for Confusion dataset

Table 8.10 presents results of sequential cuts according to the first eigenvectors on the sets resulting from the previous cuts. As before, the modularity transformation leads to three discernible clusters of numerals, {1, 4, 7}, {6, 8, 0} and {2, 3, 5, 9}. The uniform data transformation, by subtracting the mean of the similarities on the current set, additionally separates 2 from cluster {2, 3, 5, 9} – the remaining set {3,5,9} cannot be further divided because the maximum eigenvalue at that is negative, thus no positive value of the criterion at the division.

Worked example 8.4. Spectral clusters for Cockroach network

Table 8.11 presents results of the first two cuts according to the spectral clusters derived at the modularity and uniform data transformations. They differ on nodes 4

8.2 Two Cluster Case: Cut, Normalized Cut and Spectral Clustering

Table 8.10 First eigenvectors according to the modularity and uniform data preprocessing options

	Modularity			Uniform, current mean subtracted			
Set	0 – 9	2 3 5 6 8 9 0	2 3 5 9	0 – 9	2 3 5 6 8 9 0	2 3 5 9	3 5 9
1	−0.57			0.52			
2	0.08	0.06	0.5	−0.08	0.07	0.46	
3	0.07	0.51	0.5	−0.12	0.50	−0.50	0.74
4	−0.27			0.25			
5	0.21	0.22	0.5	−0.24	0.22	−0.29	−0.58
6	0.26	−0.36		−0.29	−0.36		
7	−0.53			0.49			
8	0.34	−0.43		−0.38	−0.44		
9	0.19	0.43	0.5	−0.25	0.43	−0.68	0.35
0	0.21	−0.43		−0.23	−0.44		
λ.	703.7	189.6	0.0	358.7	189.6	83.8	−46.4

Table 8.11 First eigenvectors according to the modularity and uniform data preprocessing options at Cockroach network: two cuts

	Modularity		Uniform, current mean subtracted	
Set	1–12	1–4 7–10	1–12	1–3 7–9
1	0.21	−0.30	−0.22	0.3536
2	0.32	−0.46	−0.26	0.5000
3	0.24	−0.41	−0.10	0.3536
4	0.00	−0.16	0.22	
5	−0.37		0.46	
6	−0.40		0.34	
7	0.21	0.30	−0.22	−0.3536
8	0.32	0.46	−0.26	−0.5000
9	0.24	0.41	−0.10	−0.3536
10	0.00	0.16	0.22	
11	−0.37		0.46	
12	−0.40		0.34	
λ.	1.71	1.53	1.88	1.41

and 10 at which they are ether merged with the thicker end of the network, at the uniform clustering, or not, – at the modularity clustering. At the second cut, they go to different parts, according to the network topology on Fig. 8.4.

Q.8.1. Consider an agglomerative clustering algorithm, in which the similarity between clusters is taken to be the sum of between cluster similarities. Prove that

(a) this algorithm maximizes the summary within cluster similarity criterion;
(b) the algorithm stops when all between cluster summary similarities are negative (which will happen if the similarity matrix has been preprocessed with either modularity or uniform transformation).

Worked example 8.5. Spectral clustering of affinity data

The spectral clustering approach is much successful on the affinity data for the Company dataset – the three clusters corresponding to the three products are recovered well on both data transformation options, the uniform and the modularity (see Table 8.12). The uniform version does not divide the B product cluster {4, 5, 6} in smaller parts because all components of the first eigenvector here have the same sign.

F8.2.1.2 Minimum Cut and Spectral Clustering: Formulation

Given a symmetric similarity matrix $A = (a_{ij})$ on set I, consider the issue of dividing I in two parts, S_1 and S_2, in such a way that the similarity between S_1 and S_2 is minimum while it is maximum within them. This requirement can be explicated most naturally by using the summary similarity criterion. Using indices f, $g = 1, 2$, let us denote the summary similarity "between" S_f and S_g by $a(S_f, S_g)$ so that $a(S_f, S_g) = \sum_{i \in S_f} \sum_{j \in S_g} a_{ij}$. Then, obviously, $a(S_1, S_1)$ is the summary similarity within S_1 and $a(S_1, S_2) = a(S_2, S_1)$ is the summary similarity between S_1 and S_2; the equation follows from the symmetry of A. Moreover, the sum $a(S_1, S_1) + a(S_2, S_2) + a(S_1, S_2) + a(S_2, S_1)$ is equal to the constant sum $a(I)$ of all the similarities, as Fig. 8.6 clearly demonstrates. The common value $a(S_1, S_2)$ of the between cluster similarity is referred to as the cut. Then a natural clustering criterion, the minimum cut, corresponds to the maximum of the summary within cluster similarity

$$aw(S_1, S_2) = a(S_1, S_1) + a(S_2, S_2),$$

because $aw(S_1, S_2) = a(I) - 2a(S_1, S_2)$, which shows that the minimum between-cluster summary similarity simultaneously provides for the maximum within cluster summary similarity. Yet the criterion of minimum cut usually is not considered

Table 8.12 First eigenvectors according to the modularity and uniform data preprocessing options at company affinity data set

Set	Modularity			Uniform, current mean subtracted		
	1–8	4–8	4–6	1–8	4–8	4–6
1	0.50			0.53		
2	0.51			0.55		
3	0.32			0.35		
4	−0.13	0.35	0.58	−0.09	0.36	−0.40
5	−0.29	0.25	0.58	−0.23	0.27	−0.62
6	−0.24	0.48	0.58	−0.20	0.49	−0.68
7	−0.38	−0.45		−0.33	−0.43	
8	−0.30	−0.62		−0.29	−0.61	
λ	0.41	0.21	0.00	0.41	0.21	0.01

appropriate for clustering because, at a nonnegative A, it obviously reaches the minimum when the out-of-diagonal blocks are reduced to just mere one line and column, independently of the structure of the similarities, leading thus to a most unbalanced partition: a singleton and the rest, which is not what should be considered a proper aggregation. Yet with pre-processing of the similarities by subtracting either a constant threshold or the random interactions as described in Section 8.1, the structure of the similarity matrix becomes identifiable with the minimum cut criterion. Unfortunately, when A-entries can be both positive and negative, the problem of minimum cut becomes computationally intensive, referred to as NP-complete in the theory of combinatorial optimization (see, for example, Johnsonbaugh and Schaefer 2004). This implies that local or approximate algorithms would be a welcome development for the problem.

One of such algorithms is AddRem from Section 8.1 because collecting a cluster is equivalent to splitting the set in two parts, the cluster and the rest, if the criterion of maximum summary within cluster similarity is extended to cover both clusters. Of course, the operation of addition-removal loses its one-cluster asymmetry and becomes just operation of exchange between the two clusters.

Another approach comes from the spectral theory on matrices, which is devoted to the analysis and computation of eigenvalues and corresponding eigenvectors for square matrices. Indeed, define N-dimensional vector $z = (z_i)$ such that $z_i = 1$ if $i \in S_1$ and $z_i = -1$ if $i \in S_2$. Obviously, $z_i^2 = 1$ for any $i \in I$ so that $z^T z = N$ which is constant at any given entity set I. On the other hand, $z^T A z = a(S_1, S_1) + a(S_2, S_2) - 2a(S_1, S_2) = 2(a(S_1, S_1) + a(S_2, S_2)) - a(I)$, which means that the summary criterion is maximized when $z^T A z$ is maximized, that is, the problem of finding a minimum cut is equivalent to the problem of maximization of Rayleigh quotient

$$g(z) = \frac{z^T W z}{z^T z} \qquad (8.8)$$

with respect to the unknown N-dimensional z whose components are either 1 or -1. Matrix W is A pre-processed into either B, with subtraction of a threshold, or C, with subtraction of the random interactions (see Section 8.1 for more detail) or using a different transformation.

As is well known, the maximum of (8.8) with respect to arbitrary z is equal to the maximum eigenvalue of W and it is reached at the corresponding eigenvector referred to as the first eigenvector. This brings forth the idea that is referred to as spectral clustering: Find the first eigenvector as the best solution and then approximate it with a $(1, -1)$-vector by putting 1 for positive components and -1 for non-positive components – then produce S_1 as the set of entities corresponding to 1, and S_2, corresponding to -1.

C8.2.1.3 Spectral Clustering for the Minimum Cut Problem: Computation

To find the maximum eigenvalue and corresponding eigenvector for a symmetric similarity matrix W, MatLab command [Z,L]=eig(W) should be executed.

Resulting L is a diagonal matrix with eigenvalues located on the diagonal in the ascending order, so that the last one is the maximum eigenvalue. Accordingly, the last column is the corresponding, "first", normed eigenvector. Its positive components correspond to one cluster, and the non-positive components to the other. Here is a sequence of commands to determine the split parts S_1 and S_2:

```
>> [Z,L]=eig(W);
>> [n,n]=size(L);
>> z=Z(:,n);
>> S{1}=find(z>0); S{2}=find(z<=0);
```

If W is non-negative then the first eigenvector is proven to be not negative either – no partition of I can emerge in such a situation.

8.2.2 Normalized Cut and Laplace Transformation

P8.2.2.1 Normalized Cut: Presentation

The concept of normalized cut is a relatively recent development started by Shi and Malik 2000. It belongs to a series of graph cutting criteria that balance the cluster sizes by normalizing the sums of within cluster similarities by the size-dependent values. Given a similarity matrix A, let us take the volume of $S \subseteq I$, the summary similarity in rows $i \in S$, $a(S) = \Sigma_{i \in S} a_{i+}$. Then the normalized cut over a partition $\{S_1, S_2\}$ is defined as $a(S_1, S_2)/a(S_1) + a(S_1, S_2)/a(S_2)$. This is to be minimized over all splits $\{S_1, S_2\}$ of set I. An equivalent criterion would maximize the sum of normalized within cluster similarities, $a(S_1, S_1)/a(S_1) + a(S_1) + a(S_2, S_2)/a(S_2)$ – the two criteria sum to 2.

Yet the normalized cut brought forward a less intuitive type of data preprocessing, the Laplace transformation of a similarity matrix, W, which may be in its original format A or with a constant threshold subtracted, B, or with the random interaction subtracted, C, into its Laplacian, L. In its normalized form, this transformation normalizes every similarity w_{ij} by dividing it by the square root of the product of i and j volumes, $w_{ij}/\sqrt{w_{i+}w_{i+}}$, and then subtracts the resulting matrix from the identity matrix which has all its entries zero except for unities on the diagonal (see the lower triangle matrix in Table 8.13). With this transformation we are into the realm of spectral clustering again. The Laplacian matrix is proven to have all the eigenvalues non-negative. Besides, L has a specific minimum eigenvalue – the zero. Yet the next minimum eigenvalue and the corresponding eigenvector provide for a relaxation of the minimum of normalized cut problem reformulated in terms of the Rayleigh quotient for the Laplacian matrix (see in part F of this section). Then split $\{S_1, S_2\}$ can be found by using this second minimum eigenvector in the same way as in the previous section: S_1 is defined by indices of all positive components and S_2 of all negative components.

8.2 Two Cluster Case: Cut, Normalized Cut and Spectral Clustering

Table 8.13 Affinity similarities between eight companies in the company data in Table 6.1 (upper triangle) and the result of the normalized Laplace transformation (lower triangle)

	1	2	3	4	5	6	7	8
1	1	**0.3623**	**0.1730**	0.1005	0.0123	0.0111	0.0101	0.0024
2	−0.5360	1	**0.2143**	0.0447	0.0261	0.0025	0.0224	0.0080
3	−0.2803	−0.3450	1	0.0207	0.0989	0.0252	0.0266	0.0085
4	−0.1611	−0.0712	−0.0361	1	0.1424	**0.1752**	0.0880	0.0073
5	−0.0177	−0.0372	−0.1548	−0.2207	1	**0.2248**	**0.1918**	0.0236
6	−0.0196	−0.0044	−0.0486	−0.3343	−0.3845	1	0.0347	0.0011
7	−0.0150	−0.0332	−0.0431	−0.1411	−0.2759	−0.0614	1	**0.2982**
8	0.0050	−0.0164	−0.0191	−0.0162	−0.0471	−0.0026	−0.6158	1

Worked example 8.6. Normalized cut for Company data: Laplacian and Lapin matrices

To show how this works, consider the affinity data for Company data set and its Laplacian matrix in Table 8.13. The minimum eigenvalue of the Laplacian matrix is 0 whereas the second minimum eigenvalue is 0.32. The eigenvector corresponding to the latter, as expected, well separates the first three entities, A-product companies.

Although the result is natural, there is no way to see it from the Laplacian matrix by itself. Unlike the original affinity matrix, the visible structure of the Laplacian gives no useful indications on the cluster structure underlying its entries. To make the structure visible, the Laplacian should be further transformed. The Laplacian Pseudo Inverse (Lapin, for short) transformation takes the spectral decomposition of the Laplacian, inverses the non-zero eigenvalues λ into $1/\lambda$, and returns a pseudo-inverse Laplacian which is presented in the lower triangle of Table 8.14.

One can easily see that the cluster structure is more pronounced in the Lapin matrix than it is in the original affinity matrix. First, there is no need for guessing a right threshold value to subtract: it is 0 here. Second, there is only one not-fitting entry here, (5,7), but it would not make any difference anyway because it is rather

Table 8.14 Affinity similarities between eight companies as in Table 8.13 (upper triangle; entries larger than 0.15 are highlighted in bold; those not fitting in the structure underlined) and the result of Lapin transformation (lower triangle, positive entries highlighted in bold, that not fitting underlined)

	1	2	3	4	5	6	7	8
1		**0.3623**	**0.1730**	0.1005	0.0123	0.0111	0.0101	0.0024
2	**0.4734**		**0.2143**	0.0447	0.0261	0.0025	0.0224	0.0080
3	**0.2221**	**0.2677**		0.0207	0.0989	0.0252	0.0266	0.0085
4	−0.2073	−0.2719	−0.2405		0.1424	**0.1752**	0.0880	0.0073
5	−0.4281	−0.4314	−0.2577	**0.0190**		**0.2248**	**0.1918**	0.0236
6	−0.3534	−0.3839	−0.2638	**0.1563**	**0.1984**		0.0347	0.0011
7	−0.5430	−0.5337	−0.4076	−0.1213	**0.0457**	−0.1020		**0.2982**
8	−0.4440	−0.4316	−0.3385	−0.1650	−0.0504	−0.1478	**0.6003**	

Table 8.15 Reciprocal non-zero eigenvalues of the Laplacian and Lapin matrices corresponding to the same eigenvectors

Eigenvalue labels	I	II	III	IV	V	VI	VII
Normalized Laplacian	0.32	0.59	1.11	1.35	1.40	1.55	1.67
Lapin	3.08	1.70	0.90	0.74	0.71	0.65	0.60

small in comparison with the other negative entries (6,7) and (4,7) linking item 7 to B-product cluster, or entry (5,8) linking item 5 to C-product cluster.

One more result of the Lapin transformation is that the eigenvalue to look for is the maximum one, and it is much better separated from the rest because of the inversion (see Table 8.15). The corresponding eigenvector does not change.

Indeed the three product based clusters, {1,2,3}, {4,5,6}, {7,8}, are found with both the summary clustering criterion and spectral approach applied to the Lapin transformed Company affinity data.

Worked example 8.8. Failure of spectral clustering at Cockroach network

Lapin matrix for Cockroach network Fig. 8.4 is presented in Table 8.16. It manifests a rather clear cut cluster structure embracing three clusters, {1,2,3,4}, {7, 8, 9, 10}, {5,6, 11, 12}. Indeed, the positive entries are those within the clusters, except for two positive – but rather small – entries, at (4,5) and (10,11). Yet the first eigenvector reflects none of that; it cuts through by separating six nodes {1,2,3,4,5,6} (negative components) from the rest (positive components). This is an example of a situation in which the spectral approach fails: the normalized cut criterion at the partition separating the first 6 nodes from the other 6 nodes is equal to 0.46, whereas its value at cluster {5, 6, 11, 12} cut from the rest is 0.32. The same value of the criterion,

Table 8.16 Lapin similarity data between nodes of the cockroach network in Fig. 8.3; the positive entries are highlighted in bold

	2	3	4	5	6	7	8	9	10	11	12
1	2.43	1.18	0.05	−0.52	−0.58	−0.69	−0.92	−0.75	−0.59	−0.69	−0.63
2		1.75	0.16	−0.64	−0.74	−0.92	−1.22	−0.99	−0.74	−0.88	−0.81
3			0.44	−0.36	−0.51	−0.75	−0.99	−0.76	−0.46	−0.60	−0.58
4				0.14	−0.15	−0.59	−0.74	−0.46	0.02	−0.16	−0.24
5					0.68	−0.69	−0.88	−0.60	−0.16	**0.46**	**0.44**
6						−0.63	−0.81	−0.58	−0.24	**0.44**	**0.94**
7							2.43	1.18	0.05	−0.52	−0.58
8								1.75	0.16	−0.64	−0.74
9									0.44	−0.36	−0.51
10										0.14	−0.15
11											0.68

0.32, is attained at cluster {4,5,6,10,11,12} cut from the rest. These two cuts are optimal according to the criterion, and the spectral cut is not.

Case Study 8.3. Circular Cluster Exposed by Lapin Transformation

To further demonstrate the formidable ability of the Lapin transformation in manifesting clusters according to human intuition, let us consider the 2D set presented on Fig. 8.7.

This set has been generated as follows. Three 100×2 data matrices, $a1$, $a2$ and $a3$, were generated from Gaussian distribution $N(0,1)$. Then matrix $a2$ was normed row-wise into b, so that each row in b is a 2D normed vector, after which matrix c has been defined as $c = 0.5*a3 + 8*b$. Its rows form a ring-wise shape, while rows of $a1$ fall into a heap in the circle's center as presented on Fig. 8.7. Then $a1$ and c are merged into a 200×2 matrix X, in which $a1$ takes the first 100 rows and c the next 100 rows.

The conventional data standardization methods would not change the picture, and conventional clustering procedures like K-Means clustering would not be able to separate the ring as a whole. The single link clustering will be able to separate these two clusters, which would once again reminds as of a rift between the data approximation clustering and graph theoretic approaches. Yet the Laplace transformation allows us to put this dataset into the data approximation context too.

The data is first transformed into a 200×200 affinity similarity matrix, which is then Lapin transformed into a final similarity matrix. This final matrix shows a clear-cut pattern: all similarities between the first hundred and the second hundred of points are negative whereas all the Lapin similarities within these sets are positive.

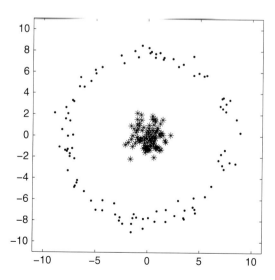

Fig. 8.7 Two intuitively obvious clusters that are difficult for conventional approaches: *stars* in the heap and *dots* in the ring

Table 8.17 Five points from Fig. 8.7, on the left, the affinity similarities between them, in the middle, and Lapin similarities, on the right

	x−axis	y−axis	2	3	4	5	2	3	4	5
1	−1.1465	0.3274	0.63	0.00	0.02	0.01	**0.04**	−0.12	−0.14	−0.10
2	0.8956	0.5529		0.00	0.00	0.00		−0.12	−0.15	−0.11
3	0.3086	7.9059			0.00	0.03			**0.16**	**0.40**
4	−7.1827	0.0625				0.01				**0.46**
5	−5.0025	5.8504								

Such a structure clearly separates the two clusters with any reasonable algorithm, the summary criterion based AddRem and the spectral approach included.

Take a look, for example, at a randomly selected 5×2 fragment from matrix X concerning 2 rows from $a1$ and 3 rows from c in the left-hand part of Table 8.17. It is not easy to cluster points 3, 4, 5 together because of the great distances between them.

This is reflected in the Gaussian affinity matrix A between 200 rows of the data matrix too, which is defined as described in Section 8.1 according to formula $a_{ij} = \exp(-d(x_i, x_j)/s)$ where $d(x,y) = \sum_{v \in V}(x_v - y_v)^2$ is the squared Euclidean distance between vectors x and y. The value of s relates to the denominator of exponent $2\sigma^2$ in the definition of the Gaussian density so that if one takes σ to be half of the range, then s should be about the same, which leads to $s = 9$ in this case. The part of affinity matrix A related to the set of five points is presented in the middle of Table 8.17. One can see indeed a high affinity value between the first two entities, which belong to the heap in the middle of Fig. 8.7 and are close to each other indeed, while the other similarities are close to zero – no visible structure. A similar pattern can be seen on the Laplacian except that all non-diagonal entries are negative there because of the definition. After the Lapin transformation, however, the similarity structure, once again, becomes clear-cut, as shown on the right part of Table 8.17 for the 5-point subset, and in fact is true for the entire dataset.

This ability of Lapin transformation in transforming elongated structures into convex clusters has been a subject of mathematical scrutiny. An analogy with electricity circuits has been found. Roughly speaking, if w_{ij} measures the conductivity of the wire between nodes i and j in a "linear electricity network", then the corresponding element of a Lapin matrix expresses the "effective resistance" between i and j in the circuit (Klein and Randic 1993). Yet there can be cases of elongated structures, as shown in Worked example 8.8, at which Lapin transformation does not work at all.

F8.2.2.2 Partition Criteria and Spectral Clustering: Formulation

Given a symmetric similarity matrix $A = (a_{ij})$ on set I, consider the issue of dividing I in two parts, S_1 and S_2, in such a way that the similarity between S_1 and S_2 is

8.2 Two Cluster Case: Cut, Normalized Cut and Spectral Clustering

minimum while it is maximum within the parts. Denote by $a(S_f, S_g)$ the summary similarity "between" S_f and S_g so that $a(S_f, S_g) = \sum_{i \in S_f} \sum_{j \in S_g} a_{ij}$.

The normalized cut utilizes the summary similarities $a_{i+} = a(i, I)$. Denote

$$a(S_k) = \sum_{i \in S_k} a_{i+}$$

Obviously, $a(S_1) = a(S_1, S_1) + a(S_1, S_2)$; a similar equation holds for $a(S_2)$. The normalized cut is defined as

$$nc(S) = \frac{a(S_1, S_2)}{a(S_1)} + \frac{a(S_2, S_1)}{a(S_2)} \quad (8.9)$$

to be minimized.

It should be noted that minimized cut (8.9), in fact, includes the requirement of maximization of the within-cluster similarities. Indeed consider the normalized within-cluster similarity

$$nt(S) = \frac{a(S_1, S_1)}{a(S_1)} + \frac{a(S_2, S_2)}{a(S_2)}, \quad (8.10)$$

scoring the tightness of clusters. These two measures are highly related: $nc(S) + nt(S) = 2$ (see Q.8.2). This latter equation warrants that minimizing the normalized cut simultaneously maximizes the normalized tightness.

It appears, the criterion of minimizing $nc(S)$ can be expressed in terms of a corresponding Rayleigh quotient – for the so-called Laplacian. Given a (pre-processed) similarity matrix $W = (w_{ij})$, let us denote its row sums, as usual, by $w_{i+} = \sum_{j \in I} w_{ij}$ ($i \in I$) and introduce diagonal matrix D in which all entries are zero except for diagonal elements (i,i) that hold w_{i+} for each $i \in I$. The so-called (normalized) Laplacian is defined as $L = E - D^{-1/2}WD^{-1/2}$ where E is identity matrix and $D^{-1/2}$ is a diagonal matrix with (i,i)-th entry equal to $1/\sqrt{w_{i+}}$. That means that L's (i,j)-th entry is $\delta_{ij} - w_{ij}/\sqrt{w_{i+}w_{j+}}$ where δ_{ij} is 1 if $i = j$ and 0, otherwise. It is not difficult to prove that $Lf_0 = 0$ where $f_0 = (\sqrt{w_{i+}}) = D^{1/2}I_N$ where I_N is N-dimensional vector whose all entries are unity. That means that 0 is an eigenvalue of L with f_0 being the corresponding eigenvector.

Moreover, it is possible to prove that for any N-dimensional f, the following equation holds:

$$f^T L f = \frac{1}{2} \sum_{i,j \in I} w_{ij} \left(\frac{f_i}{\sqrt{w_{i+}}} - \frac{f_j}{\sqrt{w_{j+}}} \right)^2 \quad (8.11)$$

This equation proves that matrix L is semipositive definite, which means that product $f^T L F$ is not negative for any vector f. It is proven in matrix theory that any semipositive definite matrix has all its eigenvalues non-negative so that 0 is the minimum eigenvalue.

Given a partition $S = \{S_1, S_2\}$ of I, let us define vector s by condition $s_i = \sqrt{w_{i+}w(S_2)/w(S_1)}$ for $i \in S_1$ and $s_i = -\sqrt{w_{i+}w(S_1)/w(S_2)}$ for $i \in S_2$. Obviously, the squared norm of this vector is constant, $\sum_{i \in I} s_i^2 = w(S_2) + w(S_1) = w_{++}$. Moreover, s is orthogonal to the trivial eigenvector $f_0 = D^{1/2}I_N$ of L. Indeed, the product of i-th components of these vectors has w_{i+} as its factor multiplied by a value which is constant within clusters. Then summation of these components over S_1 will produce $w(S_1)\sqrt{w(S_2)/w(S_1)} = \sqrt{w(S_1)w(S_2)}$, and over S_2, $-w(S_2)\sqrt{w(S_1)/w(S_2)} = -\sqrt{w(S_1)w(S_2)}$. These two sum to 0, which proves the statement.

It remains to prove that minimization of (8.9) is equivalent to minimization of $s^T L s / s^T s$ for thus defined s. Indeed, at $f = s$, the squared item in (8.11) is equal to 0 for i,j from the same set S_1 or S_2. When i and j belong to different classes of S, the squared item is equal to $w(S_1)/w(S_2) + w(S_2)/a(S_1) + 2 = [w_{++} - w(S_2)]/w(S_2) + [w_{++} - w(S_1)]/w(S_1) + 2 = w_{++}/w(S_2) + w_{++}/w(S_1)$. That means that $s^T L s = w_{++}nc(S)$ that is, $nc(S) = s^T L s / s^T s$ indeed.

We have proven that the normalized cut minimizes the Rayleigh quotient for Laplacian matrix L over specially defined vectors s that are orthogonal to the eigenvector $f_0 = (w_{i+}^{1/2})$ corresponding to the minimum eigenvalue 0 of L.

Therefore, one may consider the problem of finding the minimum non-zero eigenvalue for L along with the corresponding eigenvector as a proper relaxation of the normalized cut problem. That means that the spectral clustering approach in this case would be to grab that eigenvector and approximate it with an s-like binary vector. The simplest way to do that would be by putting all plus components to S_1 and all negative to S_2.

It remains to define the pseudo-inverse Laplacian transformation, Lapin for short, for a symmetric matrix W. Consider all non-zero eigenvalues $\lambda_1, \lambda_2, \ldots, \lambda_r$ of matrix L and corresponding eigenvectors $f_1, f_2, \ldots, f_r$. The following spectral decomposition equation holds:

$$L = \lambda_1 f_1 f_1^T + \lambda_2 f_2 f_2^T + \ldots + \lambda_r f_r f_r^T \qquad (8.12)$$

The pseudo-inverse is defined by leaving the same eigenvectors but reversing the eigenvalues, which causes no problems since they are all non-zero:

$$L^+ = \frac{1}{\lambda_1} f_1 f_1^T + \frac{1}{\lambda_2} f_2 f_2^T + \ldots + \frac{1}{\lambda_r} f_r f_r^T \qquad (8.13)$$

Q.8.2. Prove that $nc(S) + nt(S) = 2$ where the constituents are defined by Equations (8.9) and (8.10).

Q.8.3. Prove that a one cluster extension of the normalized cut criterion, maximize $ng(S) = a(S,S)/a(S)$, does not work at nonnegative similarity data because the maximum is always reached at the universal cluster $S = I$. **A.** Indeed, $ng(I) = 1$, whereas $ng(S) < 1$ at all other S unless there are all zeros outside of $A(S,S)$.

Q.8.4. What's wrong with the idea of expressing the summary similarity criterion as a Rayleigh quotient?

C8.2.2.3 Pseudo-Inverse Laplacian: Computation

Given a non-negative matrix W with none of its rows summing to 0, its Laplacian can be found with these MatLab commands:
```
>> W=(W+W')/2; % to warrant the symmetry
>> wr=sum(W);
>> D=diag(wr);
>> D=sqrt(D);
>> Di=inv(D);
>> L=eye(size(W)) – Di*W*Di;
```

Then the pseudo-inverse transformation can work like this.

```
>> L=(L+L')/2;
>>[Z,M]=eig(L);
>>ee=diag(M);
>>ind=find(ee~=0); % indices of non-zero eigenvalues;
>>Zn=Z(ind,ind);
>>Mn=M(ind,ind);
>>Mi=inv(Mn);
>>Lapin=Zn*Mi*Zn';
```

8.3 Additive Clusters

P8.3.1 Decomposing a Similarity Matrix over Clusters: Presentation

The idea behind additive clustering is this. Since the raw data are similarities measuring relations between entities, let us decode a cluster in the same relational format. That is, let us make a cluster S to assign every two entities, i and j, a similarity value: say unity if they belong to the cluster or 0 if at least one of them does not. This cluster similarity matrix s plays the role of a dummy variable – in the format of a similarity matrix. Consider, for example, subset $S = \{1,3,4\}$ of $I = \{1,2,3,4,5,6\}$: its corresponding matrices s, $2s$, and $2s\text{-}1$ are in Table 8.18.

Therefore, it is reasonable to think that a similarity matrix may reflect a number of attribute-based similarity matrices possibly taken with different weights. Consider, for example, the matrix of similarities between first five amino acids in

Table 8.18 Binary matrices for cluster $S = \{1, 3, 4\}$ in a 6-element set.

	Matrix s						Matrix $2s$						Matrix $2s-1$					
	1	2	3	4	5	6	1	2	3	4	5	6	1	2	3	4	5	6
1	1	0	1	1	0	0	2	0	2	2	0	0	1	-1	1	1	-1	-1
2	0	1	0	0	0	0	0	2	0	0	0	0	-1	1	-1	-1	-1	-1
3	1	0	1	1	0	0	2	0	2	2	0	0	1	-1	1	1	-1	-1
4	1	0	1	1	0	0	2	0	2	2	0	0	1	-1	1	1	-1	-1
5	0	0	0	0	1	0	0	0	0	0	2	0	-1	-1	-1	-1	1	-1
6	0	0	0	0	0	1	0	0	0	0	0	2	-1	-1	-1	-1	-1	1

Table 1.8 (B is omitted from the list because it is synonymous to D, see Table 1.9) as presented in Table 8.19; that on the right is obtained by subtracting the minimum, -4, from all entries to make it non-negative.

Table 8.20 presents similarity matrices between these amino-acids according to attributes from Table 8.23 on p. 346. The attributes reflect popular molecular properties of amino acids related to their size (Small or not), electricity charge (Polar or not) and the propensity to keep inside of the molecules (Hydrophobic or not). That on the right represents a weighted sum of the three with an added intercept to mimic the matrix of similarities (Table 8.19).

The two similarity matrices are compared in Table 8.21. Overall, the result does not look too bad: there are only two significant differences, in similarities between amino acids A and E, and C and D, that probably require taking into account more attributes. If we go for regression of the observed similarity over attribute-based

Table 8.19 Part of matrix Table 1.8 related to amino acids A, C, D, E, F: original on the left, rearranged in the middle, and with 4 added to all entries on the right

	A	C	D	E	F		C	D	E	F		C	D	E	F
A	4	0	-2	-1	-2		0	-2	-1	-2		4	2	3	2
C	0	9	-3	-4	-2			-3	-4	-2			1	0	2
D	-2	-3	6	2	-3				2	-3				6	1
E	-1	-4	2	5	-3					-3					1
F	-2	-2	-3	-3	6										

Table 8.20 Similarity matrices between five amino acids according to attributes Small, Polar and Hydrophobic from Table 8.23 on p. 346

				Sm:				Po:				Hy:				
	Sm	Po	Hy	C	D	E	F	C	D	E	F	C	D	E	F	2Sm+6Po+2Hy+1
A	+			1	1	0	0	0	0	0	0	0	0	0	0	3 3 1 1
C	+		+		1	0	0		0	0	0		0	0	1	3 1 2
D	+	+				0	0			1	0			0	0	6 1
E		+					0				0				0	1
F			+													

8.3 Additive Clusters

Table 8.21 Comparison of two similarity matrices between five amino acids, one taken from observations (Table 8.19), and the other additively composed using attribute clusters (Table 8.20)

	BLOSUM62				2Sm+6Po+2Hy+1				Difference			
	C	D	E	F	C	D	E	F	C	D	E	F
A	4	2	3	2	3	3	1	1	1	−1	2	0
C		1	0	2		3	1	2		−2	−1	0
D			6	1			6	1			0	0
E				1				1				0

similarity we could get slightly better results. This idea is pursued in Project 8.1 on the whole set of amino acids.

There are situations, though, in which the user prefers to find clusters underlying the observed similarities, according to the additive model, by the matrix itself, without much bothering of trying to obtain related attributes. This is the realm of additive clustering model analyzed further in Section 8.3.2. This model can be considered as an extension of the spectral decomposition of similarity matrices to the case when the vectors to be found are constrained to be 1/0 binary. Assuming the conventional least-squares criterion for this specification of the summarization problem, a natural idea coming to mind is to mimic the one-by-one approach of the Principal component analysis. The other idea, just working on all clusters in parallel, is not considered in this text.

Yet even at the restricted, one cluster, model, there can be a number of different approaches to minimizing the least-squares criterion or, equivalently, maximizing the Rayleigh quotient. Those two, tried at the maximum tightness criteria in Sections 8.1 and 8.2, should be considered first:

(i) **Spectrum of similarity matrix**
   Let us drop the constraint of vectors being binary and find the optimal solution among arbitrary vectors, that is, the maximum eigenvalue and corresponding eigenvector, and then adjust somehow its components to the zero-one setting. It seems reasonable that the larger components of the eigenvector are to be changed for unity while those smaller ones are changed for zero. If true, this would drastically reduce computation.

(ii) **Hill-climb clustering**
   The strategy of finding a cluster by adding/removing entities in a best possible way implemented, for the summary similarity criterion, in Section 8.1 can be applied here too. At least, it leads to provably tight clusters. This is the strategy pursued further in this text with AddRem algorithm formulated in Section 8.3.3.

The one-cluster model assumes, rather boldly, that all observed similarities can be explained by a summary action of just two constant-level causes and noise:

(i) general associations between all entities at a constant level;
(ii) specific associations between members of a hidden cluster, also on a constant level, though not necessarily the same as the general one.

This is a much simplified model, but it brings in a nice clustering criterion to implement the least-squares approach: the underlying cluster S must maximize the product of the average within-cluster similarity $a(S)$ and the number of elements in S, $|S|$: $g(S) = |S| a(S)$. The greater the within-cluster similarity, the better, and the larger the cluster, the better too. These two objectives do not necessarily go along. In fact, they are at odds in most cases: the greater the number of elements in in a cluster, the smaller the within-cluster similarities are. That is, criterion $g(S)$ is a compromise between the two. When S is small, an increase in its size would dominate the unavoidable fall in similarities. But later in the addition process, when S becomes larger, the relative size change diminishes and cannot dominate the fall in within-cluster similarities – the process of generating S stops. This can be put, in terms of the attraction function, as follows: the cluster S found using algorithm AddRemA has all its elements positively attractive, whereas each entity outside S is negatively attracted to S. The attraction of entity i to S is defined as its average similarity to S minus half the within cluster average, $a(S)/2$.

The pre-specified level of between-entity associations (i) is captured by using the concept of similarity shift illustrated on Fig. 8.2 above.

Worked example 8.9. Additive clusters at Confusion dataset

Consider the symmetrised Confusion data set in Table 8.2 and apply algorithm AddRemA at different levels of similarity shift starting at different entities (Table 8.22).

Table 8.22 Non-singleton clusters at symmetrised, no diagonal, Confusion matrix found at different similarity shift values; the average out-of-diagonal similarity value is Av = 33.46

Similarity shift	Cluster lists						Intensity	Contribution
0	(i)	2	3	5	8	10	45.67	37.14
	(ii)	1	4	7			91.83	21.46
Av/2 = 16.72	(i)	1	4	7			75.11	21.11
	(ii)	3	5	9			71.94	19.37
	(iii)	6	8	0			71.44	19.10
Av = 33.46	(i)	1	7				131.04	25.42
	(ii)	3	5	9			55.21	13.54
	(iii)	6	8	0			54.71	13.29
3Av/2 = 50.18	(i)	1	7				114.32	16.31
	(ii)	3	9				81.32	8.25
	(iii)	6	8	0			37.98	5.40
2Av = 66.91	(i)	1	7				97.59	8.08
	(ii)	3	9				64.59	3.54
	(iii)	8	0				54.59	2.53
	(iv)	6	8				45.59	1.76

8.3 Additive Clusters

Table 8.22 presents each approximate cluster with all three characteristics implied by the additive clustering model:

(1) The cluster list S of its entities;
(2) The cluster-specific intensity $\lambda = a(S)$, the average within cluster similarity;
(3) The cluster contribution to the data scatter, $g^2(S) = \lambda^2|S|^2$.

The cluster sizes decrease when the similarity threshold grows, as illustrated on Fig. 8.2 and stated in Q.8.7. The corresponding intensity changes reflect the ever increasing shift values subtracted from the similarities. The table also shows that there is no point in making the similarity shift values greater than the average similarity value. In fact, setting the similarity shift value equal to the average can be seen as a step of the one-by-one cluster extracting strategy: subtracting the average from all the similarities is equivalent to extracting the universal cluster with its optimal intensity value – provided the cluster is considered on its own, without the presence of other clusters. At the similarity shift equal to the average, cluster {1,4,7} loses digit 4 because of its weak connections. The results best matching those of Figure 7.5 in case study 7.3 are found at the similarity shift equal to Av/2.

Project 8.1. Analysis of structure of amino acid substitution rates

Let us consider the data of substitution between amino acids in Table 1.8 and try explaining them in terms of properties of amino acids. An amino acid molecule can be considered as consisting of three groups of atoms: (i) an amine group, (ii) a carboxylic acid group, and (iii) a side chain. The side chain varies between different amino acids, thus affecting their biochemical properties. Among important features of side chains are the size and polarity, the latter affecting the interaction of proteins with solutions in which the life processes act: the polar amino acids tend to be on protein surfaces, i.e., hydrophilic, whereas other amino acids hide within membranes (hydrophobicity). There are also so-called aromatic amino acids, containing a stable ring, and aliphatic amino acids whose side chains contain only hydrogen or carbon atoms. These are presented in Table 8.23. As can be easily seen, these five attributes cover all amino acids but only once or twice.

A natural idea would be to check what relation these features have to the substitutions between amino acids. To explore the idea one needs to represent the features in the format of the matrix of substitutions, that is, in the similarity matrix format. Such a format is readily available as the adjacency matrix format. That is, a feature, say, "Small" corresponds to a subset S of entities, amino acids, that fall in it. The subset generates a binary relation "i and j belong to S" expressed by the Cartesian product $S \times S$ or, equivalently, by the $N \times N$ binary entity-to-entity similarity matrix $s = (s_{ij})$ such that $s_{ij} = 1$ if both i and j belong to S, and $s_{ij} = 0$, otherwise. For example, on the set of first five entities $I = \{A,C,D,E,F\}$ in Table 8.23, the binary similarity matrices for attributes Small, Polar and Hydrophobic are presented in Table 8.20.

Table 8.23 Attributes of twenty amino acids

Amino acid		Small	Polar	Hydrophobic	Aliphatic	Aromatic
A	Ala	+			+	
C	Cys	+		+		
D	Asp	+	+			
E	Glu		+			
F	Phe			+		+
G	Gly	+		+		
H	His					+
I	Ile			+	+	
K	Lys		+			
L	Leu			+	+	
M	Met			+		
N	Asn	+	+			
P	Pro	+				
Q	Gln		+			
R	Arg		+			
S	Ser	+				
T	Thr	+				
V	Val			+	+	
W	Trp			+		+
Y	Tyr					+

To analyze contributions of the attributes to the substitution rate data A one can use a linear regression model (see Section 4.3)

$$A = \lambda_1 Sm + \lambda_2 Po + \lambda_3 Hy + \lambda_4 Al + \lambda_5 Ar + \lambda_0$$

which in this context suggests that the similarity matrix A (after the intercept λ_0 is subtracted from it) can be decomposed, up to a minimized residual matrix, according to features in such a way that each coefficient $\lambda_1, \ldots, \lambda_5$, expresses the intensity level supplied by it to the overall similarity. The intercept λ_0, as usual, sums shifts in the individual attribute similarity scales.

To fit the regression model, let us utilize upper parts of the matrices only. In this way, we

(i) take into account the similarity symmetry and
(ii) make the diagonal substitution rates, that is, similarity to itself, not affecting the results.

As one can see from Table 8.24, the estimates of the slope regression coefficients are all positive, giving them the meaning of the weights or similarity intensities indeed, of which dummies representing categories Small, Polar, and Aromatic are the most contributing, according to the last line in Table 8.24. The intercept, though, is negative.

8.3 Additive Clusters

Table 8.24 Least-squares regression results. The last line entries (standardized intensities) are products of the corresponding entries in the first and second lines

	Sm	Po	Hy	Al	Ar	Intercept
Intensity λ	2.46	1.48	1.02	0.81	2.65	−2.06
Standard deviation	0.27	0.31	0.36	0.22	0.18	
Standardized Intensities	0.66	0.47	0.36	0.18	0.46	

Unfortunately, the five attributes cannot explain the pattern of amino acid substitution: the determination coefficient is just 37.3%, less than a half. That means one needs to find different attributes for explaining the amino acid substitution patterns.

Then the idea of additive clustering comes (see also Shepard and Arabie 1979 and Mirkin 1996). Why cannot we find attributes to fit in the similarity matrix from the matrix itself rather than by trying to search the amino acid feature databases? That is, let us consider unknown subsets $S_1, S_2, \ldots, S_K$ of the entity set along with the corresponding binary membership vectors $s_1, s_2, \ldots, s_K$ such that $s_{ik} = 1$ if $i \in S_k$, and $s_{ik} = 0$, otherwise, $k = 1, 2, \ldots, K$, and find them according to model

$$a_{ij} = \lambda_1 s_{i1} s_{j1} + \lambda_2 s_{i2} s_{j2} + \ldots + \lambda_K s_{iK} s_{jK} + \lambda_0 + e_{ij} \quad (8.14)$$

According to this model, each of the similarities a_{ij} is equal to a weighted sum of the corresponding cluster similarities $s_{ik} s_{jk}$, up to small residuals, $e_{ij} (i, j \in I)$.

Unfortunately, there are too many items to find, given the similarity matrix $A = (a_{ij})$: the number of clusters K, the clusters $S_1, S_2, \ldots, S_K$ themselves as well as their intensity weights, $\lambda_1, \lambda_2, \ldots, \lambda_K$, and the intercept, λ_0. This makes the solution much dependent on the starting point, as it is with the general mixture of distributions model.

If, however, we rewrite the model by moving the intercept to the left as

$$a_{ij} - \lambda_0 = \lambda_1 s_{i1} s_{j1} + \lambda_2 s_{i2} s_{j2} + \ldots + \lambda_K s_{iK} s_{jK} + e_{ij}, \quad (8.15)$$

the model reminds the equation for the Principal Component Analysis very much, especially as expressed in terms of the square matrices, see 5.2.2.3 – the $a_{ij} - \lambda_0$ plays the role of the covariance values, s_{ik}, the role of the loading values, that is, k-th eigenvector, and λ_k, the role of the k-th eigenvalue, the only difference being that the binarity constraints are imposed on the values s_{ik} that must be either 1 or 0. In (8.14), the intercept value λ_0 is the intensity of the universal cluster $S_0 = I$, which is assumed to be part of the solution. In (8.15), however, this is just a similarity shift, with the shifted similarity matrix $A^s = (a_{ij}^s)$ defined by $a_{ij}^s = a_{ij} - \lambda_0$ which is akin to the uniform data transformation in Section 8.1. Most important is that the value of λ_0 in model (8.15) ought to come from external considerations rather than from inside of the model as it is in (8.14).

The machinery for identifying additive clusters one-by-one developed further on leads to the following clusters found at different scale shift value λ_0 (see Table 8.25).

Table 8.25 Non-singleton clusters at Amino acid substitution data found at different similarity shift values; the average out-of-diagonal similarity value is Av $=-1.43$

Similarity shift	Cluster lists	Intensity	Contribution
0	(i) ILMV	1.67	2.04
	(ii) FWY	2.00	1.47
	(iii) EKQR	1.17	1.00
	(iv) DEQ	1.33	0.65
	(v) AST	0.67	0.16
Av/2 $=-0.71$	(i) ILMV	2.38	6.47
	(ii) DEKNQRS	1.05	4.38
	(iii) FWY	2.71	4.21
	(iv) AST	1.38	1.09
Av $=-1.43$	(i) DEHKNQRS	1.60	16.83
	(ii) FILMVY	1.96	13.44
	(iii) FWY	3.43	8.22

At the similarity shift equal to the average, there are three clusters covering 38.5% of the variance of the data. These concern three features of those considered above: Polar (cluster i), Hydrophobic (cluster ii), and Aromatic (cluster iii). The clusters slightly differ from those presented in Table 8.22, which can be well justified by the physical and chemical properties of amino acids. In particular, cluster (i) adds to Polar group two more amino acids: H (Histidine) and S (Serine). These two, in fact, are frequently considered polar too. Cluster (ii) differs from the Hydrophobic group by the absence of C (Cysteine) and W (Tryptophan) and the presence of Y (Tyrosine). This corresponds to a specific aspect of hydrophobicity, the so-called octanol scale that does exclude C and include Y (for some most recent measurements, see, for example, http://blanco.biomol.uci.edu). The absence of Tryptophan from the cluster is probably due to the fact that it is not easily substituted by the others because it is by far the most hydrophobic of the pack. Cluster (iii) consists of hydrophobic aromatic amino acids which excludes F (Phenylalanine) because it is not hydrophobic.

F8.3.2 Additive Clusters One-by-One: Formulation

Let us reformulate the additive cluster model from Project 8.1, keeping the same equation labels. Let I be a set of entities under consideration and $A = (a_{ij})$ a symmetric similarity matrix $i, j \in I$. The additive clustering model assumes that the similarities in A are generated by a set of additive clusters $S_k \subseteq I$ together with their intensities $\lambda_k (k = 0, 1, \ldots, K)$ in such a way that each a_{ij} is approximated by the sum of the intensities of those clusters that contain both i and j:

$$a_{ij} = \lambda_1 s_{i1} s_{j1} + \lambda_2 s_{i2} s_{j2} + \ldots + \lambda_K s_{iK} s_{jK} + \lambda_0 + e_{ij} \quad (8.14)$$

8.3 Additive Clusters

where $s_k = (s_{ik})$ are the membership vectors of unknown clusters S_k, and λ_k are their positive intensity values, $k = 1, 2, \ldots, K$. Residuals e_{ij} are to be minimized.

The zero's cluster S_0 is assumed to coincide with the entire set I so that its intensity λ_0 is the intercept in (8.14). On the other hand, λ_0 has a meaning of the similarity shift, with the shifted similarity matrix $A' = (a'_{ij})$ defined by $a'_{ij} = a_{ij} - \lambda_0$. Equation (8.14) for the shifted model can be rewritten as

$$a'_{ij} = \lambda_1 s_{i1} s_{j1} + \lambda_2 s_{i2} s_{j2} + \ldots + \lambda_K s_{iK} s_{jK} + e_{ij}, \tag{8.15}$$

so the shifted similarity matrix $a'_{ij} = a_{ij} - \lambda_0$ is the sum of cluster binary matrices weighted by their intensities. The role of the intercept λ_0 in (8.15) as a "soft" similarity threshold is of a special interest when λ_0 is user specified, because the shifted similarity matrix a'_{ij} may lead to different clusters at different λ_0 values, as Fig. 8.2 and Table 8.22 clearly demonstrate.

Model (8.15) can be considered using two different assumptions of the underlying cluster structure:

A. Overlapping additive clusters
B. Non-overlapping clusters

In the latter case, the summation in model (8.14) – (8.15) hides the fact that no summation of intensities goes on. Every similarity a'_{ij} is assumed to be approximately equal to the intensity value of that cluster that contains both i and j, or 0 if no cluster contains both of the entities.

The Equations in (8.15) coincide with those in the spectral decomposition (8.12) up to the condition that vectors s's in (8.15) are bound to be 1/0 binary, whereas no constraint is imposed on f's in (8.12). That means that the additive clustering model is an extension of the spectral decomposition onto the case when vectors are binary. This type of decomposition, with additional constraints such as say non-negativity of the elements of the solution is becoming increasingly popular in data analysis. Assuming the conventional least-squares criterion for this specification of the summarization problem, a natural idea coming to mind is to imitate the one-by-one approach of the Principal component analysis. The other idea, just working on all clusters in parallel, is not considered in this text.

Therefore, we turn to a simplest version of (8.14)–(8.15) model which is a single cluster model:

$$w_{ij} = \lambda s_i s_j + e_{ij}, \tag{8.16}$$

where w_{ij} are not necessarily the original similarities but rather any similarities including the shifted a'_{ij}, and $s = (s_i)$ is an N-dimensional zero-one vector of the memberships to cluster S to be found and λ its intensity.

To fit the model (8.16), we minimize the square error criterion

$$L^2(\lambda, s) = \sum_{i,j \in I} (w_{ij} - \lambda s_i s_j)^2 \tag{8.17}$$

We first note that, with no loss of generality, the similarity matrix W can always be considered symmetric, because otherwise W can be equivalently changed for a symmetric matrix $\hat{W} = (W + W^T)/2$.

Indeed, the part of criterion (8.17) related to a particular pair $i,j \in I$ is $(w_{ij} - \lambda s_i s_j)^2 + (w_{ji} - \lambda s_j s_i)^2$ which is equal to $w_{ij}^2 + w_{ji}^2 - 2\lambda(w_{ij} + w_{ji})s_i s_j + 2\lambda^2 s_i s_j$. The $s_i s_j$ on right are not squared because they are 0 or 1, thus do not change under this operation. The same part at matrix $\hat{W} = (\hat{w}_{ij})$ reads as $(w_{ij}^2 + w_{ji}^2 + 2w_{ij}w_{ji})/2 - 2\lambda(w_{ij} + w_{ji})s_i s_j + 2\lambda^2 s_i s_j$ so that the only parts affected are constant while those depending on the cluster to be found are identical, which proves the statement. Thus, the assumption that the similarity matrix is symmetric does not change a thing: it can always be transformed to a symmetric form $\hat{W} = (W + W^T)/2$.

For the sake of simplicity we assume that the matrix W comes with no diagonal entries, or that the diagonal entries w_{ii} are all zero.

Let us take a look at criterion (8.17) under each of two assumptions (Mirkin et al. 2010):

(a) Cluster intensity λ is pre-specified by the user
(b) Cluster intensity λ is to be found according to the criterion.

We first analyze the case (a) of λ pre-specified. Let us slightly rewrite criterion (8.17):

$$L^2(\lambda, s) = \sum_{i,j \in I}(w_{ij} - \lambda s_i s_j)^2 = \sum_{i,j \in I} w_{ij}^2 - 2\lambda \sum_{i,j \in I}\left(w_{ij} - \frac{\lambda}{2}\right)s_i s_j \quad (8.17')$$

Assume that λ is positive. Then minimizing (8.17) is equivalent to maximizing the sum on the right, which is just the summary uniform criterion (8.2) at $\pi = \lambda/2$ that has been described and utilized in Section 8.1. Indeed, the equation $\sum_{i,j \in I}(w_{ij} - \lambda/2)s_i s_j = \sum_{i,j \in S}(w_{ij} - \lambda/2)$ easily follows from the fact that $s_i = 1$ if and only if $i \in S$. That means that the algorithm AddRem from 8.1.3 is applicable here to produce $\lambda/2$-tight clusters.

Consider now case (b), when intensity λ in (8.17) is to be adjusted to further minimize the criterion. It is easy to prove that, given an S, the optimal λ is just the average of within cluster similarities, $\lambda = \lambda(S)$, where

$$\lambda(S) = \sum_{\substack{i,j \in I \\ i \neq j}} w_{ij} s_i s_j \Big/ \sum_{\substack{i,j \in I \\ i \neq j}} s_i s_j = \frac{\sum_{\substack{i,j \in S \\ i \neq j}} w_{ij}}{|S|(|S| - 1)} \quad (8.18)$$

as it is always the case for the least-squares approximation of a series of numbers by a central value (see Section 2.2).

That means that, again, the criterion is equivalent to the summary uniform criterion (8.2), but this time with a variable value of the threshold $\pi = \lambda(S)/2$ that

8.3 Additive Clusters

depends on S. In particular, a locally optimal cluster is $\lambda(S)/2$-tight: the average similarities of entities $i \in I$ to S are greater than $\lambda(S)/2$ for those i in S and smaller than $\lambda(S)/2$ for i's out of S.

If one puts the optimal $\lambda = \lambda(S)$ in (8.17), the least squares criterion is decomposed as follows

$$L^2(\lambda(S), s) = \sum_{i,j \in I}(w_{ij} - \lambda(S) s_i s_j)^2 = \sum_{i,j \in I} w_{ij}^2 - \lambda^2(S) \sum_{\substack{i,j \in I \\ i \neq j}} s_i s_j = T - \lambda^2(S)|S|(|S|-1)$$

where T is the data scatter, the sum of all the similarities squared, so that a Pythagorean decomposition of the data scatter holds:

$$T = \sum_{i,j \in I} w_{ij}^2 = \lambda(S)^2 |S|(|S| - 1) + L^2 \qquad (8.19)$$

where L^2 is the unexplained minimized part (8.17) whereas the item in the middle is the explained part of the data scatter.

The decomposition (8.19) looks more elegant with the diagonal similarities $\Sigma_{i,j \in I} s_i s_j$ is equal to $|S|^2$ and $\lambda(S) = s^T W s / (s^T s)^2$ so that

$$\sum_{i,j \in I} w_{ij}^2 = \lambda(S)^2 |S|^2 + L^2 \qquad (8.19')$$

The explained part in (8.19'), which is to be maximized to minimize L^2 because the scatter T is constant, is

$$g(S)^2 = [\lambda(S)|S|]^2 = \left[\frac{s^T W s}{s^T s}\right]^2 \qquad (8.20)$$

which is but the square of the Rayleigh quotient

$$g(S) = \frac{s^T W s}{s^T s} = \lambda(S)|S| \qquad (8.21)$$

Since it is assumed that at least some of the similarities in A are positive, the maximum of (8.11) over all binary s's is positive as well. Indeed, take a positive w_{ij} and a vector s with all components equal to zero except for just i-th and j-th components that are unities. Obviously (8.21) is positive on that, the more so the maximum. If, however, all the similarities between entities are negative, then no non singleton cluster can make (8.21) positive – that is, no nontrivial cluster can come up with the criterion.

That means that a version of AddRem(i) algorithm with a variant threshold π, AddRemAdd(i) in Section 8.3.3, in fact (locally) optimizes the Rayleigh quotient (8.21).

Now we can return to the case of the original model with multiple clusters. The situation will slightly differ depending on whether clusters are assumed non-overlapping (A) or possibly overlapping (B).

Consider, first, the case of model (8.15) with the restriction that clusters to be found must not overlap. The fact that clusters S_f and S_g do not overlap can be equivalently stated in terms of their binary membership vectors s_f and s_g: these must be orthogonal so that $<s_f, s_g> = 0$. This implies that the shifted data scatter admits the following decomposition:

$$<A',A'> = \sum_{k=1}^{K} [s_k^T A s_k / s_k^T s_k]^2 + <E,E> \qquad (8.22)$$

which extends Equation (8.19) to the multiple cluster case. In (8.22), the inner products $<A',A'>$ and $<E,E>$ denote the sums of the squared elements of the corresponding matrices. To derive (8.22), one can take the inner product of Equation (8.15) by itself, considering all matrices as $N \times N$ vectors, and taking into account the fact that matrices $s_k s_k^T$ and $s_l s_l^T$ are orthogonal as $N \times N$ vectors at $k \neq l$, because the corresponding vectors s_k and s_l are orthogonal.

Equation (8.22) means that each of the optimal non-overlapping clusters indeed contributes the squared Rayleigh quotient (8.20) to the shifted data scatter, and, moreover, the optimal intensity value λ_k of cluster S_k is, in fact, the within cluster average $\lambda_k = \lambda(S_k)$. The sum in the middle represents the part of the data scatter $<A',A'>$ "explained" by the model, whereas $<E,E>$ relates to the "unexplained" part. Both can be expressed in percentages of the data scatter. Obviously, the greater the explained part the better the fit.

Assuming that the cluster contributions differ significantly, one can apply the one-by-one principal component analysis strategy to the cluster case as well – though, in this case, the process does not necessarily lead to an optimal solution. This strategy can be put as follows. First, a cluster S is found at the entire data set to maximize the Rayleigh quotient (8.21). It is denoted by S_1 along with its intensity value $\lambda_1 = \lambda(S_1)$ and the contribution $g^2(S_1)$ in (8.20) and removed from the entity set I. The next cluster S_2 is found in the same way over the remaining entity set, and removed as well. The process of sequential extraction iterates until no positive entries in A' over the remaining entities can be found. This would mean the remaining entities are all to remain singletons. In general, the process yields suboptimal, not necessarily optimal, clusters.

Let us turn now to the case of overlapping clusters B.

To fit the model (8.15), the one-by-one cluster extracting strategy will require minimizing, at each step $k = 1, 2, \ldots, K$ the criterion (8.17) applied to a corresponding residual similarity matrix A_k (Mirkin 1987, 1996). Specifically, A_1 is taken to coincide with the shifted similarity matrix, $A_1 = A'$. At k-th step, a (locally) optimal cluster maximizing (8.20) over $W = A_k$ is found to be set as S_k along with its intensity value λ_k, equal to the average of the residual similarities within S_k. Its contribution to the data scatter is equal to the optimized criterion (8.20). The residual similarities are updated after each step k by subtracting the found $\lambda_k s_{ik} s_{jk}$:

8.3 Additive Clusters

$$a_{ij,k+1} = a_{ij,k} - \lambda_k s_{ik} s_{jk} \tag{8.23}$$

In spite of the fact that thus found clusters may and frequently do overlap, this one-by-one strategy leads to a decomposition of the data scatter into the contributions of the extracted clusters (S_k, λ_k) and the minimized residual square error, which is analogous to (8.22):

$$<A', A'> = \sum_{k=1}^{K} [s_k^T A_k s_k / s_k^T s_k]^2 + <E, E> \tag{8.24}$$

except that it is residual similarity matrix A_k stands in the middle rather than the original matrix A'.

To prove (8.24), one needs just the Equation (8.19′) applied to $W = A_k$,

$$\sum_{i,j \in I} a_{ij,k}^2 = \left(s_k^T A_k s_k / s_k^T s_k\right)^2 + L_k^2 \tag{8.25}$$

Since $L_k^2 = \sum_{i,j \in I} a_{ij,k+1}^2$, (8.24) can be obtained by summing the Equation (8.25) over all $k = 1, 2, \ldots, K$.

Q.8.5. What happens if $\lambda < 0$ in criterion (8.17)? **A.** According to formula (8.17′), that would mean that the summary uniform similarity $\sum_{i,j \in S} (w_{ij} - \lambda/2)$ must be minimized rather than maximized. An optimal set S would consist of most dissimilar entities. Such a set sometimes is referred to as an anti-cluster.

Q.8.6. Can you think of a real world problem that would amount to the goal of finding anti-clusters rather than clusters?

Q.8.7. Consider the uniform summary criterion $u(S, \pi)$ and two values of threshold, $\pi_1 < \pi_2$. Prove that the size of optimal cluster at π_2 cannot be greater than that at π_1, thus supporting the intuition illustrated on Fig. 8.2.

C8.3.3 Finding (Sub)Optimal Additive Clusters: Computation

Before starting computation of additive clusters, the similarity matrix should be made symmetric, by averaging it with its transpose, and shifted by a scale shift value λ_0 which is to be user defined. A default value for λ_0 can be the average value of the similarity matrix if similarity values vary across the matrix or $\lambda_0 = 1/2$ if the similarity matrix is the flat zero-one matrix of an ordinary graph.

We consider here only one cluster based additive clustering algorithms.

Given a matrix $W = (w_{ij})$, consider an additive clustering analogue to AddRem(i) algorithm from 8.1.3 to adjust the cluster intensity, that is, the threshold value, according to criterion (8.21). Again vector $z = 2s - 1$ is used to hold the information of cluster S being built. Its components are: $z_i = 1$ if $i \in S$ and $z_i = -1$, otherwise. This allows for the same action of changing the sign of z_i to express both addition of i into S if $i \notin S$ and removal of i out of S if $i \in S$.

8.3.3.1 AddRemAdd(j) Algorithm

Input: matrix $W = (w_{ij})$; Output: cluster S containing j, its intensity λ and contribution g^2 to the original A' matrix scatter.

1. *Initialization.* Set N-dimensional z to have all its entries equal to -1 except for $z_j = 1$, the number of elements $n = 1$, intensity $\lambda = 0$, and contribution $g^2 = 0$. For each entity $i \in I$, compute its average similarity to S, $w(i,S) = w_{ij}$.
2. *Selection.* Find i^* maximizing $w(i,S)$.
3. *Test.*

    a. If $w(i^*,S) > \lambda/2$

        i. Change the sign of z_{i^*} in vector z, $z_{i^*} \Leftarrow -z_{i^*}$
        ii. Update: $n \Leftarrow n + z_i^*$ (the number of elements in S), $\lambda \Leftarrow (n-2)[\lambda + z_{i^*}2w(i^*S)/(n-2)]/n$ (the average similarity within S), $w(i,S) \Leftarrow [(n-1)a(i,S) + z_{i^*}w_{ii^*}]/n$ (the average similarities of all entities to S), and $g^2 = \lambda^2 n^2$ (the contribution), and go to 2.

    b. Else

        i. Output S, λ and g^2.

    c. End

The general step is justified by the fact that indeed Equation (8.5) imply that maximizing $w(i,S)$ over all $i \in I$ does maximize the increment of $g(S)$ among all sets that can be obtained from S by either adding an entity to S or removing an entity from S. Updating formulas can be derived from the definitions of the concepts involved.

The algorithm AddRemAdd(j) utilizes no ad hoc parameters, except for the similarity shift value, so the cluster sizes are determined by the process of clustering itself. Yet, changing the similarity shift λ_0 may affect the clustering results indeed, which can be of an advantage when one needs to contrast within- and between-cluster similarities.

To use AddRemAdd algorithm for the case of non-overlapping clusters, one needs to perform a set of repetitive steps arranged as the algorithm ADN (ADditive clusters Non-overlapping) as follows.

8.3.3.2 ADN Algorithm

Input: matrix $A' = (a'_{ij})$; Output: a set of non-overlapping clusters $S_1, S_2, \ldots, S_K$ where

(i) number of clusters K is not pre-specified and
(ii) they do not necessarily cover all the entity set,

8.3 Additive Clusters

together with their intensities λ_k and contributions g_k^2 to the A' matrix scatter.

0. *Initialization.* Set $k = 1$, $I_k = I$ and $A_k = A'$.
1. *Stopping test.* Check whether I_k contains more than one entity and whether A_k contains positive values. If either is not true, the computation stops and those clusters found so far are output.
2. *Cluster.* Apply AddRemAdd(j) for every $j \in I_k$. Select that of the results maximizing the contribution and put is as S_k along with the corresponding intensity λ_k and contribution g_k^2.
3. *Update.* Set $I_k = I_k - S_k$, $k = k + 1$, and A_k the part of matrix A' related to elements of I_k only.

The number of clusters is not pre-specified by the user with ADN nor the subset of entities remaining unclustered. Yet both are predetermined by the choice of the scale shift parameter λ_0 leading to matrix A'. This choice, in fact, defines the granularity of clustering as illustrated on Fig. 8.2.

A similar algorithm, ADO (ADditive clusters Overlapping) can be drawn for the case when clusters are not necessarily non-overlapping.

8.3.3.3 ADO Algorithm

Input: matrix $A' = (a'_{ij})$ and parameters for halting the computation: (i) threshold of contribution of individual clusters ς, say $\varsigma = 5\%$, (ii) threshold of explained contribution η, say $\eta = 50\%$; Output: a set of possibly overlapping clusters $S_1, S_2, \ldots, S_K$ where

(i) number of clusters K is not pre-specified,
(ii) they do not necessarily cover all the entity set, and
(iii) they may overlap,

together with their intensities λ_k and contributions g_k^2 to the A' matrix scatter.

0. *Initialization.* Compute the data scatter $D = <A', A'>$. Set $k = 1$ and $A_k = A'$.
1. *Cluster.* Apply AddRemAdd(j) to A_k for every $j \in I$. Select of the results that maximizing the contribution and put is as S_k along with the corresponding intensity λ_k and contribution g_k^2.
2. *Stopping test.* Check whether $g_k^2/D > \zeta$ and $\Sigma_{f \le k} g_f^2/D \le \eta$. If either is not true, the computation stops and only clusters found at the previous iterations are output.
3. *Update.* Set $A_k = A_k - \lambda_k s_k s_k^T$, $k = k + 1$.
4. *Similarity positivity test.* Check whether A_k contains positive values. If yes, go to 1. If not, the computation stops and all clusters found so far are output.

Algorithm ADO extracts clusters from the similarity matrix one by one so that the residual elements are getting smaller at each step overall (Mirkin 1996). A drawback of ADO is that any cluster, once extracted, is never updated, so that a version of the algorithm should be developed with an inbuilt mechanism for updating the extracted clusters.

8.4 Summary

This chapter is an attempt to make a unified teaching material from diverse approaches to finding clusters in networks. The unifying theme is the summary within-cluster similarity criterion that, first, embraces the uniform and modularity approaches to confront the data with background noise, and then runs in the spectral clustering approach and the additive clustering approach. These two latter approaches represent two different pathways for extending the theory of spectral matrix decomposition to clustering. The spectral clustering does it by finding such combinatorial clustering criteria and such data transformations at which the spectral problem becomes an unconstrained relaxation of the combinatorial task. The additive clustering does just the opposite: it formulates a clustering problem as an extension of the spectral decomposition and tries to solve it by using combinatorial methods. Both of the approaches are effective; they do find good clusters, although there are specifics such as, for example, that the uniform criterion is better fitting to flat ordinary graph structures while the modularity criterion is better fitting at the data reflecting the diversity of individual entities. It is clear however that this part of data analysis technology is quickly moving forward to further developments.

References

Guattery, S., Miller, G.: On the quality of spectral separators. SIAM J. Matrix Anal. Appl. **19**(3), 701–719 (1998).
Johnsonbaugh, R., Schaefer, M.: Algorithms. Pearson Prentice Hall, Upper Saddle River (2004). ISBN 0-13-122853-6.
Klein, C., Randic, M.: Resistance distance. J. Mathematical Chem. **12**, 81–95 (1993).
Luxburg, U.: A tutorial on spectral clustering. Stat. Comput. **17**, 395–416 (2007).
Mirkin, B.: Additive clustering and qualitative factor analysis methods for similarity matrices. J. Classif. **4**, 7–31 (1987); Erratum (1989), **6**, 271–272.
Mirkin, B.: Mathematical Classification and Clustering. Kluwer Academic Press, Boston-Dordrecht (1996).
Mirkin, B., Camargo, R., Fenner, T., Loizou, G., Kellam, P.: Similarity clustering of proteins using substantive knowledge and reconstruction of evolutionary gene histories in herpesvirus. Theor. Chem. Acc.: Theory, Comput. Mod. **125**(3–6), 569–582 (2010).
Newman, M.E.J.: Modularity and community structure in networks. PNAS. **103**(23), 8577–8582 (2006).
Newman, M., Girvan, M.: Finding and evaluating community structure in networks. Phys. Rev. E. **69**, 026113 (2004).
Shepard, R.N., Arabie, P.: Additive clustering: Representation of similarities as combinations of discrete overlapping properties. Psychol. Rev. **86**, 87–123 (1979).
Shi, J., Malik, J.: Normalized cuts and image segmentation. IEEE Trans. Pattern Anal. Machine Intelligence. **22**(8), 888–905 (2000).

Appendix

A1 Basic Linear Algebra

Table A.1 presents data matrix from Table 5.9. It has 8 rows and 7 columns, that is, it is 8×7 matrix.

Table A.1 Company data standardized

	v1	v2	v3	v4	v5	v6	V7
e1	−0.20	0.23	−0.33	−0.63	0.36	−0.22	−0.14
e2	0.40	0.05	0	−0.63	0.36	−0.22	−0.14
e3	0.08	0.09	0	−0.63	−0.22	0.36	−0.14
e4	−0.23	−0.15	−0.33	0.38	0.36	−0.22	−0.14
e5	0.19	−0.29	0	0.38	−0.22	0.36	−0.14
e6	−0.60	−0.42	−0.33	0.38	−0.22	0.36	−0.14
e7	0.08	−0.10	0.33	0.38	−0.22	−0.22	0.43
e8	0.27	0.58	0.67	0.38	−0.22	−0.22	0.43

A1.1 Inner Product and Distance

Every row in data matrix Table A.1 represents an entity as a 7-dimensional vector, or point, such as e1=(−0.20, 0.23, −0.33, −0.63, 0.36, −0.22, −0.14) which is simultaneously a 1×7 matrix. Similarly, every column represents a feature or category as an 8-dimensional vector, or a 8×1 matrix, such as

v1
−0.20
0.40
0.08
−0.23
0.19
−0.60
0.08
0.27

or, its transpose, a 1×8 row
$v1^T = (-0.20, 0.40, 0.08, -0.23, 0.19, -0.60, 0.08, 0.27)^T.$

B. Mirkin, *Core Concepts in Data Analysis: Summarization, Correlation and Visualization*, Undergraduate Topics in Computer Science,
DOI 10.1007/978-0-85729-287-2, © Springer-Verlag London Limited 2011

Elements of vectors are referred to as their components. Operations of summation and subtraction are defined component-wise:

$$e1 = (-0.20, 0.23, -0.33, -0.63, 0.36, -0.22, -0.14)$$
$$+$$
$$e2 = (0.40, 0.05, 0, -0.63, 0.36, -0.22, -0.14)$$
$$e1+e2 = (0.20, 0.28, -0.33, -1.26, 0.72, -0.44, -0.28)$$

and

$$e1 = (-0.20, 0.23, -0.33, -0.63, 0.36, -0.22, -0.14)$$
$$-$$
$$e2 = (0.40, 0.05, 0, -0.63, 0.36, -0.22, -0.14)$$
$$e1-e2 = (-0.60, 0.18, -0.33, 0, 0, 0, 0)$$

The second important operation is multiplication of a vector by a real defined as multiplication of all components simultaneously:

$$3*e1 = (-0.60, 0.69, -0.99, -1.89, 1.08, -0.66, -0.42),$$
$$10*e1 = (-2.00, 2.30, -3.30, -6.30, 3.60, -2.20, -1.40)$$

To get some intuition, let us consider Cartesian plane representation of 2D vectors obtained by cutting off all components of the rows except for the first two (Figure A.1(a)).

Figure A.1 illustrates two geometric facts: (a) the sum of two vectors sits in the fourth node of the parallelogram formed by connecting 0 and the vectors; (b) given vector x, all vectors ax at the constant a taking any value form the line through the origin 0 and x.

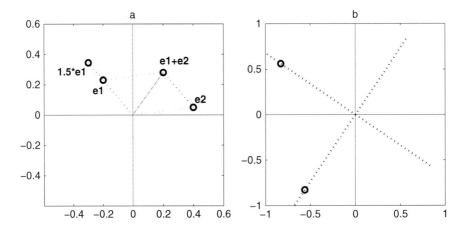

Fig. A.1 Plane geometry representation of 2D vectors on (**a**) and eigenvector lines for symmetric matrix A (**b**)

Appendix

The third important operation over vectors is inner product. The inner, or scalar, product is defined for every pair of vectors x and y of the same dimension and it is equal to – not a vector – but just a number equal to the sum of the products of the corresponding components and denoted by $<x.y>$. For example, for 2D parts of vectors $e1 = (-0.20, 0.23)$ and $e2 = (0.40, 0.05)$, the inner product is $<e1, e2> = -0.20*0.40 + 0.23*0.05 = -0.08 + 0.01 = -0.07$. A full computation $<e1, e2> :=$ sum (e1.* e2) is below:

$$e1 = (-0.20, 0.23, -0.33, -0.63, 0.36, -0.22, -0.14)$$
$$e2 = (0.40, 0.05, 0, -0.63, 0.36, -0.22, -0.14)$$
$$e1.*\, e2 = (-0.08, 0.01, 0, 0.39, 0.13, 0.05, 0.02)$$
$$<e1,e2> = \text{sum}(e1.*\, e2) = -0.08 + 0.01 + 0 + 0.39 + 0.13 + 0.05 + 0.02 = 0.52$$

The inner product is a linear operation so that, for example, $<e1, 2*e1+3*e2> = 2*<e1,e1> + 3*<e1,e2>$, which can be proven in this case straightforwardly by computation.

The inner square, that is, the product of a vector by itself, like $<e1, e1> = -0.20*(-0.20) + 0.23*0.23 = 0.040 + 0.053 = 0.093$, is the sum of squares of its components, which is the square length of the line connecting the origin 0 and the point on Cartesian plane such as Fig. A.1a. This follows from the Pythagoras theorem illustrated on Fig. A.2. The theorem states that the square of hypothenuse's length in any right-angled triangle is equal to the sum of squares of the sides' lengths, $c^2 = a^2 + b^2$. By extending this property to multidimensional points and vectors, the square root of the inner square $<x, x>$ is referred to as the norm of x and denoted $\|x\|$.

This allows us to introduce Euclidean distance between any two vectors/points x and y as the norm of their difference, $r(x, y) = \|x - y\|$. In MatLab, this can be expressed as r(x,y)= sqrt(sum((x-y).*(x-y)). For example, the distance between e1 and e2 as rows of Table A.1 can be computed as follows:

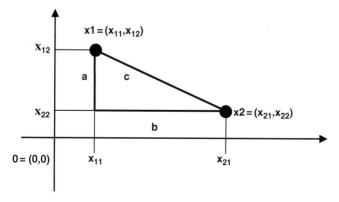

Fig. A.2 Pythagoras' theorem: the squared Euclidean distance between x1 and x2 is $d(x1, x2) = (x_{11} - x_{21})^2 + (x_{12} - x_{22})^2$

e1−e2 = (− 0.60, 0.18, −0.33, 0, 0, 0, 0)
(e1−e2).*(e1−e2) = (0.36, 0.03, 0.11, 0, 0, 0, 0)
d(e1,e2) = sum((e1−e2).*(e1−e2)) = 0.36 +0.03+0.11+0+0+0+0=0.50
r(e1,e2) = sqrt(d(e1,e2)) = sqrt(0.50) = 0.71

An important function in this computation is the squared Euclidean distance *d(e1,e2)* – this is the base of the least-squares approach in data analysis.

Some other distances are popular too. Among them: Manhattan or City-block distance defined as $m(x1,x2) = |x_{11} - x_{21}| + |x_{12} - x_{22}| + \ldots + |x_{1V} - x_{2V}|$ and Chebyshev or L_∞ distance defined as $c(x1,x2) = max(|x_{11} - x_{21}|, |x_{12} - x_{22}|, \ldots, |x_{1V} - x_{2V}|)$. A popular exercise in getting intuition about the distances is drawing sets of points that are equidistant to origin 0: this is a circle in the case of Euclidean distance, rhomb in the case of city-block distance, and square in the case of Chebyshev distance.

An important relation between (Euclidean squared) distance and inner product is this:

$$d(x,y) = <x-y, x-y> = <x,x> + <y,y> - 2<x,y>,$$

which is especially simple if $<x,y> = 0$:

$$d(x,y) = <x,x> + <y,y>$$

just like in Pythagoras' theorem. This is why vectors/points *x* and *y* satisfying <x,y>=0 are referred to as orthogonal. This property underlies the decompositions of data scatter presented in the text.

A1.2 Matrix Algebra

A general denotation for a matrix *A* is like this:

$$A = \begin{pmatrix} a_{11} & a_{12} & \cdots & a_{1V} \\ \vdots & & \ddots & \vdots \\ a_{N1} & a_{N2} & \cdots & a_{NV} \end{pmatrix}$$

so that *A* has *N* rows and *V* columns which is denoted as $N \times V$ size, and a common element is a_{iv} ($i = 1,\ldots,N$, $v = 1,\ldots,V$) – the row's index always goes first. The transpose A^T of matrix *A* is defined by switching the rows and columns so that $A^T = (a_{vi})$ is of $V \times N$ size:

$$A^T = \begin{pmatrix} a_{11} & a_{21} & \cdots & a_{V1} \\ \vdots & & \ddots & \vdots \\ a_{1N} & a_{2N} & \cdots & a_{VN} \end{pmatrix}$$

Appendix

A matrix of $N \times V$ size is referred to as a square matrix if $N = V$. A square matrix A is referred to as symmetric if $A = A^T$. The set of elements a_{ii} with coinciding indices is referred to as diagonal of matrix A. The symmetry then literally is over the diagonal.

Operations of summation, subtraction and multiplication by a number are defined for matrices component-wise exactly as it is for vectors. Matrices of different sizes cannot be summed with or subtracted from each other. Here is an example:

$$\begin{vmatrix} -0.20 & 0.23 \\ -0.40 & 0.05 \\ 0.08 & 0.09 \\ -0.23 & -0.15 \end{vmatrix} + 2 \begin{vmatrix} 0.19 & -0.29 \\ 0.60 & -0.42 \\ 0.08 & -0.10 \\ 0.27 & -0.58 \end{vmatrix} = \begin{vmatrix} 0.18 & -0.35 \\ 0.80 & -0.79 \\ 0.24 & -0.11 \\ 0.31 & 1.01 \end{vmatrix}$$

An $N \times V$ matrix A can be multiplied by a column vector b of the size $V \times 1$ to produce an $N \times 1$ vector $c = Ab$ – note that the number of components in b must be equal to the number of columns in A. This is just the sum of A columns weighted by the corresponding components of b (hence is the rule of the size of b). Here is an example:

$$\begin{pmatrix} -0.20 & 0.23 \\ 0.40 & 0.05 \\ 0.08 & 0.09 \\ -0.23 & -0.15 \end{pmatrix} \begin{pmatrix} 3 \\ 2 \end{pmatrix} = 3 \begin{pmatrix} -0.20 \\ 0.40 \\ 0.08 \\ -0.23 \end{pmatrix} + 2 \begin{pmatrix} 0.23 \\ 0.05 \\ 0.09 \\ -0.15 \end{pmatrix} = \begin{pmatrix} -0.14 \\ 1.30 \\ 0.42 \\ -0.99 \end{pmatrix}$$

This definition can be reformulated using the inner product: in fact, each component of Ab is the inner product of the corresponding row of A and b. Using the same example,

$$\begin{pmatrix} -0.20 & 0.23 \\ 0.40 & 0.05 \\ 0.08 & 0.09 \\ -0.23 & -0.15 \end{pmatrix} \begin{pmatrix} 3 \\ 2 \end{pmatrix} - \begin{pmatrix} <(-0.20\ 0.23),\ (3\ 2)> \\ <(0.40\ 0.05),\ (3\ 2)> \\ <(0.08\ 0.09),\ (3\ 2)> \\ <(-0.23 - 0.15),\ (3\ 2)> \end{pmatrix} - \begin{pmatrix} -0.14 \\ 1.30 \\ 0.42 \\ -0.99 \end{pmatrix}$$

Based on this, matrix product AB is defined for matrices A of size $N \times V$ and B of size $V \times M$ as a matrix of size $N \times M$ whose columns are products of A and corresponding columns of B. Let us extend our example to this case:

$$\begin{pmatrix} -0.20 & 0.23 \\ 0.40 & 0.05 \\ 0.08 & 0.09 \\ -0.23 & -0.15 \end{pmatrix} \begin{pmatrix} 3 & 1 \\ 2 & 0 \end{pmatrix} = \begin{pmatrix} -0.14 & -0.20 \\ 1.30 & 0.40 \\ 0.42 & 0.08 \\ -0.99 & -0.23 \end{pmatrix}$$

Given a square $n \times n$ matrix A and an $n \times 1$ vector b, the product $c = Ab$ is again an $n \times 1$ vector. A vector b is of a special interest if c lies on the line drawn through 0 and b, that is, if equation $Ab = \lambda b$ holds for some number λ. Such a number is referred to as an eigenvalue of A and b the corresponding eigenvector. The set of eigenvalues is not too large – the number of eigenvalues cannot exceed the matrix size n. If A is symmetric, then all its eigenvalues are real numbers and the eigenvectors corresponding to different eigenvalues are orthogonal to each other. In data analysis, it is usually assumed that all the eigenvalues are different indeed if the matrices are based on observations of quantitative variables because of random errors. Then the eigenvectors of A represent "inner" directions for Cartesian axes that follow the structure of A. Geometrically speaking, matrix multiplication transforms lines into lines. Then it would be correct to say that A transforms axes of the Cartesian space into its inner axes specified by the eigenvectors. Figure A.1b represents the eigenvector-defined axes for matrix $A = e + e^T$ where e is the matrix composed of two-dimensional row-vectors $e1$ and $e2$ considered above.

A2 Basic Optimization

Given a function $f(x)$ for $x \in X$, it is natural to look for points x in X at which $f(x)$ takes extreme values, ether maximum or minimum, hence is the problem of optimization, that is, finding a point that either minimizes f or maximizes it. Let us focus on minimization for certainty. There are two approaches to optimization: one is the classical one and the other of nature-inspired computational intelligence.

The classical approach is informed by calculus.

This approach has been first developed for one-dimensional functions $f(x)$ like the one whose graph is on Fig. A.3. In the point of minimum, like A or D, or maximum, like C, or change in the orientation of convexity, like B, the first derivative $f'(x)$ which expresses the tangent of the curve $f(x)$ in the point is 0 – this is what is referred to as the first-order necessary condition of minimum. It is possible to separate the minima from the rest by using the second order derivatives, but there is no way to tell one local minimum from the other unless reaching each of them, and to add to the misery, there is not much usually known of how to find them all or just the global minimum either. Sometimes the calculus is not of much help – a case in hand is the curve on Fig. A.3: its global minimum is at the very left point of the graph, and the first-order condition cannot help because it is valid only in interior points of the admissible set X.

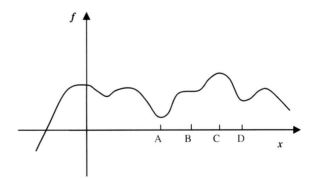

Fig. A.3 Graph of a typical multi-optimum function

Fig. A.4 New point taken in the direction opposite to the tangent

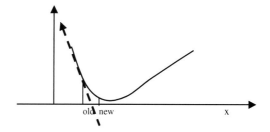

Yet to reach a local minimum satisfying the first-order minimum condition, a most universal method is of steepest descent. This method relies on the derivative of the function in any given point. This shows the direction of the steepest ascent over the optimized function, so that the opposite direction makes the steepest descent. Given an x and values $f(x)$ and $f'(x)$, this method finds another point x_{new} by subtracting the derivative scaled by a step factor, $x_{new} = x - \mu f'(x)$, where μ is the step factor (see Fig. A.4). The closer the point to the minimum, the smaller is the value of the derivative, thus the smaller the change. Of course, the method can converge to a minimum point, not necessarily the global minimum.

The situation when x is multidimensional is even more complex. The mathematics have made a good progress on the theory of optimization when only one minimum can exist – such is the case of so called convex or linear programming when both, function $f(x)$ and set of admissible points X, are convex or linear. In the more general situation, though, the steepest descent frequently remains the only tool available, even in spite of the fact it finds a local minimum with no estimates on the global one. Here, however, the concept of gradient is involved rather that of the derivative. For a function of n-dimensional vectors, $f(x_1, x_2, \ldots, x_n)$, the gradient is an n-dimensional vector grad(f(x)) whose k-th component is partial derivative $\frac{\partial y}{\partial x_k}$ ($k = 1, \ldots, n$). The different term is used because there are examples of functions that have no derivatives at some points but still have gradients in those points. The gradient, in n-dimensional space, shows direction of the steepest ascent. So, by taking the opposite direction, the process is supposed to go in the direction of steepest descent. That makes the method of steepest descent to work iterations. Each iteration takes in a point $x = (x_1, x_2, \ldots, x_n)$ and outputs a new point in the direction opposite to gradient:

$$x(new) = x(old) - \mu^* grad(f(x(old)))$$

where μ is the step size. This new point is taken then by the next iteration. By changing the step-size from iteration to iteration, one may achieve a better rate of convergence.

In the case when the set of arguments can be naturally partitioned in two or more parts such that the function is easy to minimize over each part taken separately, an iterative process applies to involve steps optimizing each part at pre-specified values of the other parts. This process is referred to as alternating minimization. Consider that $x = (y, z)$ so that $f(x) = f(y, z)$ and, at any given y^* and z^*, the minimum of $f(y^*, z)$ with respect to z can be found easily, as well as minimum $f(y, z^*)$ over y. Then, starting from some y^0 the alternating minimization process would produce a sequence $y^0, z^1, y^1, z^2, z^2, \ldots$ in which z^t is a minimizer of $f(y^{t-1}, z)$ and y^t a minimizer of $f(y, z^t)$ at each $t = 1, 2, \ldots$. This sequence would provide for an ever decreasing sequence of values $f(y^t, z^t)$. In a situation when there is a bound on them from below, this would warrant that the sequence converges to a local minimum. If either y or z can have only a finite number of values, the process of alternating minimization would converge in a finite number of steps.

Q.A.1. What is gradient of function: (i) $f(x_1, x_2) = x_1^2 + x_2^2$, (ii) $f(x_1, x_2) = (x_1 - 1)^2 + 3^*(x_2 - 4)^2$, (iii) $f(z_1, z_2) = 3^*z_1^2 + (1 - z_2)^4$? **A:** (i) $(2x_1, 2x_2)$, (ii) $[2^*(x_1 - 1), 3^*(x_2 - 4)]$, (iii) $(6^*z_1, -4^*(1 - z_2)^3)$.

In contrast to classical approaches, a nature inspired optimization approach does not try to reach a minimum by improving and updating a single solution point. Just the opposite. According to this approach, a population of admissible solutions is thrown in randomly and all the attention is given not to an individual solution but the population as a whole. Probabilistic rules are defined to generate the next generation of the population, usually in the same numbers, so that a process of evolution of the population from generation to generation is defined and executed computationally. Because of its probabilistic rules, each instance of the process may differ from the others. To warrant that the population improves in the process of evolution, a special "elite maintenance" policy is defined so that the elite – which is the best solution or a set of best solutions reached so far in the process – is used as an improvement device. The presence of the probabilistic component is considered an important device to warrant that the population does not stuck in a local optimum but rather roam over the entire area of admissible solutions.

A3 Basic MatLab

A3.1 Introduction

The working place within a processor's memory is up to the user. A recommended option:

– a folder with user-made MatLab codes, termed say Code and two or more subfolders, Data and Result, in which data and results, respectively, are to be stored.

MatLab's icon then is clicked on, after which MatLab opens as a set of windows of which the working area is referred to as Command Window. MatLab can be brought in to the working folder/directory with traditional MSDOS or UNIX based

Appendix   365

commands such as: cd <Path_To_Working_Directory> in its Command Window. MatLab remembers then this path; and it is available to the user in a tiny window on top of the Command Window.

MatLab is organized as a set of packages, each in its own directory, consisting of program files with extension .m each. Character "%" symbolizes a comment for humans till the end of the line.

Help can be invoked Windows-wise or within the working area. In the latter, "help" command allows seeing names of the packages as well as of individual program files; the latter are operations that can be executed within MatLab. Example: Command "help" shows a bunch of packages, "matlab\datafun" among them; command "help datafun" displays a number of operations such as "max – largest component"; command "help max" explains the operation in detail.

A3.2 Loading and Storing Files

A numeric data file should be organized as an entity-to-feature data table: rows correspond to entities, columns to features (see studn.dat and studn.var). Such a data structure, with all entries numerical, is referred to as a 2D array, corresponding to a matrix in mathematics; 1d arrays correspond to solitary entities or columns (features) or rows (entity records). Array is a most important MatLab data format to hold numeric data. It works on the principle of a chess-board: its (i,k)-th entry arr(i,k) is the element in i-th row and k-th column. An Excel file has a similar structure but it is interlaced with strings. A 2D array's defining feature is that every row has the same number of digits.

To load such a file one may use a command from package "iofun". A simple one is "load" to load a numeric array, organized as described, into the current MatLab processor memory:

>> arr=load('Data\stud.dat');
% symbol "%" is used for comments:
% MatLab interpreter doesn't read lines beginning with "%".
% "arr" is a place in computer's memory to put the data (variable);
% semicolon ";" should stand at the end of an instruction;
% if it does not, then the result will be printed to the screen,
% which can be very useful for the user for checking the process of computation
% studn.dat is a 100x8 file of 100 part-time students with 8 features:
% 3 binary for Occupation categories; then Age, NumberChildren,
% and scores over three disciplines.
% All feature names are in file studn.var stored in Data folder.

An 1D array can be put into the workspace with a command like

>> a=[3 4 7 0];

which is a 4×1 array, which can be transposed into a 1×4 array with a "transpose" command

>> b=a'

Since no semicolon is put in the end, b will be displayed on screen as

    3
    4
    7
    0

To get its 2d entry, a command

>> c=b(2)

can be utilized. Similarly, command

>> d=arr(7,8)

puts the value in arr's 7th row and 8th column into workspace as variable d.

If a numeric array in working memory is to be stored, one may use MatLab command "save" which admits a number of storage formats including internal .mat format (see more with "help save"). To store array X into file Result\good.res in ASCII format (which is a text format covering characters in a standard keypad set), one may use command

>> save Result\good.res X –ascii

If you need to check, before saving, what files and variables are currently in the workspace, you may use the upper-left part of the MatLab window or command

>> whos

that produces the list on the screen.

Names are handled as *strings*, with ' ' symbol. The entity/feature name sizes may vary, thus cannot be handled in the array format.

To do this, another data format is used: the *cell*. Cells involve curly braces rather than round brackets (parentheses) utilized for arrays. See the difference: arr(i) is 1D array arr's i-th element, whereas brr{i} is cell brr's i-th element, which can be not only a number or character, as in arrays, but also a string, an array, or even another cell.

There are other data structures as well in MatLab (video, audio, internet) which are not covered here.

MatLab supports several data formats, including Excel which is popular among scientists and practitioners alike (see more in help iofun). An Excel file with extension .xls can be dealt with in MatLab by using commands xlsread ans xlswrite. Straightforward as they are, the user should not expect a comfortable switch between Excel and MatLab with these commands. Take a look, for example, onto Excel data file of several students in the Table A.2 below.

Table A.2 An Excel spreadsheet with data of five students over four features (age in years, number of children, occupation [information technology IT or business administration BA or other AN], and mark over a range of 0–100%)

Feature Student	Age	#Children	Occup	Cl_Mark
John	35	0	IT	94
Peggy	28	2	BA	67
Fred	27	1	BA	85
Chris	28	0	OT	48
Liz	25	0	IT	87

The xlsread command produces three data structures from an xls file: one for numeric part, the other for text part, and the third for all data in the file. Specifically, if the table above is stored in Data subfolder as student.xls file, this works as follows:

>> [nn,tt,rr]=xlsread('Data\student.xls');
% nn is array of numeric values, tt – is cell of text,
% and rr is cell covering all the data in file

to produce a numeric 5×4 array nn:

$$nn = \begin{matrix} 35 & 0 & NaN & 94 \\ 28 & 2 & NaN & 67 \\ 27 & 1 & NaN & 85 \\ 28 & 0 & NaN & 48 \\ 25 & 0 & NaN & 87 \end{matrix}$$

and a text 8×5 cell tt:

$$tt = \begin{matrix} \text{'Feature'} & \text{'Age'} & \text{'\#Children'} & \text{'Occup'} & \text{'ML\_Mark'} \\ \text{'Student'} & \text{``} & \text{``} & \text{``} & \text{``} \\ \text{``} & \text{``} & \text{``} & \text{``} & \text{``} \\ \text{'John'} & \text{``} & \text{``} & \text{'IT'} & \text{``} \\ \text{'Peggy'} & \text{``} & \text{``} & \text{'BA'} & \text{``} \\ \text{'Fred'} & \text{``} & \text{``} & \text{'BA'} & \text{``} \\ \text{'Chris'} & \text{``} & \text{``} & \text{'OT'} & \text{``} \\ \text{'Liz'} & \text{``} & \text{``} & \text{'IT'} & \text{``} \end{matrix}$$

The full dataset is in 8×5 cell rr:

```
          'Feature'   'Age'    '#Children'  'Occup'   'CI_Mark'
          'Student'   [NaN]  [   NaN]   [ NaN]   [  NaN]
          [  NaN]    [NaN]  [   NaN]   [ NaN]   [  NaN]
   rr =  'John'     [ 35]   [    0]    'IT'     [   94]
          'Peggy'    [ 28]   [    2]    'BA'     [   67]
          'Fred'     [ 27]   [    1]    'BA'     [   85]
          'Chris'    [ 28]   [    0]    'OT'     [   48]
          'Liz'      [ 25]   [    0]    'IT'     [   87]
```

The NaN symbol applies in MatLab to undefined numeric values such as emerge from division by zero and the like.

As one can see these are not exactly clean-cut structures to work with. The numerical array nn contains an incomprehensible column of NaN values, and the text file tt mixes up names of students and features.

A3.3 Using Subsets of Entities and Features

If one wants working with only three of the six features, say "Age", "Children" and "OOProgramming_Score", one must put together their indices into a named 1D array:

```
>> ii=[4 5 7]
% no semicolon in the end to display ii on screen as a row;
```

Then commands to reduce the dataset and the set of feature names over ii columns can be like these:
```
>> newa=arr(:,ii); %new data array
>> newb=b(ii);
 % newb is new feature set: to set it, one uses round braces rather than curly ones,
 % in spite of the fact that cells are involved here, not arrays
```

A similar command makes it to a subset of entities. If, for instance, we want to limit our attention to only those students who received 60 or more at "OOProgramming", we first find their indices with command "find":

```
>> jj=find(arr(:,7)>=60);
% jj is the set of the students defined in find()
% arr(:,7) is the seventh column of arr
```

Now we can apply "arr" to "jj":

```
>> al=arr(jj,:); % partial data of better off students
```

Appendix

The size of the data file al can be found with command

\>\>size(al)
% note: no semicolon to see the size on the screen

to produce a screen output:

ans =

  55 8

meaning that al consists of 55 rows and 8 columns. If one needs to maintain these in the workspace, use command

\>\>[n,m]=size(al)
that will put 55 into n and 8 into m.

Now we are ready to discern meaningful data from numerical array nn and text cell tt in workspace for Table A.2 on p. 367. As shown on that page, array nn's meaningless column is 3. Thus we can remove it like this:

\>\> [rnn,cnn]=size(nn);
% thus, the number of columns is cnn

\>\> vv=setdiff([1:cnn],3);
% operation setdiff(x,y) removes from x all elements of array y occurring in x
% [1:cnn] is an array of all integers from 1 to cnn inclusive, e.g., [1:4] is [1 2 3 4]
% thus, vv consists of all indices but 3

\>\> nnr=nn(:,vv);
% this puts all nn, except for column 3, into nnr:

$$nnr = \begin{matrix} 35 & 0 & 94 \\ 28 & 2 & 67 \\ 27 & 1 & 85 \\ 28 & 0 & 48 \\ 25 & 0 & 87 \end{matrix}$$

To create a cell containing the corresponding feature set, we need first to have a cell with all features. These constitute the final fragment of the first row of cell tt, without the very first string, "Feature", as can be seen from the tt contents shown above. Thus command

\>\> fe=tt(1,2:5);
% only first row in tt concerning its four columns, 2–5, goes to cell fe

leads to cell fe of size 1×4 containing of four features. To remove feature 3, we apply the array vv produced above:

>>fer=fe(vv);

Cell fer contains strings "Age" , "#Children", "ML_Mark" indexed by 1, 2 and 3 and corresponding, in respect, to columns of array nnr.

Many additional operations of MatLab are introduced and utilized in projects, worked examples and case studies, especially in Chapters 2, 3, 4.

A4 MatLab Program Codes

A4.1 Minkowski's Center: Evolutionary Algorithm

```
%cm.m, computing Minkowski p-distance central point c of a series x
%along with the average distance and its proportion in the sum

function [c,d,pe]=cm(x,p)

n=length(x);
lb=min(x);
rb=max(x); %----------------lb, rb are boundaries of the area (i)---
de=0;
for ik=1:n
   de=de+(abs(x(ik)))^p;
end
de=de/n;%--------------------------average p-th power of the data

%------------population setting (ii), setting the limit, iter, to iterations
pp=15; %population size
feas=(rb-lb)*rand(pp,1)+lb; %  generated population of p c values within the range
flag=1;
count=0;
iter=5000;
%--------- evaluation of the initially generated population (iii) ---
funp=0;
for ii=1:pp
   vv(ii)=mink(p,x,feas(ii));
end
[funi, ini]=min(vv);
soli=feas(ini) %initial best c value
```

Appendix

```
funi %initial error
si=1;%0.5; %step of change
%------------evolution of the population (iv) ----------------
while flag==1
   count=count+1;
   feas=feas+si*randn(pp,1); % Gaussian mutation added with step si
   for ii=1:pp
      feas(ii)=max(lb, feas(ii));
      feas(ii,:)=min(rb,feas(ii));% keeping the population within the range
      vec(ii)=mink(p,x,feas(ii)); %evaluation
   end
%-------------- elite maintenance (v) ---------------
   [fun, in]=min(vec); %best distance value
   sol=feas(in,:);%corresponding c value
   [wf,wi]=max(vec);
   wun=feas(wi); %worst c
   if wf>funi
      feas(wi)=soli;
      vec(wi)=funi; % changing the worst for the elite
   end
   if fun < funi
      soli=sol;
      funi=fun;
   end
   if (count>=iter)
      flag=0;
   end
   pe=funi/de;
%------------ screen the results of every 1000th iteration
   if rem(count,1000)==0
      %funp=funi;
      disp([soli pe]);
   end
end
c=soli;
d=funi;
pe=d/de;

return

%--------computing the quality of ce, the average deviation in p-th power
function dis=mink(p,x,ce)

nn=length(x);
dis=0;
```

```
for ik=1:nn
   dis=dis+(abs(x(ik)-ce))^p;
end
dis=dis/nn;

return
```

A4.2 Fitting Power Law: Non-linear Evolutionary and Linearization

```
% plan.m, power law analysis assuming the predictor x and target y are
% available as variables in matlab
% the power law is a function:    y=ax^b          (1)
% its linearized form:            log(y)=log(a)+b*log(x)  (2)

function plan(x,y)

%-----linear analysis of log(x) and log(y)
for ii=1:length(x);xc(ii)=max(.05,x(ii));yc(ii)=max(0.05,y(ii));end;
%0.05 instead of 0 to make logarithms possible
xll=log(xc);
yll=log(yc);
[all,bll,cll, rvll]=lr(xll,yll);

%all the slope, bll the intercept of the linear regression
%cll the correlation coefficient, rvll the residual variance of the linear regression
yle=all*xll+bll;% linear-regression estimated yll
cd=cll^2;%determination coefficient, it should be cd=1-rvll
cd
rvll
%figure(1);plot(xll,yll,'k.',xll,yle,'rp');

%-----linearized: fitting equation (1) by first fitting equation(2)

[al,bl, rl]=llr(x,y);
% al the estimate of a, bl the estimate of b and
% rl the proportion of the residual variance in the variance of y

% ylr - the linearized rule estimate for the power law
for ii=1:length(x);ylr(ii)=al*x(ii)^bl;end;

%------as is: fitting equation (1) by straightforwardly minimizing the
%------residual variance with an evolutionary algorithm
```

Appendix

```
[an,bn,f, rn]=nlr(x,y);
% an the estimate of a, bn the estimate of b and
% rn the proportion of the residual variance in the variance of y
for ii=1:length(x);yn(ii)=an*x(ii)^bn;end; %estimated power law

%-----------output: two-plot figure, real on the left, log on the right
%figure(2);
subplot(1,2,1);
plot(x,y,'k.',x,ylr,'b.',x,yn,'r.');%data scatter with two estimated power laws,
% blue-linearized, red- as is
subplot(1,2,2);plot(xll,yll,'k.',xll,yle,'rp');

%-----------output: text file of the results
saveplan('rep', cll, al, bl, rl, an, bn, rn,cd);

return

% llr.m, fitting a nonlinear regression function y=ax^b
% using linearization
% x is predictor, y is target, a,b -regression parameters to be fitted

function [a,b, residvar]=llr(xt,yt);

% regression is power law y=a*x^b as reflected in the procedure
% residvar is the average square error's proportion to the variance of y;
% xt, yt are predictor and target

%-----an elementary check of length compatibility--------
ll=length(xt);
if ll~=length(yt)
   disp('Something wrong is with the data');
   pause;
end

%--------- calculating a and b using the linearization
for ii=1:ll;xc(ii)=max(.05,xt(ii));yc(ii)=max(0.05,yt(ii));end;
%putting 0.05 instead of zero to make possible logarithms of the data
xl=log(xc); %taking log of x and y
yl=log(yc);

[al,bl,dl]=lr(xl,yl);
b=al;
a=exp(bl);
ab=[a b];
residvar=delta(ab,xt,yt)/var(yt,1);
```

return

%-------- computing the quality of the approximation y=a*(x^b)
%which is the residual variance

function esq=delta(tt,x,y)%tt=[a, b]; x predictor, y target
a=tt(1);
b=tt(2);
esq=0;
for ii=1:length(x)
   yp(ii)=a*(x(ii)^b); %this power law function can be changed
   esq=esq+(y(ii)-yp(ii))^2;
end
esq=esq/length(x);
return;

% nlr.m, evolutionary fitting of a nonlinear regression function y=f(x,a,b)
% x is predictor, y is target, a,b -regression prameters to be fitted

function [a,b, funi,residvar]=nlr(xt,yt);

% in this version the regression equation is power law y=a*x^b which is
% reflected only in the subroutine 'delta' in the bottom for computing the
% value of the average error squared;
% funi is the average square error's best value;
% residvar is its proportion to the variance of y;
% xt, yt are predictor and target

%-----an elementary check of length compatibility--------
ll=length(xt);
if ll~=length(yt)
   disp('Something is wrong with the data');
   pause;
end
%--------------- determine rectangle at which (a,b)-populations fluctuate
[ab,bb]=ddr(xt,yt);

lb=[ab(1) bb(1)];
rb=[ab(2) bb(2)];
lb
rb
disp('Hit ENTER if you wish to proceed. ');
pause;
%------------organisation of the iterations, iter the limit to their number
p=15; %population size

Appendix

```
for ii=1:p;feas(ii,:)=(rb-lb).*rand(1,2)+lb;end; %  generated population of p pairs
coefficients within the range
flag=1;
count=0;
iter=10000;%5000;
%--------- evaluation of the initially generated population
funp=0;
for ii=1:p
   vv(ii)=delta(feas(ii,:),xt,yt);
end
[funi, ini]=min(vv);
soli=feas(ini,:) %initial coeffts
funi %initial error
si=1;%0.5; %step of change
%------------evolution of the population
while flag==1
   count=count+1;
   feas=feas+si*randn(p,2); %mutation added with step si
   for ii=1:p
     feas(ii,:)=max([lb;feas(ii,:)]);
     feas(ii,:)=min([rb;feas(ii,:)]);% keeping the population within the range
     vec(ii)=delta(feas(ii,:),xt,yt); %evaluation
   end

   [fun, in]=min(vec); %best approximation value
   sol=feas(in,:);%corresponding parameters
   [wf,wi]=max(vec);
   wun=feas(wi,:); %worst case
   if wf>funi
     feas(wi,:)=soli;
     vec(wi)=funi;
%changing the worst for the best of the previous generation
   end
   if fun < funi
     soli=sol;
     funi=fun;
   end
   if (count>=iter)
      flag=0;
   end
 residvar=funi/var(yt,1);
%------------ screen the results of every 500th iteration
   if rem(count,500)==0
      %funp=funi;
      disp([soli residvar]);
```

```
   end
end
a=soli(1);
b=soli(2);
return

%-------- computing the quality of the approximation y==a*(x^b)
function esq=delta(tt,x,y)%tt=[a, b]; x predictor, y target
a=tt(1);
b=tt(2);
esq=0;
for ii=1:length(x)
   yp(ii)=a*(x(ii)^b); %this is a power law function
   esq=esq+(y(ii)-yp(ii))^2;
end
esq=esq/length(x); %the average difference squared
return;

% ddr.m, determination of the domain for power law y=a*x^b with b
% restricted
function [ab,bb]=ddr(x,y)
n=length(x);
bm=(log(y(1))-log(y(2)))/(log(x(1))-log(x(2)));
am=y(1)/(x(1)^bm);
ab=[am am];
bb=[bm bm];
%------------finding extreme values for a and b using pairwise equations
bs=0;as=0; bsq=0;asq=0;
count=0;
for ii=1:(n-1);
   if min(x(ii),y(ii))>.25
     for jj=(ii+1):n
       if min(x(jj),y(jj))>.25
       if (x(ii)/x(jj)<0.75)|(x(ii)/x(jj)>1.25)
          count=count+1;
          bt=(log(y(ii))-log(y(jj)))/(log(x(ii))-log(x(jj)));
          aij=y(ii)/(x(ii)^bt);
          aij=min(aij,100);%restriction
          %if (aij>100)
          %   disp([ii jj]); aij
          %end;
          bs=bs+bt;
          bsq=bsq+bt*bt;
          as=as+aij;
          asq=asq+aij*aij;
```

Appendix

```
            if bt>bb(2)
                bb(2)=bt;
            end;
            if bt<bb(1)
                bb(1)=bt;
            end;
            if aij>ab(2)
                ab(2)=aij;
            end;
            if aij<ab(1)
                ab(1)=aij;
            end;
        end;
        end;
    end;
    end;
end;
as=as/count
asq=asq/count;
sas=sqrt(asq-as^2)
bs=bs/count
bsq=bsq/count;
sbs=sqrt(bsq-bs^2)
ab(1)=as-4*sas;ab(2)=as+4*sas;
bb(1)=bs-4*sbs;bb(2)=bs+4*sbs;
count
return


% saveplan.m, saving results of the power-law analysis in plan.m

function saveplan(file, cc, al, bl, rl, an, bn, rn,cd);

ct =num2str(cc);
first=['Results of the power-law analysis y=ax^b' ];
alla=[ 'On the level of logarithms, the correlation is ' num2str(cc)];
alex=['Explained proportion of log(y)-variance is ' num2str(100*cd) '%'];
nt=[ ];
lt1=['Linearized estimate parameter values are a= ' num2str(al) ', b= ' num2str(bl)];
lt2=['Explained proportion of y-variance is r= ' num2str(100*(1-rl)) '%' ];
nt1=['"As is" estimate parameter values are a= ' num2str(an) ', b= ' num2str(bn)];
nt2=['Explained proportion of y-variance is r= ' num2str(100*(1-rn)) '%'];
alltext=strvcat(alla, lt1,lt2,nt1,nt2);

oul=[' These are visualized on the Figure produced:']
our=[' The power-law estimates on the left, the logarithms, on the right'];
```

```
    alltt=strvcat(alltext, oul, our);
    alltt
Filename=[ file '.out'];
fid= fopen(Filename, 'at');
if fid~=-1
    fprintf(fid, '%s\n', first);
    fprintf(fid, '%s\n', '    ');
    fprintf(fid, '%s\n', alla);
    fprintf(fid, '%s\n', alex);
    fprintf(fid, '%s\n', '    ');
    fprintf(fid, '%s\n', lt1);
    fprintf(fid, '%s\n', lt2);
     fprintf(fid, '%s\n', '    ');
     fprintf(fid, '%s\n', nt1);
    fprintf(fid, '%s\n', nt2);
     fprintf(fid, '%s\n', '    ');
     fprintf(fid, '%s\n', oul);
    fprintf(fid, '%s\n', our);
    fprintf(fid, '%s\n', '    ');
    fprintf(fid, '%s\n', '    ');
    fclose(fid);
end;
return
```

A4.3 Training Neuron Network with One Hidden Layer

```
% nnn.m for learning a set of features from a data set
% with a neural net with a single hidden layer
% with the symmetric sigmoid (hyperbolic tangent) in the hidden layer
% and data normalisation to [−10,10] interval

function [V,W, mede]=nnn(hiddenn,muin)

% hiddenn - number of neurons in the hidden layer
% muin - the learning rate, should be of order of 0.0001 or less
% V, W - wiring coefficients learnt
% mede - vector of absolute values of errors in output features

%--------------1.loading data ---------------------
da=load('Data\studn.dat'); %this is where the data file is put!!!
% da=load('Data\iris.dat'); %this will be for iris data
[n,m]=size(da);
```

```
%------2.normalizing to [-10,10] scale--------------------
mr=max(da);
ml=min(da);
ra=mr-ml;
ba=mr+ml;
tda=2*da-ones(n,1)*ba;
dan=tda./(ones(n,1)*ra);
dan=10*dan;
%------------3. preparing input and output target)--------
ip=[1:5]; % here is list of indexes of input features!!!
%ip=[1:2];%only two input features in the case of iris
ic=length(ip);
op=[6:8]; % here is list of indexes of output features!!!
%op=[3:4];% output iris features
oc=length(op);
output=dan(:,op); %target features file
input=dan(:,ip); %input features file
input(:,ic+1)=10;    %bias component
%-----------------4.initialising the network --------------------
h=hiddenn;     %the number of hidden neurons!!!
W=randn(ic+1,h); %initialising w weights
V=randn(h,oc);   %initialising v weights
W0=W;
V0=V;
count=0; %counter of epochs
stopp=0; %stop-condition to change
%pause(3);

while(stopp==0)
mede=zeros(1,oc); % mean errors after an epoch
%---------------5. cycling over entities in a random order
   ror=randperm(n);
   for ii=1:n
     x=input(ror(ii),:); %current instance's input
     u=output(ror(ii),:);% current instance's output
%---------------6. forward pass (to calculate response ru)------
     ow=x*W;
     o1=1+exp(-ow);
     oow=ones(1,h)./o1;
     oow=2*oow-1;% symmetric sigmoid output of the hidden layer
     ov=oow*V; %output of the output layer
     err=u-ov; %the error
     mede=mede+abs(err)/n;
%------------ 7. error back-propagation-------------------------
     gV=-oow'*err;    % gradient vector for matrix V
```

```
    t1=V*err'; % error propagated to the hidden layer
    t2=(1-oow).*(1+oow)/2; %the derivative
    t3=t2.*t1';% error multiplied by the th's derivative
    gW=-x'*t3;  % gradient vector for matrix W
%---------------8. weights update----------------------
    mu=muin; %the learning rate from the input!!!
    V=V-mu*gV;
    W=W-mu*gW;
  end;
%-----------------. stop-condition ------------------------
  count=count+1;
  ss=mean(mede);
  if ss<0.01|count>=10000
    stopp=1;
  end;
  mede;
  if rem(count,500)==0
    count
    mede
  end
end;
```

A4.4 Building Classification Trees

```
% clatree.m a program for building a decision tree over quantitative data,
% according to a method, 'gini', 'chi' or 'ing' in 3.5
% and specified stopping conditions: (a) number of entities,
% (b) prevailing feature; Inputs: data matrix X, partition as cell s, method
% variables untouched

function Clusters=clatree(X, s, method)

[n,mm]=size(X)
TS=10; %cluster size threshold
ee=0.8;%threshold to an s-class contents in a cluster
tin=0; %threshold on the scoring function to be set
switch method
        case 'gini'
            tin=0.03;
        case 'chi'
            tin=0.08;
        case 'ing'
            tin=0.15;
```

```
        otherwise
            disp('The method is wrong');
            pause(10);
        end
for ik=1:length(s);
   ds(ik)=length(s{ik});
end
ds=ds/sum(ds);
%distribution of s
ss=1; %cluster counter
bb=ss; %the last cluster's index
Clusters{ss,1}=[1:n];%entity set to cluster
%Clusters{ss,2}=[1:m];%features to be used
if max(ds)<ee
   Clusters{ss,2}=1; %should be split further
else
   Clusters{ss,2}=0; %should not be split further
end
Clusters{ss,3}=[0]; %parent's index
Clusters{ss,4}=[]; %characteristics
Clusters{ss,5}=ds;%distribution of s
tt=0;%counter of clusters to split taking into account added clusters
while ~(tt==ss),
   for uu=(tt+1):ss
      uu
      realnum=Clusters{uu,1}; %cluster to be split
      flag=Clusters{uu,2};
      if (flag==1)
        ma=-1;%starting gain value
        vv=0;%starting feature
        yy=-1000;%starting value
        for v0=1:mm
           xs=X(realnum,v0); %variable to be used
           [g,res,y]=msplit(xs,s,method);%producing split
        disp(['var ' num2str(v0) ' val ' num2str(y) ' ' num2str(res)])
        % this line is to see action of each feature at each cluster
           if res>=ma
              ma=res;
              yy=y;
              vv=v0;
           end;
        end
        if ma>tin
           xt=X(realnum,vv);
           g{1}=realnum(find(xt<=yy));
```

```
            g{2}=realnum(find(xt>yy));
            l1=length(g{1});
            l2=length(g{2});
            if (l1*l2)==0
               Clusters{uu,2}=0;
            else
               if (l1>TS & l2>TS)
                  cc=clfil(g{1},s,ee,vv,uu,-1,yy,ma);
                  for il=1:5
                     Clusters{bb+1,il}=cc{il};
                  end
                  cc=clfil(g{2},s,ee,vv,uu,1,yy,ma);
                  for il=1:5
                     Clusters{bb+2,il}=cc{il};
                  end
                  bb=bb+2;
               elseif l1>TS
                  cc=clfil(g{1},s,ee,vv,uu,-1,yy,ma);
                  for il=1:5
                     Clusters{bb+1,il}=cc{il};
                  end
                  bb=bb+1;
               elseif l2>TS
                  cc=clfil(g{2},s,ee,vv,uu,1,yy,ma);
                  for il=1:5
                     Clusters{bb+1,il}=cc{il};
                  end
                  bb=bb+1;
               end;
               Clusters{uu,2}=0;
            end
         end;
      end;
   end;
  tt=ss;
  ss=bb;
end;
 %savrdnew(file,Clusters,CC,B,yent);
return
%-------------- assigning a cluster object
function cc=clfil(gg,s,ee,vv,uu,t,y,ma)
%t=-1 for 1-split, 1 for 2-split
 cc{1}=gg;
 for ik=1:length(s)
    ds(ik)=length(intersect(s{ik},gg));
```

```
end;
ds=ds/sum(ds);%distribution of s in gg
if (max(ds)>ee)
   cc{2}=0;
else
   cc{2}=1;
end
cc{3}=uu; %parent
cc{4}=[vv t y ma];
% variable, less/more than, split y, gain
cc{5}=ds;
return
```

A5 Two Random Samples

A5.1 Short.dat

Short.dat is a dataset of random samples from three different distributions in Table A.3.

Table A.3 Three columns from three different distributions

8	20	1,512
12	21	50
11	23	48
10	21	206
9	9	12
7	20	199
10	22	51
12	18	50
9	20	198
13	21	843
9	5	12
13	13	8
10	10	7
11	14	9
9	18	39
9	13	12
7	21	51
11	20	46
11	21	50
9	18	54
8	20	1,391
10	19	49
10	19	41
13	24	35
12	23	45
10	13	11
12	9	9
10	21	49

Table A.3 (continued)

7	10	10
8	17	52
12	8	8
11	20	48
12	17	199
8	11	9
8	11	13
9	20	978
12	17	51
9	20	6, 233
13	19	23
10	21	47
11	11	8
11	20	973
11	7	43
13	20	201
9	18	200
10	19	49
9	10	7
14	20	36
9	10	8
11	21	203

A5.2 A Sample of 280 N(0,10) Values

Here is the sample from the Gaussian distribution N(0,10). The sample has been sorted in the ascending order in Table A.4.

Table A.4 Sample of 280 values from N(0,1) sorted in the ascending order

−30.29	−12.48	−7.01	−2.99	1.76	5.58	10.35
−25.57	−12.29	−6.94	−2.91	1.98	5.59	10.50
−25.34	−12.27	−6.83	−2.83	1.98	5.63	10.94
−23.79	−11.89	−6.79	−2.78	2.07	5.65	10.98
−23.34	−11.61	−6.65	−2.75	2.08	5.65	11.08
−22.38	−11.50	−6.64	−2.66	2.14	5.74	11.13
−22.37	−11.33	−6.11	−2.66	2.14	5.74	11.64
−21.78	−11.10	−6.02	−2.58	2.18	5.81	12.28
−21.05	−10.78	−5.98	−2.52	2.21	5.82	12.33
−20.89	−10.57	−5.87	−2.23	2.27	5.89	12.59
−20.65	−10.52	−5.53	−2.07	2.28	6.13	12.79
−19.10	−10.44	−5.35	−2.06	2.29	6.26	12.93
−18.16	−10.13	−5.33	−1.91	2.36	6.29	13.15
−17.95	−10.09	−5.22	−1.90	2.37	6.51	13.24
−17.79	−10.08	−5.17	−1.74	2.56	6.55	13.42
−17.58	−10.06	−4.91	−1.60	2.71	6.59	13.44
−16.47	−9.79	−4.82	−1.51	2.79	6.59	13.48
−16.43	−9.11	−4.62	−1.44	2.85	6.65	13.56
−16.31	−9.08	−4.58	−1.42	2.91	7.00	13.99
−16.19	−9.01	−4.53	−1.28	2.94	7.09	14.27

Table A.4 (continued)

−16.15	−8.95	−4.43	−1.26	2.98	7.16	14.69
−16.14	−8.93	−4.26	−0.80	3.16	7.30	14.95
−15.90	−8.71	−4.18	−0.79	3.21	7.58	15.35
−15.89	−8.53	−4.17	−0.73	3.27	7.99	15.74
−15.67	−8.49	−4.08	−0.50	3.27	8.34	15.82
−15.56	−8.01	−4.01	−0.49	3.46	8.57	15.84
−15.50	−7.98	−3.98	−0.23	3.66	8.58	15.99
−15.04	−7.97	−3.95	−0.21	3.74	8.70	16.03
−15.00	−7.75	−3.84	−0.08	3.80	8.85	16.84
−14.91	−7.67	−3.78	−0.02	4.29	8.87	16.87
−14.16	−7.48	−3.74	0.03	4.39	8.97	17.29
−14.14	−7.46	−3.65	0.33	4.41	9.02	17.62
−14.04	−7.44	−3.61	0.65	4.42	9.08	18.43
−13.88	−7.37	−3.59	0.70	4.48	9.12	19.57
−13.84	−7.37	−3.47	0.78	4.60	9.39	19.58
−13.72	−7.35	−3.46	0.80	4.78	9.57	20.80
−13.58	−7.27	−3.39	1.10	4.94	9.83	22.38
−13.33	−7.24	−3.14	1.20	5.28	10.02	22.66
−12.98	−7.20	−3.02	1.38	5.41	10.08	29.50
−12.68	−7.03	−3.01	1.58	5.54	10.09	32.03

Index

A

Absolute deviation, 41
Absolute Quetelet index, 111
Activation function, 170
Additive clustering, 341, 356
Additive non-overlapping clusters, 354
AddRemAdd algorithm, 354
AddRem algorithm, 330, 343
ADN algorithm, 354
Affinity data, 335
Agglomerative clustering, 291, 312
Alternating minimization, 234, 363
Anomalous pattern, 250
Anti-cluster, 353
Artificial neuron, 159
Attraction of an entity, 327, 344

B

Background similarity pattern, 316
"Bag-of-words" model, 122
Baire distance, 285
Batch K-Means, 254
Bayes decision rule, 118
Bayesian Information Criterion (BIC), 277
Bernoulli model, 48
Binary scale, 9, 44
Binary tree, 307
Bisecting K-Means, 293
Bit, 45
Bootstrapping, 59
Box-plot, 91
Build algorithm, 248

C

Category utility function, 151
Cell, 370
Central fuzzy set, 52
Centroid, 222

Chi-square contingency coefficient, 106, 109
  decomposition of, 106
Chi-square distance, 215
Chi-square distribution, 109
Classification tree, 141
Classifier, 113
Clustering, 221
Cluster representative, 262
Cluster-specific intensity, 345, 349
Coefficient of determination, 75, 132
Conceptual association, 101
Conceptual clustering, 293
Conceptual description of clusters, 267
Confusion table, 124
Connected component, 302
Contingency table, 100
Contribution of the PCA model to the data scatter, 201
Conventional formulation for PCA, 202
Convex function, 166
Correlation, iii
Correlation coefficient, 75
Correlation ratio, 95
Correspondence analysis, 212
Cosine, 209
Covariance matrix, 203
Cross-over, 242
Cut, 329

D

Data analysis perspective, 39
Data recovery approach, 6, 40, 174
Data scatter, 177
Data scatter decomposed over clusters and features, 268
Data standardization, 177
Data visualization, 21, 195
Decision rule, 113
Decision tree, 141

387

Decoder, 173
Decomposition of the scatter, 94
Density function, 35
Derivative, 380
Diagonal, 361
Discriminant function, 133
Distribution, 31
Divisive clustering, 292
Dual optimization problem, 139
Dummy, 48

E
Edge, 311
Eigenvalue, 202, 361
Eigenvector, 202, 362
Ellipsoid, 204
EM algorithm, 277
Entropy, 45
Epoch, 164
Error back-propagation, 168
Euclidean distance, 360
Evolutionary approach, 244
Explained proportion of data scatter, 268

F
F-measure, 126
Falsifability principle, 115
Feature-cluster contributions, 268
Feature contribution to data scatter, 180
Feature weights, 182
First eigenvector, 202, 340
First-order condition of minimum, 362
Fisher's criterion, 134
Four-fold contingency table, 110
FP rate, 127
Fuzzy cluster, 271
Fuzzy set, 50
Fuzzy K-Means, 271

G
Gaussian cluster, 277
Gaussian distribution, 34, 42, 119
Gaussian kernel, 140
Genetic algorithm, 244
Genuine similarity data, 317
Gini index, 49
Goodman-Kruskal tau, 111
Goodness-of-split criterion, 145, 293
Gradient, 363
Gradient method, 163
Grand mean, 91, 178, 195
Graph, 309

H
Hartigan index, 248
Hierarchical clustering, 283
Hierarchy, 321
Histogram, 38

I
IK-Means, 253
Impurity function, 145
Incremental K-Means, 238
Information gain, 148
  ratio, 148
Information retrieval, 209
Initialization, 247
Initialization of K-Means, 246
Inner product, 361
Instance processing procedure, 169
Intercept, 69
Interpretation of principal components, 203
Interval predicate, 98
  classifier, 98

K
Kernel function, 140
K-fold cross validation, 60
K-Means, 222

L
Lance-Williams coefficients, 292
Lapin transformation, 340
Laplacian, 339
Latent semantic indexing, 208
Leaf cluster, 294
Learning correlation, 117
Learning rate, 164
Least squares criterion, 175
Lift chart, 156
Linear classifier, 133
Linearization, 77, 84
Linear regression, 78, 133
Logistic regression, 120

M
Manhattan distance, 360
Margin in SVM method, 139
Margin of a point, 134
Matrix, 360
Matrix product, 361
Matthew's effect, 34
Maximum likelihood, 42, 276
Maximum spanning tree (MST), 304
Mean, 34
Meaning of fuzzy K-Means criterion, 273
Median, 40

Index 389

Medoid, 245
Metadata, 3
Method of steepest descent, 165, 363
Midrange, 40
Minimum cut, 333
Minimum distance rule, 223
Minkowski's criterion, 40
Minkowski's metric center, 52
Mixture of distributions clustering, 275
Modularity attraction, 327
Modularity criterion, 323
Multiple correlation coefficient, 132
Multiplicative decoder, 188, 198
Mutation, 244
Mutual information, 148

N
Naïve Bayes classifier, 120
Nature inspired approach, 240
Nearest-neighbor classifier, 95
Nearest neighbor clustering, 308
Network data, 322
Neural network, 159
Node height, 284
Nominal scale, 10, 46
Non-pivotal validation, 60
Norm, 199
Normal, or Gaussian, law, 42
Normalization by the range, 178
Normalized cut, 334, 340

O
Occam's razor, 115
One cluster model, 349
One cluster summary criterion, 328
One-by-one cluster extraction, 352
Orthogonal, 362
    projection operator, 132
Overfitting, 115

P
Particle swarm, 245
Partitioning around medoids (PAM), 245
Pearson chi-squared normalized, 270
Pearson index, 110, 214
Pie-chart, 33
Pivotal validation method, 60
Poisson distribution, 48, 269
Population, 42, 54, 240, 277
Power law, 34
P-quantile, 38
Precision, 124, 126
Predictor, 74, 95, 113

Prevalence, 125
Prim's algorithm, 311
Principal component analysis, 188, 202
Principle of maximum likelihood, 42
Principle of maximum parsimony, 115
Probabilistic statistics perspective, 42
Pseudo-inverse matrix, 132
Pythagoras' theorem, 175

Q
Quantitative scale, 11, 31
Quetelet index, 103, 108

R
Rank of matrix, 191
Rayleigh quotient, 202, 334, 352
Receiver operating characteristics (ROC)
    graph, 127
Reference point, 250
Regression-wise clustering, 274
Reject option, 155
Rescaling of dummy features, 186
Residual similarities, 352

S
Scatter plot, 68
Self-organizing map (SOM), 278
Semipositive definite matrix, 340
Sensitivity, 126
Shifting the origin, 178
Sigmoid function, 137, 170
Similarity shift, 310
Single link clustering, 343
Singular triplet, 199
Singular value, 199
Singular value decomposition
    (SVD), 201
Slope, 69
Space's origin, 181
Span, 132
Spectral clustering, 333
Spectral decomposition, 340
Splitting criterion, 293
Square-error criterion, 237
Square matrix, 361
Standardization of mixed scale
    data, 182
Statistical independence, 108
String, 370
Subplot, 36
Summarization, iii
Summarization problem, 180
Summary Gini index, 300

Support vector, 135
Support Vector Machine (SVM), 138

T
Tabular regression, 90, 98
Target feature, 57, 113
Tf-idf normalization, 208
Three sigma rule, 34
Threshold graph, 323
Tight cluster, 326
TP rate, 124, 127
Trace of matrix, 191
Training protocol, 117
Transition equations, 215
Transpose, 360
Two-splitting, 293, 301

U
Ultrametric, 284
Uniform criterion, 325

Uniform distribution, 45
Union of fuzzy sets, 51

V
Variance, 34, 39
VC-complexity, 116
Vector, 358
Visualization, iii
Volume of subset, 327

W
Ward agglomeration, 287
Ward distance, 290
Ward-like divisive clustering, 300
Weighted data scatter, 216
Weighted graph, 309

Z
Z-scoring, 35, 76, 178